Optimization and Energy Management in Smart Grids

Optimization and Energy Management in Smart Grids

Editor

Germano Lambert-Torres

MDPI

Basel • Beijing • Wuhan • Barcelona • Belgrade • Novi Sad • Cluj • Manchester

Germano Lambert-Torres
Smart Energy Master
Gnarus Institute
Itajuba
Brazil

Editorial Office
MDPI AG
Grosspeteranlage 5
4052 Basel, Switzerland

This is a reprint of the Special Issue, published open access by the journal *Energies* (ISSN 1996-1073), freely accessible at: www.mdpi.com/journal/energies/special_issues/optimization_and_energy_management.

For citation purposes, cite each article independently as indicated on the article page online and using the guide below:

Lastname, A.A.; Lastname, B.B. Article Title. *Journal Name* **Year**, *Volume Number*, Page Range.

ISBN 978-3-7258-1624-8 (Hbk)
ISBN 978-3-7258-1623-1 (PDF)
https://doi.org/10.3390/books978-3-7258-1623-1

Contents

About the Editor

Germano Lambert-Torres

Germano Lambert-Torres received his Ph.D. in Electrical Engineering from the École Polytechnique de Montreal, Canada, in 1990. Currently, he is the director of research and development at PS Solutions and a member of the Scientific–Technical Council at Gnarus Institute, Brazil. Dr. Lambert-Torres has led research in intelligent systems, maintenance management, and power system operation for many power industries in Brazil and South America. He has also taught seminars and short courses in many countries worldwide at the Institute of Electrical and Electronics Engineers (IEEE), International Council on Large Electric Systems (CIGRÉ), and International Federation of Automatic Control (IFAC) conferences, universities, and research centers. Moreover, he has supervised over 120 M.Sc. and Ph.D. theses and published over 800 journal and technical conference papers. He is also the author/editor of ten books. Dr. Lambert-Torres is a fellow of the IEEE.

Preface

Smart Grids (SGs) have significantly impacted how electrical networks operate. They involve elements and techniques of telecommunications, information/data computing, and the Internet, among others. A Smart Grid is a combination of hardware, management, and reporting software built atop an intelligent communications infrastructure.

SGs must consider all sides of the power grid, making it more innovative and flexible. This idea can be represented by the following four pillars: generation optimization, transmission optimization, distribution optimization, and demand-side management. In generation optimization, topics such as (but not limited to) wind farms and other non-conventional sources, dispatching scheduling optimization, the use of distributed generation, and fuel cells are welcome in this Special Issue. In the transmission optimization area, possible topics include the phasor measurement technique, comprehensive area measurement, flexible AC transmission system, adoptive islanding, self-healing grids, probabilistic and dynamic stability assessment, substation monitoring, and control, among others. In the distribution optimization area, topics such as (but not limited to) distribution automation, demand optimization, selective load control, and islanding of microgrids are of interest. Finally, the demand side management area includes (but is not limited to) distributed generation, on-site generation, district energy, decentralized energy, distributed energy resources (DER), and distributed energy storage systems. Topics such as energy efficiency and demand response are also very welcome to this Special Issue.

Descriptions of numerical, statistical, intelligent, or hybrid methodologies are also welcome. These include situations where market barriers are solved, such as extensive field testing before widespread integration into the power grid, new technologies being too expensive, a risk-averse mentality, long-scale cycles for equipment, regulatory questions, and financial constraints.

This Special Issue will focus on optimizing the electrical grid process and contributing to better practices in SGs. It will explain how the process could benefit the more effective management of SGs, such as reduced energy costs, reduced stress on the electric energy infrastructure, and improved energy efficiency.

Germano Lambert-Torres
Editor

energies

MDPI

Communication

GA-Based Voltage Optimization of Distribution Feeder with High-Penetration of DERs Using Megawatt-Scale Units

Aswad Adib, Joao Onofre Pereira Pinto * and Madhu S. Chinthavali

Oak Ridge National Laboratory, Oak Ridge, TN 37830, USA; adiba@ornl.gov (A.A.); chinthavalim@ornl.gov (M.S.C.)

* Correspondence: pintoj@ornl.gov

Abstract: In this paper, genetic algorithm (GA)-based voltage optimization of a modified IEEE-34 node distribution feeder with high penetration of distributed energy resources (DERs) is proposed using two megawatt-scale reactive power sources. Traditional voltage support units present in distribution grids are not suitable for DER-rich feeders, while voltage support using small-scale DERs present in the feeder requires considerable communication effort to reach a global solution. In this work, two megawatt-scale units are placed to improve the voltage profile across the IEEE 34-node feeder, which has been modified to include several PV units and an energy storage unit. The megawatt-scale units are optimized using GA for fast and accurate operation. The performance of the proposed scheme is verified using simulation results with a multi-platform setup where the modified IEEE-34 node feeder is modeled in OpenDSS while the GA optimization scheme is programmed in MATLAB.

Keywords: high-penetration of DERs; GA-based optimization; megawatt-scale reactive power units; multi-platform simulation

check for updates

Citation: Adib, A.; Pinto, J.O.P.; Chinthavali, M.S. GA-Based Voltage Optimization of Distribution Feeder with High-Penetration of DERs Using Megawatt-Scale Units. *Energies* **2023**, *16*, 4842. https://doi.org/10.3390/en16134842

Academic Editor: Germano Lambert-Torres

Received: 28 April 2023
Revised: 13 June 2023
Accepted: 15 June 2023
Published: 21 June 2023

1. Introduction

The landscape of the traditional distribution grid is rapidly changing with the increasing penetration of distributed energy resources (DERs) such as PVs and energy storage (ES) units. Although the addition of DER units presents several benefits, such as the reduction of fossil-fuel-based energy usage, and more potential benefits, such as improved grid resiliency, the current grid infrastructure is not equipped to deal with the stochastic nature and the bidirectional power flow in feeders associated with the high penetration of such units [1–3]. Therefore, more research is necessary regarding the high penetration of DER-based distribution feeders to fulfill their potential.

Traditionally, voltage support is provided in distribution grids through on-load tap-changing (OLTC) transformers and shunt capacitors. However, the tap changers and mechanical switches associated with such devices cannot respond fast enough to regulate voltage in high DER penetration distribution grids [4], making them unsuitable for such feeders. Flexible AC transmission systems (FACTS) devices scaled for the distribution grid level, referred to as distributed-FACTS (D-FACTS), can provide improved grid support. D-STATCOMs (distributed static synchronous compensators), which are the most widely utilized D-FACTS device due to their simple shunt connection, can only provide reactive power support, while grid support at the distribution level requires both active and reactive power support [5]. D-FACTS devices capable of providing both active and reactive power, e.g., UPFCs (unified power flow controllers), are less common due to their complicated architecture requiring both a shunt and series connection [6]. On the other hand, large-scale DER units can fulfill the functionalities of D-FACTS devices while being capable of providing additional support functions such as black start capability. Proper coordination is necessary between such large-scale DERs to improve performance across the feeder.

Several state-of-the-art optimization and communication techniques are present in the literature to improve distribution feeder performance with high penetration of DERs. The techniques presented can broadly be categorized as decentralized, distributed, and centralized techniques [7]. Under decentralized schemes, each DER is locally controlled and cannot guarantee feeder-level improvement. Therefore, the remaining two techniques are generally discussed in the literature. A two-stage distributed optimization process is presented in [8], where the traditional voltage support devices are optimized hourly and the individual small-scale DERs present are optimized every few minutes. The feeder is divided across several zones, and each zone is equipped with an agent that is primarily focused on optimizing the particular zone while sharing a global optimization goal with other zones. A similar zone-based distributed scheme is proposed in [9], where the optimization method is enhanced through the integration of the alternative direction multiplier method for inter-regional coordination and the branch and bound method for handling nonconvexity introduced by traditional devices. A distributed approach is also proposed in [10], where three microgrids optimize themselves independently while the distribution system operator (DSO) sets global boundaries for power exchange between the microgrids and the remaining DGs in the feeder. The distributed approaches presented in [11,12] display good performance and require less communication infrastructure than centralized options; however, an optimal global solution to the optimization problem cannot be guaranteed. Several centralized optimization techniques can also be found in the literature. In [11], a centralized energy management scheme is presented that puts emphasis on consumer electricity costs. In [12], a distributionally robust chance-constrained optimization scheme is presented that centrally optimizes the dispatch of PVs and ES units, while the DER uncertainties are modeled through a set of probability distributions. However, the work considering a centralized scheme considers simplified feeders with a small number of DER units to reduce the complexity associated with centralized techniques. Moreover, the works referenced thus far perform optimization based on traditional power system tools such as mixed integer linear programming, which require significant computation time and are not suitable for the stochastic nature of DERs. Furthermore, forecasting data and probability distributions generally used for DERs to realize real-time optimization will inevitably result in inaccuracies. One solution can be to use meta-heuristic techniques for optimization. Meta-heuristic algorithms have been extensively used for allocating resources such as shunt capacitors and DERs in distribution feeders [13,14]. In [15], a global multidimensional constrained optimization technique is used for developing energy-harvesting gravity-based devices for wind applications. The presented algorithm shows improved results over the state-of-the-art; however, its applicability to larger systems such as distribution feeders is not evident. In [16], a chaotic differential evolution algorithm is used to optimally allocate and size DERs, considering economic loss, power loss, and voltage deviation in feeders. In [17], an improved differential search algorithm is used to optimally allocate DERs in a radial distribution grid to minimize real power loss, network operational cost, and voltage profile. In [18], both DERs and D-STATCOMs are optimally sized and allocated to minimize real power loss using a genetic algorithm (GA). Similarly, DERs and capacitor banks are sized and placed using a combination of GA, DE, and the strength pareto evolutionary algorithm [19]. Power loss minimization and voltage deviation optimization are performed in [20] by controlling DERs, shunt capacitors, OLTC transformers, and flexible loads using normal distribution crossover-based non-dominated sorting GA. However, the issue regarding communication infrastructure associated with centrally controlling a large number of units still persists. Therefore, more research is necessary in real-time optimization of feeders using such techniques.

In this paper, a GA-based optimization scheme is presented to improve the voltage profile of a modified IEEE-34 node feeder with high penetration of DERs using two megawatt-scale DER units. The small-scale DER units are controlled locally independently of the proposed scheme, while the large-scale ones are centrally optimized. Central optimization allows for a globally optimal solution, and since only two units are introduced, the communication infrastructure required is minimal. On the other hand, using GA allows for fast and accurate real-time optimization.

The rest of the paper is organized as follows. In Section 2, the modified IEEE-34 node feeder is described. In Section 3, the presented optimization scheme is discussed in detail. Results are presented in Section 4 to validate the proposed scheme. Finally, conclusions are drawn in Section 5.

2. Description of the Feeder

The IEEE 34-node feeder considered for this work is shown in Figure 1. It retains all the characteristics of the existing IEEE 34-node feeder, including load values, load placements, line lengths, and the traditional voltage support devices, namely OLTC transformers and shunt capacitors [21]. However, the feeder has been modified from its traditional counterpart by adding eight PV resources and an energy storage resource [22]. The ratings of the added PVs are shown in Table 1. The ES is rated at 250 kVA. The nominal load of the presented feeder is 2.05 MVA, which is lower than the total power of the DERs, making the feeder susceptible to the stochastic nature of the DERs. The MVA rating of the substation transformer has also been increased to account for the added resources in the feeder.

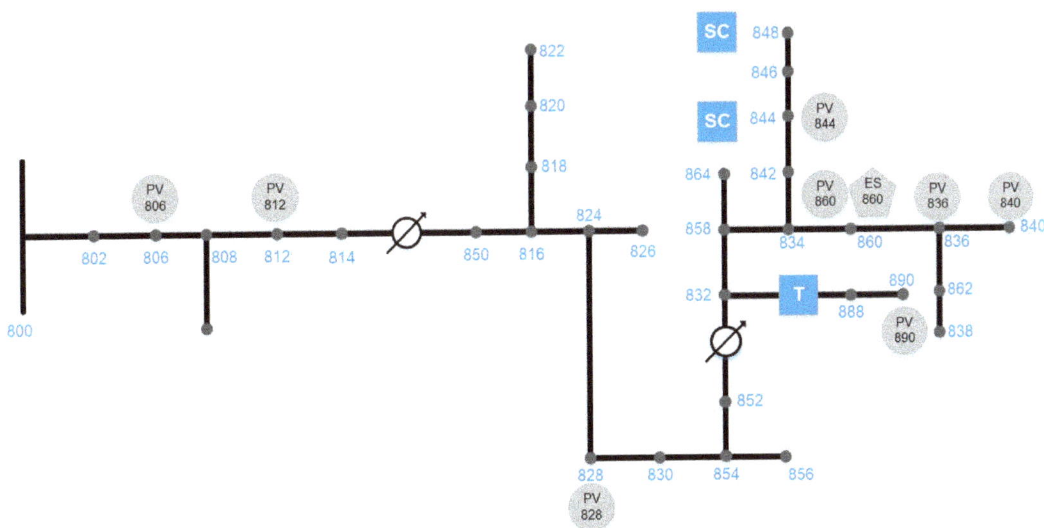

Figure 1. Schematic of the modified IEEE-34 node feeder.

Table 1. PV Ratings.

PVs	Apparent Power (kVA)	MPP (kW)
844	1000	900
890	750	500
860	1250	1000
828	200	150
806	100	100
836	150	150
840	250	200
812	250	225

The PVs added to the feeder are actively engaging in voltage support operations using the volt-var method. However, the amount of reactive power available for voltage support is constantly changing based on the maximum power point (MPP) of the PVs. Therefore, the PVs are locally controlled and act only on the voltage profile of their local nodes.

On the other hand, the reactive power resources that are being added will act depending on the feeder-level voltage profile. The voltage profile of the feeder without the reactive power resources is shown in Figure 2. Figure 2 was generated with the power generation and consumption units at their nominal values. In other words, all the loads were operating at their rated values, and all the PVs were operating at their respective MPPs. As can be seen from Figure 2, under nominal conditions, the voltage at certain nodes can swell well beyond 1 p.u. due to DER power output. It should be noted that the traditional voltage support devices are active in this scenario, which justifies the use of additional reactive power resources. Two particular points of interest in the feeder voltage shown in Figure 2 are the ones with the lowest and highest voltage values of all the three-phase nodes, which correspond to nodes 814 and 840, respectively. Two reactive power resources are therefore added to these nodes to regulate the feeder voltage. The added resources are rated for 2 MVA. Their placements are highlighted in Figure 3. A more thorough analysis regarding the placement of reactive power resources will be considered in the future.

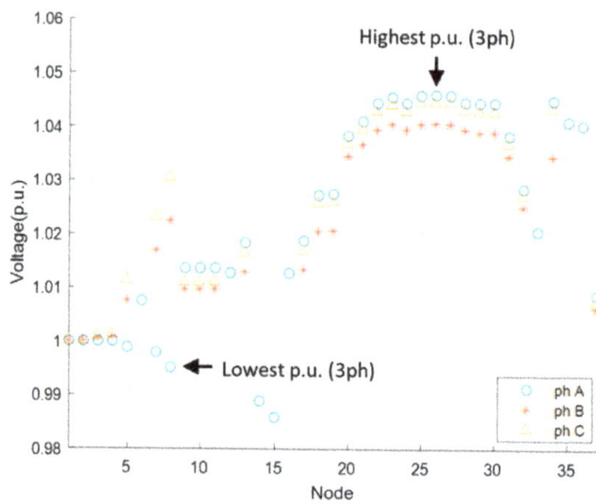

Figure 2. Voltage profile of the feeder under rated conditions.

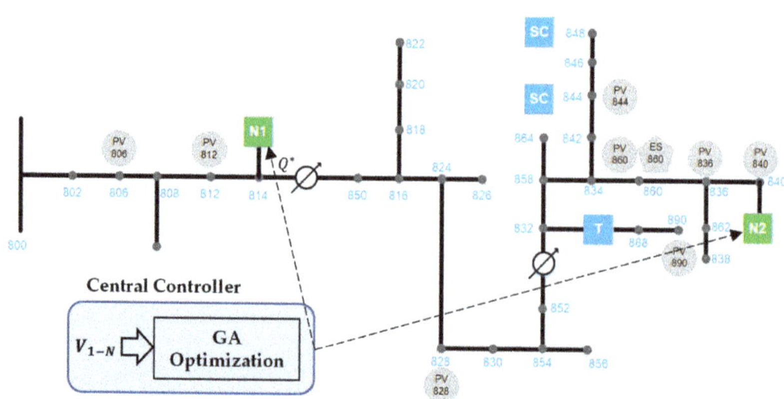

Figure 3. Schematic of the feeder with reactive power resources and the optimization framework.

3. Proposed Optimization Scheme

The entire system considered in this work is divided across two platforms. The feeder model is housed in OpenDSS, while the optimization algorithm is in MATLAB. GA has been used to optimize the reactive power settings of the reactive power resources, while the infinity norm function has been employed as the optimization function.

3.1. GA Optimization

The goal of this paper is to regulate the voltage of all nodes to not violate the ±5% limit, which is required by most of the distribution grids in the world, using the minimum amount of reactive power from the large DERs. Therefore, this problem can be addressed as an optimum reactive power flow (ORPF) type, resulting in a fairly large, non-linear, multimodal optimization problem with constraints. Furthermore, the resulting objective function is non-continuous, non-differentiable, and rugous, besides possibly being non-convex. The deterministic optimization methods have difficulties dealing with this type of problem; they usually require large computational times and, when applicable, are very likely to get trapped in local minimums. On the other hand, many evolutionary algorithms have features that make them feasible for the type of problem in scope. Among the evolutionary methods, GA is a bio-inspired technique that mimics the evolution theory of Darwin by modeling natural operators such as natural selection, crossover, and mutation. Among the powerful features of GA are intrinsic parallelism and inherent global search; no requirement to explicitly model stands out. Hybrid techniques and combinations of different evolutionary methods can also be used [13], but the simplicity of GA allows for a natural expansion of this approach for more complex systems, e.g., larger feeders and simultaneous optimization of discrete (OLTCs) and continuous variables (DERs) [23,24].

The optimization function used in this work, which is the same as the fitness function (FF) of the GA scheme, is based on the infinity norm of the feeder voltage deviations as expressed in (1),

$$FF = \max_{i\epsilon\{1,N\}}(|v_i - v_i^*|), \tag{1}$$

where, N is the total number of node voltages, v_i is the actual voltage at node i, and v_i^* is the desired voltage at node i, which for this problem was chosen as 1 p.u. The goal is to minimize *FF*, which means to minimize the largest voltage deviation throughout the whole feeder. Therefore, the optimization problem can be expressed as,

$$\min\left(\max_{i\epsilon\{1,N\}}(|v_i - v_i^*|)\right). \tag{2}$$

Based on the largest voltage deviation, the reactive power commands for the optimized units are derived in an iterative process, as will be described in the following subsection. The upper and lower bounds of the reactive powers are set at their ratings. In other words,

$$|Q|_1^*, |Q|_2^* \leq 2\text{Mvar}. \tag{3}$$

The iterations will continue until the GA encounters one of its stopping criteria, ideally the maximum value of the fitness function, which will be set by the user. The maximum deviation can be set as the largest p.u. deviation in voltage that is acceptable. It should be noted that each phase of all the node voltages is considered for the fitness function.

3.2. OpenDSS MATLAB Co-Simulation Scheme

As mentioned previously, the system under consideration is divided across two platforms, as shown in Figure 3. The platforms communicate with each other using the COM interface. The modified IEEE 34-node feeder is modeled in OpenDSS. The control scheme for all the locally controlled elements, i.e., OLTC transformers, shunt capacitors, and PVs, is also modeled in OpenDSS. The added reactive power resources are modeled as energy storage devices in the OpenDSS model, while their controls are housed in MATLAB.

The optimization in MATLAB is performed using the feeder node voltages. Therefore, at each GA generation, all the p.u. node voltages are sent from OpenDSS to MATLAB, while the reactive power commands are sent from MATLAB to OpenDSS. The stopping criteria for the algorithm is set at a fitness of 5%, i.e., when the largest voltage deviation across all nodes drops to less than 5%. The whole process is depicted as a schematic in Figure 4 and a flowchart in Figure 5. A schematic diagram depicting the real-time implementation of the proposed scheme is illustrated in Figure 6. The update time of the reactive power commands is set at three minutes to account for the model update process, although the GA optimization can operate much faster, as will be shown in the following section. Note that the feeder model can be executed on several simulation platforms, while OpenDSS is used for this work.

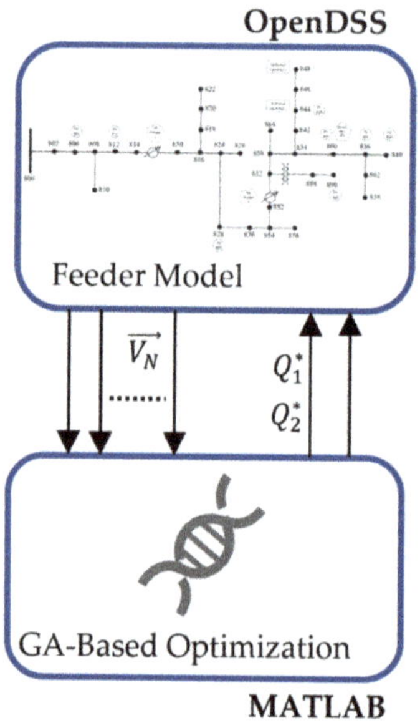

Figure 4. Schematic of the co-simulation process.

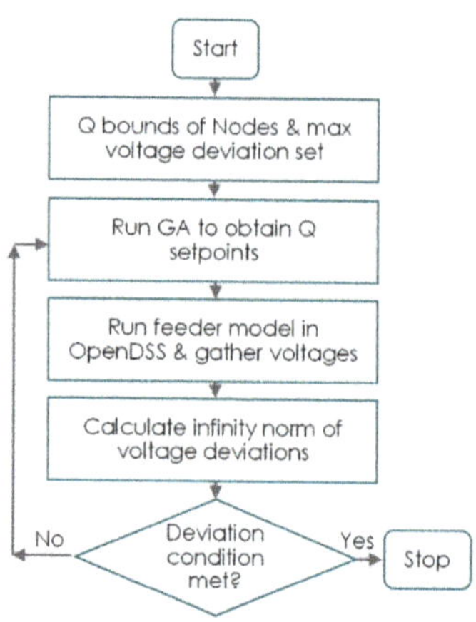

Figure 5. Flowchart of the co-simulation process.

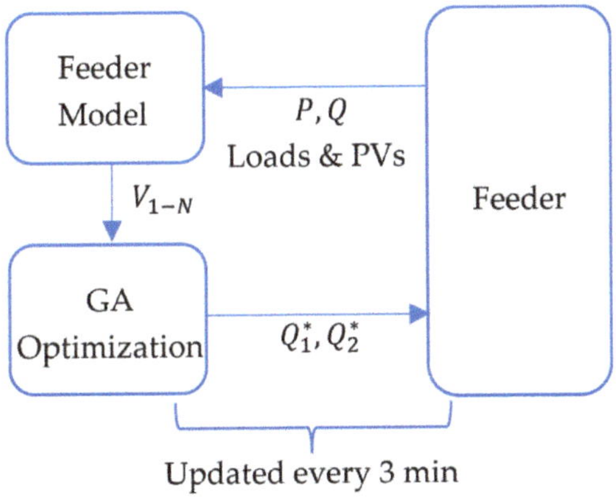

Figure 6. Real-time implementation of the proposed algorithm.

4. Results

In this section, the proposed optimization scheme is verified using use cases performed in a simulation setup. As mentioned previously, the simulation setup is built across two platforms. The modified IEEE-34 node feeder is modeled in OpenDSS, while the GA optimization scheme is housed in MATLAB. The parameters used for GA optimization are shown in Table 2. The parameters are chosen through trial and error to strike a balance between performance and computational time.

Table 2. GA Parameters.

Parameter	Value
Fitness Limit	0.04
Function Tolerance	0.001
Selection Function	Uniform
Fitness Scaling	Top
Selection Function	Uniform
Crossover	Scattered
Crossover Fraction	0.9
Mutation	Gaussian

The voltage profile of the feeder with and without GA optimization is shown in Figure 7. This profile is generated under 50% overload and a total PV capacity of 50%. The capacity of the PVs is altered by changing the irradiation input to the units. Irradiation inputs are generated in MATLAB and sent to the OpenDSS model through the COM interface. Note that all three phases of each node are shown separately. It should also be noted that the OLTC transformers and shunt capacitors are operational in these use cases. The presence of PVs causes voltage to swell in certain nodes, while overload conditions cause voltage to sag in other nodes. As one can see from Figure 7, at one particular node (Node 20 in the figure), the voltage goes below 0.9 p.u. when the GA optimization is not enabled, while six other nodes also violate the ±5% limit. When GA optimization is enabled, after 21 iterations, the voltage across the entire feeder is spread much more uniformly with respect to the rated voltage, with the number of node voltage violations reduced to two. The reactive power setpoints generated by the optimization are −1.87 Mvar and 0.63 Mvar at nodes 888 and 814, respectively. The optimization converged in around 40.5 s, making it viable for real-time operations.

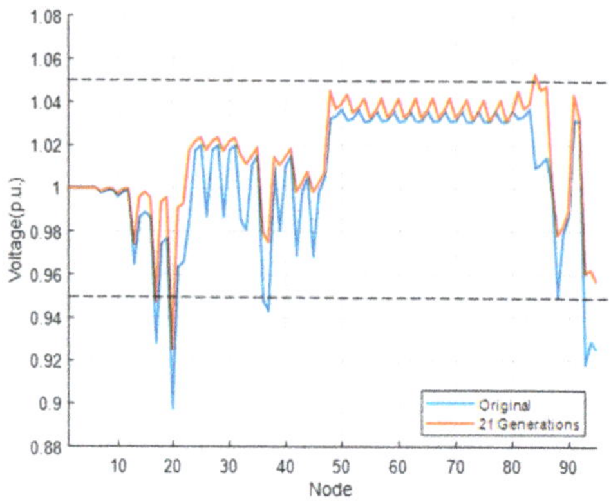

Figure 7. Voltage profile of the modified IEEE-34 node feeder with and without GA under 50% overload and half the PV capacity.

In the use case depicted in Figure 8, the feeder is operating in light load conditions with 50% of the nominal load active, while the PVs are operating in 50% overrated conditions. This causes higher voltage swells across the feeder when the optimization is not enabled. With the GA optimization enabled, a more uniform voltage profile with respect to the rated voltage can be seen in Figure 8, with the entire profile residing within the ±5% limit. A large difference in the same node voltages (nodes 84–86) can be seen in the figure with and without optimization, since these nodes correspond to Node 888, which is one of the nodes where the additional reactive power unit is added. Although the voltage deviation at that particular node is higher, the overall profile has improved. The reactive power setpoints generated in this case are −1.58 Mvar and 0.89 Mvar, while the optimization converged in about 52.8 s. It should be noted that the maximum number of stall generations was set at 20, which is why the GA results for both use cases are for 21 generations. As a future expansion of this work, the use cases will be recreated on a controller hardware-in-the-loop platform.

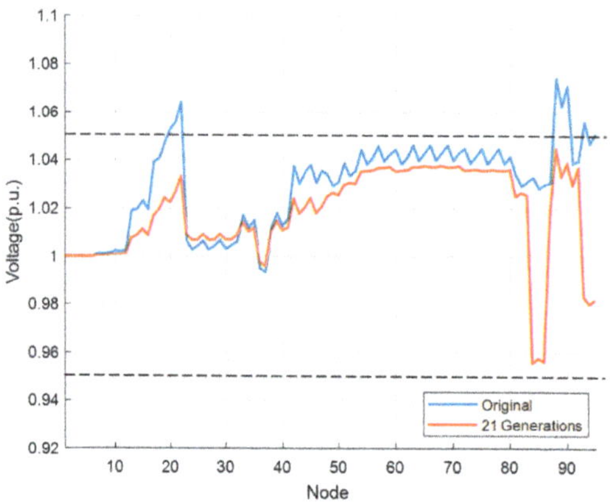

Figure 8. Voltage profile of the modified IEEE-34 node feeder with and without GA under 50% load and 50% over PV rated capacity.

5. Conclusions

In this paper, a GA-based scheme has been proposed to improve the voltage profile of a high DER penetration distribution feeder by centrally optimizing two megawatt-scale units. The proposed scheme has been implemented in a modified IEEE-34 node feeder, which was converted to a high DER penetration feeder through the introduction of several PV and ES units, the total power of which was comparable to the total load of the feeder. The utilization of two large-scale units for voltage optimization instead of the existing small-scale DERs in the feeder allows for much less communication infrastructure, therefore mitigating the biggest disadvantage of central optimization while inheriting all its benefits. On the other hand, a GA-based scheme allows for quick and accurate real-time optimization of the resources. The proposed scheme has been validated through simulation scenarios carried out in an OpenDSS-MATLAB setup.

Author Contributions: Conceptualization, J.O.P.P. and M.S.C.; methodology, A.A. and J.O.P.P.; software, A.A. and J.O.P.P.; validation, A.A. and J.O.P.P.; formal analysis, A.A.; investigation, J.O.P.P.; resources, J.O.P.P. and M.S.C.; data curation, A.A. and J.O.P.P.; writing—original draft preparation, A.A.; writing—review and editing, J.O.P.P. and M.S.C.; visualization, A.A. and J.O.P.P.; supervision, J.O.P.P. and M.S.C.; project administration, M.S.C.; funding acquisition, J.O.P.P. and M.S.C. All authors have read and agreed to the published version of the manuscript.

Funding: This research was funded by the U.S. Department of Energy, Office of Electricity, TRAC under contract number DE-AC05-00OR22725. Notice: This manuscript has been authored by UT-Battelle, LLC under Contract No. DE-AC05-00OR22725 with the U.S. Department of Energy. The United States Government retains and the publisher, by accepting the article for publication, acknowledges that the United States Government retains a non-exclusive, paid-up, irrevocable, world-wide license to publish or reproduce the published form of this manuscript, or allow others to do so, for United States Government purposes. The Department of Energy will provide public access to these results of federally sponsored research in accordance with the DOE Public Access Plan (http://energy.gov/downloads/doe-public-access-plan).

Data Availability Statement: Not applicable.

Conflicts of Interest: The authors declare no conflict of interest. The funders had no role in the design of the study; in the collection, analysis, or interpretation of data; in the writing of the manuscript; or in the decision to publish the results.

References

1. Wu, W.; Chou, S.-C.; Viswanathan, K. Optimal Dispatching of Smart Hybrid Energy Systems for Addressing a Low-Carbon Community. *Energies* **2023**, *16*, 3698. [CrossRef]
2. Laribi, O.; Rudion, K. Optimized Planning of Distribution Grids Considering Grid Expansion, Battery Systems and Dynamic Curtailment. *Energies* **2021**, *14*, 5242. [CrossRef]
3. Riaz, M.; Ahmad, S.; Hussain, I.; Naeem, M.; Mihet-Popa, L. Probabilistic Optimization Techniques in Smart Power System. *Energies* **2022**, *15*, 825. [CrossRef]
4. Peng, F.Z. Flexible AC Transmission Systems (FACTS) and Resilient AC Distribution Systems (RACDS) in Smart Grid. *Proc. IEEE* **2017**, *105*, 2099–2115. [CrossRef]
5. Bloemink, J.M.; Green, T.C. Benefits of Distribution-Level Power Electronics for Supporting Distributed Generation Growth. *IEEE Trans. Power Deliv.* **2013**, *28*, 911–919. [CrossRef]
6. Chawda, G.S.; Shaik, A.G.; Mahela, O.P.; Padmanaban, S.; Holm-Nielsen, J.B. Comprehensive Review of Distributed FACTS Control Algorithms for Power Quality Enhancement in Utility Grid with Renewable Energy Penetration. *IEEE Access* **2020**, *8*, 107614–107634. [CrossRef]
7. Zhang, Z.; Zhang, Y.; Yue, D.; Dou, C.; Ding, L.; Tan, D. Voltage Regulation with High Penetration of Low-Carbon Energy in Distribution Networks: A Source–Grid–Load-Collaboration-Based Perspective. *IEEE Trans. Ind. Inform.* **2022**, *18*, 3987–3999. [CrossRef]
8. Fazio, A.R.D.; Risi, C.; Russo, M.; Santis, M.D. Distributed Coordinated Reactive Power Control for Voltage Regulation in Distribution Networks. *IEEE Trans. Smart Grid* **2020**, *12*, 312–323.
9. Tang, Z.; Hill, D.J.; Liu, T. Coordinated Optimization for Zone-Based Voltage Control in Distribution Grids. *IEEE Trans. Ind. Appl.* **2022**, *58*, 173–184.
10. Liu, Y.; Guo, L.; Lu, C.; Chai, Y.; Gao, S.; Xu, B. A Fully Distributed Voltage Optimization Method for Distribution Networks Considering Integer Constraints of Step Voltage Regulators. *IEEE Access* **2019**, *7*, 60055–60066. [CrossRef]
11. Tziovani, L.; Hadjidemetriou, L.; Kolios, P.; Astolfi, A.; Kyriakides, E.; Timotheou, S. Energy Management and Control of Photovoltaic and Storage Systems in Active Distribution Grids. *IEEE Trans. Power Syst.* **2022**, *37*, 1956–1968. [CrossRef]
12. Rayati, M.; Bozorg, M.; Cherkaoui, R.; Carpita, M. Distributionally Robust Chance Constrained Optimization for Providing Flexibility in an Active Distribution Network. *IEEE Trans. Smart Grid* **2022**, *13*, 2920–2934. [CrossRef]

13. Adetunji, K.E.; Hofsajer, I.W.; Abu-Mahfouz, A.M.; Cheng, L. A Review of Metaheuristic Techniques for Optimal Integration of Electrical Units in Distribution Networks. *IEEE Access* **2021**, *9*, 5046–5068. [CrossRef]
14. Menesy, A.S.; Sultan, H.M.; Habiballah, I.O.; Masrur, H.; Khan, K.R.; Khalid, M. Optimal Configuration of a Hybrid Photovoltaic/Wind Turbine/Biomass/Hydro-Pumped Storage-Based Energy System Using a Heap-Based Optimization Algorithm. *Energies* **2023**, *16*, 3648. [CrossRef]
15. Yurchenko, D.; Machado, L.Q.; Wang, J.; Bowen, C.; Sharkh, S.; Moshrefi-Torbati, M.; Val, D.V. Global optimization approach for designing high-efficiency piezoelectric beam-based energy harvesting devices. *Nano Energy* **2022**, *93*, 106684. [CrossRef]
16. Kumar, S.; Mandal, K.K.; Chakraborty, N. Optimal DG placement by multi-objective opposition based chaotic differential evolution for technoeconomic analysis. *Appl. Soft Comput.* **2019**, *78*, 70–83. [CrossRef]
17. Injeti, S.K. A Pareto Optimal Approach for Allocation of Distributed Generators in Radial Distribution Systems Using Improved Differential Search Algorithm. *J. Electr. Syst. Inf. Technol.* **2018**, *5*, 908–927. [CrossRef]
18. Singh, B.; Singh, S. GA-Based Optimization for Integration of DGs, STATCOM and PHEVs In Distribution Systems. *Energy Rep.* **2019**, *5*, 84–103. [CrossRef]
19. Huiling, T.; Jiekang, W.; Fan, W.; Lingmin, C.; Zhijun, L.; Haoran, Y. An Optimization Framework for Collaborative Control of Power Loss and Voltage in Distribution Systems with DGs and EVs Using Stochastic Fuzzy Chance Constrained Programming. *IEEE Access* **2020**, *8*, 49013–49027. [CrossRef]
20. Zeynali, S.; Rostami, N.; Feyzi, M.R. Multi-Objective Optimal Short-term Planning of Renewable Distributed Generations and Capacitor Banks in Power System Considering Different Uncertainties Including Plug-In Electric Vehicles. *Int. J. Electr. Power Energy Syst.* **2020**, *119*, 105885. [CrossRef]
21. IEEE PES Test Feeder. Available online: https://cmte.ieee.org/pes-testfeeders/resources/ (accessed on 27 April 2023).
22. Dharmawardena, H.; Venayagamoorthy, G.K. A Distribution System Test Feeder for DER Integration Studies. In Proceedings of the Clemson University Power Systems Conference (PSC), Charleston, SC, USA, 4–7 September 2018.
23. Malachi, Y.; Singer, S. A genetic algorithm for the corrective control of voltage and reactive power. *IEEE Trans. Power Syst.* **2006**, *21*, 295–300. [CrossRef]
24. Rousis, A.O.; Konstantelos, I.; Strbac, G. A planning model for a hybrid AC–DC microgrid using a novel GA/AC OPF algorithm. *IEEE Trans. Power Syst.* **2020**, *35*, 227–237. [CrossRef]

energies

Hypothesis

Stochastic Security-Constrained Economic Dispatch of Load-Following and Contingency Reserves Ancillary Service Using a Grid-Connected Microgrid during Uncertainty

Kalyani Makarand Kurundkar * and Geetanjali Abhijit Vaidya

Electrical Engineering Department, PVGs COET and GKPIM, Pune 411009, India
* Correspondence: kmk_elect@pvgcoet.ac.in

Abstract: In the context of the growing penetration of renewable power sources in power systems causing probabilistic contingency conditions, a suitable economic dispatch model is decisively needed. There is a lack of research in the field of probabilistic mathematical formulation considering the uncertainties due to the stochastic nature of renewables and contingency occurrence, as it is a very complex problem to be solved. The most appropriate model is the stochastic security-constrained economic dispatch (SSCED) model for optimized economic dispatch decisions during uncertainty. However, because of its complexity, it is rarely employed. This paper attempts to solve the complex SSCED problem in the presence of the uncertainty of resources and probabilistic contingency conditions, which is a novel effort in this regard. The SSCED is carried out over multiple periods to provide the load-following or contingency reserves. In the proposed SSCED, the uncertainty problem is addressed by modeling the stochastic wind energy power source by using "probability transition scenarios". The uncertainty caused by probabilistic contingency conditions in the dispatch schedule is approximated using a "state-specific transition matrix". The frequency control reserves in contingency conditions are co-optimized with energy, and stochastic security-constrained economic dispatch is achieved. The efforts are put forward to suggest a new market model in the presence of the uncertainty of renewable energy availability. Case studies are examined to show the potential technical and financial advantages of the proposed SSCED through co-optimization. Grid-connected microgrid owners offer frequency control ancillary services by providing load-following ramping reserves in the normal state and contingency reserves in the state of contingency. The probabilistic contingencies considered are generator failure and an underloading condition. A modified "IEEE 30 bus system" is considered a grid-connected microgrid for testing the proposed SSCED. The results show that the greater the flexibility of the resources, the greater the technical and economic benefits. The increase in ramping flexibility of a wind source results in almost an 8.1% reduction in operational costs compared to the base case. The contingency condition analysis shows that the presence of ramping reserves in the system enhances the power system performance, avoiding the cascading effects that ultimately cause a power system failure.

Keywords: multiperiod; stochastic; uncertainty; load-following; contingency; security-constrained

Citation: Kurundkar, K.M.; Vaidya, G.A. Stochastic Security-Constrained Economic Dispatch of Load-Following and Contingency Reserves Ancillary Service Using a Grid-Connected Microgrid during Uncertainty. *Energies* **2023**, *16*, 2607. https://doi.org/10.3390/en16062607

Academic Editor: Germano Lambert-Torres

Received: 27 December 2022
Revised: 3 March 2023
Accepted: 5 March 2023
Published: 9 March 2023

1. Introduction

In order to maintain the voltage profile, stability, and security of the power system as well as to support the main grid, microgrids may offer possible ancillary services. Due to the increased prevalence of unreliable renewable energy sources, ancillary services have become crucial at the distribution level [1].

It is found from the literature that researchers have placed a greater emphasis on estimating spinning reserves from renewable sources and co-optimizing them with energy while taking into account frequency control ancillary services provided by microgrids [2–5]. The co-optimization of energy and reserves in presence of converter-controlled DERs by blockchain technology is discussed in [6]. The stochastic optimization of energy and reserves are considered in [7]; using scenarios, the model takes into account and incorporates uncertainties linked to the production of renewable energy, day-ahead market pricing, and costs associated with unbalanced energy. When comparing the outcomes of stochastic and deterministic models, the cost of stochastic planning is 5% less than that of deterministic planning. A deterministic co-optimization considers the bi-level coordinated dispatch model that fully takes into account the informational interactions between the main grid and active distribution network [8]. To reduce operating expenses, a day-ahead active power scheduling approach that takes into account the errors in forecasting for renewable energy is suggested [9]. Profit maximization through the co-optimization of energy and spinning reserves given by renewable sources and energy storage has been studied [10]. Security-Constrained Economic Dispatch (SCED) in the reserve allocation electricity market is discussed [11].

It has been observed that until now, the research work mainly focused on the co-optimization of energy and reserves considering the deterministic approach for economic dispatch. Much less research work has been carried out considering the stochastic co-optimization of energy and reserves under uncertainty.

Further, as inferred from the literature survey on the stochastic co-optimization of energy and reserves, it is found that although stochastic co-optimization is carried out by few researchers, addressing uncertainty due to renewables, the uncertainty due to probabilistic contingencies, is rarely considered. The research handling of both uncertainties, i.e., the uncertainty due to renewables and the uncertainty due to probabilistic contingency while performing stochastic co-optimization of energy and reserves, is hardly found as it is very complex. Worldwide, system operators are facing challenges in handling the complex problem of the uncertainty modeling of renewable resources and contingencies while co-optimizing the dispatch of energy with ramp reserves in normal and contingency conditions [12]. This issue can be solved by a systematic solution to the stochastic security-constrained economic dispatch which considers the uncertainty.

This research work contributes by addressing the issue of the stochastic security-constrained co-optimization of energy and reserves in the presence of uncertainty of resources along with probabilistic contingency occurrence, which is decisively needed.

This need is addressed in this research work, and it is novel since it focuses on:

- Solving multiperiod SSCED by the co-optimization of energy and ramping reserves in both normal and contingency conditions.
- Modeling the uncertainty of wind power output and solving the aforementioned complex problem using "probability-weighted scenarios and transition matrices" [13].
- Modeling probabilistic contingency conditions.
- Demonstrating the management of Fast Ramping (FR) resources, such as wind energy, and using its negative characteristic of power ramping into an advantageous one.
- Validating the impact of change in the flexibility (ramping rate) of generators on active power costs in normal and contingency conditions.

The structure of this paper is as follows. The typical structure of the proposed SSCED multiperiod problem structure is described in Section 2, and Section 3 describes the proposed methodology for solving the multiperiod SSCED problem. The solution of case studies, outcomes, and results are discussed in Section 4. Section 5 concludes this investigation and suggests potential directions for future research work.

2. Typical Structure of Proposed SSCED Multiperiod Problem Structure

The basic active power dispatch in this work is represented as follows for the single and multiperiod problem structure of the SSCED. The active power dispatch P, from a resource i, renewable energy generation scenario s, and contingency state c, in time period t, has the indices as shown in Figure 1.

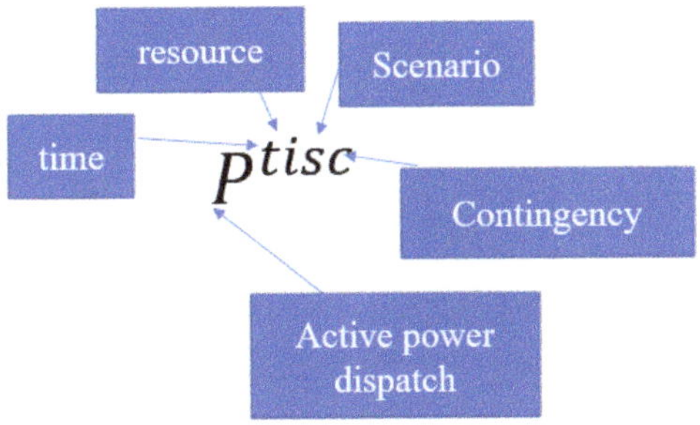

Figure 1. Nomenclature for active power dispatch in each scenario occurring in time period *t*.

2.1. Single Base Scenario (Single Renewable Generation Profile)

The structure of the proposed single-period SSCED problem with a single base generation scenario under the proposed strategy is illustrated by the diagram in Figure 2. for the time period $t = 1$, the yellow-colored box is the power flow scenario for the base case P^{1i10}, i.e., in the high-probability case in scenario $s = 1$, the dashed box shows the reference dispatch set P^{1ir}. The red circles represent the power flow scenario for "low probability contingency states 1, 2 . . . c" at time period $t = 1$, in generation scenario $s = 1$, for generator *i*, and are correspondingly P^{1i11}, P^{1i12} . . . . P^{1i1c}.

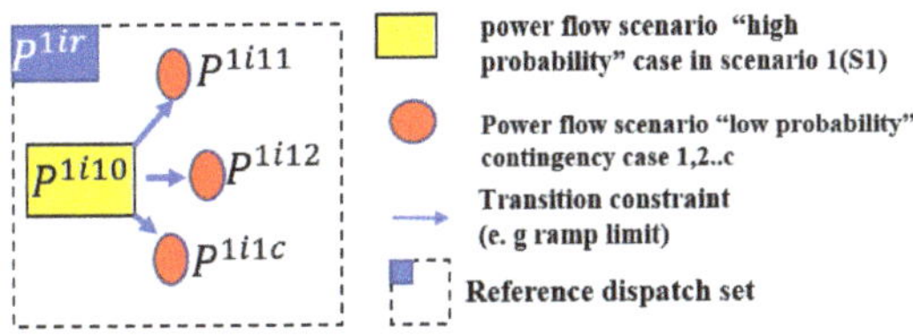

Figure 2. Typical problem structure of a single base scenario SSCED.

The ramp limits that prevent contingency state dispatches from deviating from the base case dispatch are shown by the blue colored arrows when contingencies 1, 2 . . . , and *c*, occur.

2.2. Multiple Base Scenario (Multiple Renewable Generation Profile)

Similarly, for stochastic security-constrained economic dispatch, there are multiple renewable energy generation profiles considered. This creates multiple base case scenarios in a single time period $t = 1$, as shown in Figure 3. Two generation profiles are creating two generation scenarios and three probabilistic contingencies, i.e., $c = 1, 2$, and 3.

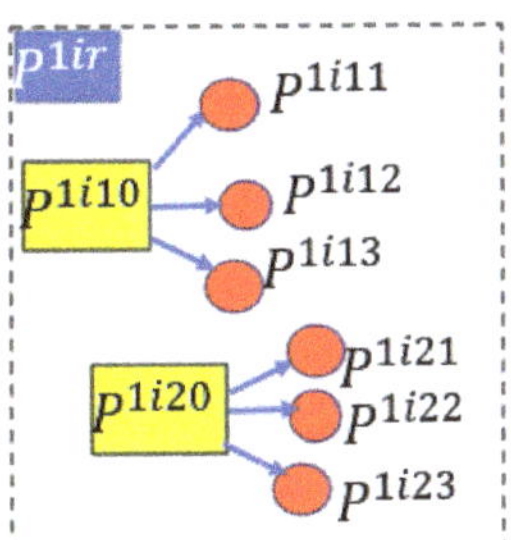

Figure 3. Problem structure for a single-period stochastic economic dispatch with multiple base case scenarios.

In the proposed stochastic multiperiod problem structure, there is a commitment schedule that is allotted to all states for simulating a variety of scenarios in the base case and contingency states for multiple periods, as shown in Figure 4.

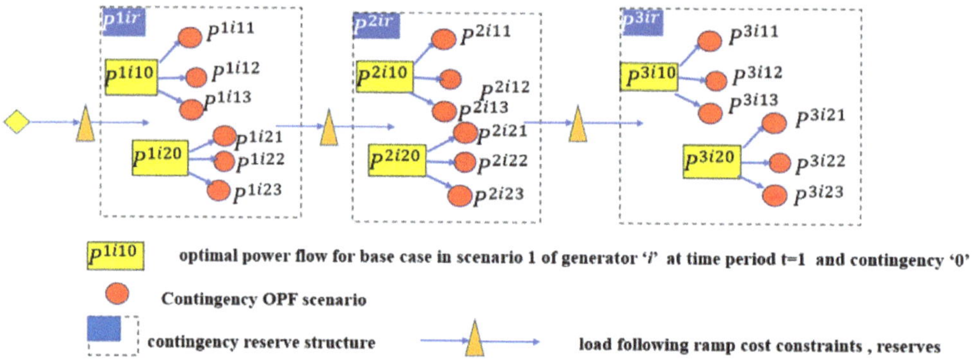

Figure 4. Stochastic security-constrained economic dispatch for multiple periods: its problem structure.

It is more challenging to extend a problem over multiple periods. In order to ensure that the contingency state dispatches, the economic dispatch decisions are made before the occurrence of contingencies from the base case power flow scenario while adhering to ramp rate limits (blue colored arrows), as seen in Figure 4. Since the uncertainty is actually revealed period by period, this issue can be resolved by adopting a "Markovian structure" [14], where a stochastic model is characterized as a succession of potential events, where each event's probability is solely dependent on the state that was attained in the previous event. Transitions describe how the status of the system changes. "Transition probabilities" are the probabilities of odds events (contingencies if any) attached to different state transitions. The process is characterized by a state space, a "transition matrix" describing the probabilities of particular transitions, and an initial state (or initial distribution) across the state space.

2.3. *Load-Following Reserve Ramping under Uncertainty Using 'Central Path'*

In normal conditions, ramping feasibility is only enforced on this high-probability 'central path' (Blue colored), as shown in Figure 5, in which all possible transitions are constrained to be feasible with respect to the physical ramping capabilities of generators as well as any load-following reserve capacity. For a high-probability central path, transitions from a small set of base scenarios in one period to a small set of base scenarios in the next period are described by a transition probability matrix.

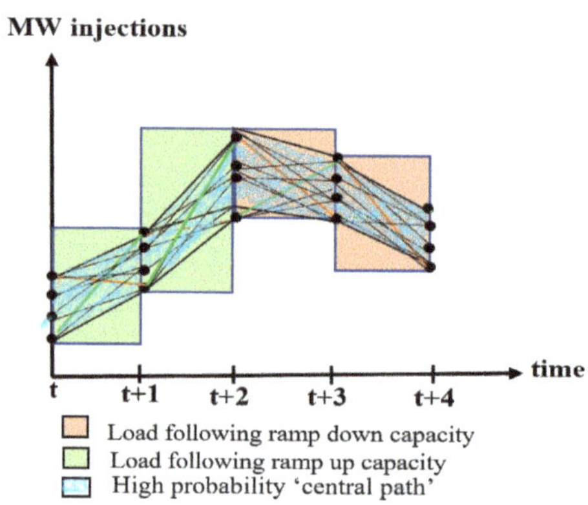

Figure 5. Load-following ramping reserves and ramping in the normal state [15].

A simple "wear and tear" cost is defined for load-following reserve/contingency reserve dispatch from one period to another, applied as a probability-weighted cost to each possible transition, as well as ramp-up and ramp-down load-following ramping reserve costs. These reserve costs apply to maximum upward and downward transitions included in the central path scenarios, as illustrated in Figure 5.

2.4. Contingency Condition Reserve Ramping during Uncertainty

In the proposed SSCED problem, as shown in Figure 6, the uncertainty of generation from renewables is modeled as generation scenarios $s = 1$, $s = 2$, using transition probability matrices. The reserves are defined by the maximum redispatch deviations across all uncertain generation scenarios and contingencies having the physical ramp rates limit as (PR^i_+) and (PR^i_-). The maximum upward deviations (cr^{ti}_+) and downward deviations (cr^{ti}_-) of contingency reserves in contingency cases (c = 1,2 . . . .) from the base case within each scenario (s) at time instant t is given by $P^{tis_1}_-$, $P^{tis_2}_+ \ldots$, etc., as seen in Figure 6.

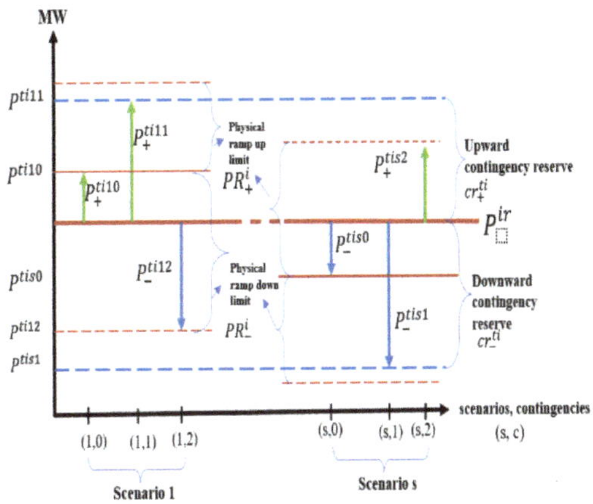

Figure 6. Reserve structure for generator i, for several contingencies c, in presence of uncertainty scenarios s.

For the multiperiod problems, the reserve structure is as shown in Figure 6; the variables include a t index as well since the same structure is used at each period t in the multiperiod problems.

3. Proposed Methodology for Solving Multiperiod SSCED Problem

3.1. Proposed Methodology for Solving SSCED considering the Uncertainty of Renewable Generation and the Uncertainty of Contingency Occurrence

The problem structure is solved using the proposed methodology for SSCED. This methodology follows the procedure shown in the flowchart in Figure 7.

The flowchart in Figure 7 shows the methodology proposed for multiperiod stochastic security-constrained economic dispatch, for a 24 h time period.

The flowchart shows that first, for the base case (without contingency), the initialization of time period $t = 1$, generation scenario $s = 1$, and as there is no contingency, i.e., $c = 0$ is carried out.

All network data are read as it includes generator data, generator ramping limits, renewable energy generation profiles, contingency data (if any), associated generation cost data, and reserve cost data. Then, system data are read for $t < 24$, and the initial base case probability for each generation profile (scenario) is set to 1. The maximum time period (T) of analysis considered in this work is T = 24. The maximum number of generation scenarios (S^{tmax}) in time period t of the analysis are initialized, the maximum number of probable contingencies in time period t and scenario s is C^{tsmax} is initialized. This methodology is based on solving the master problem, solved at each time step t.

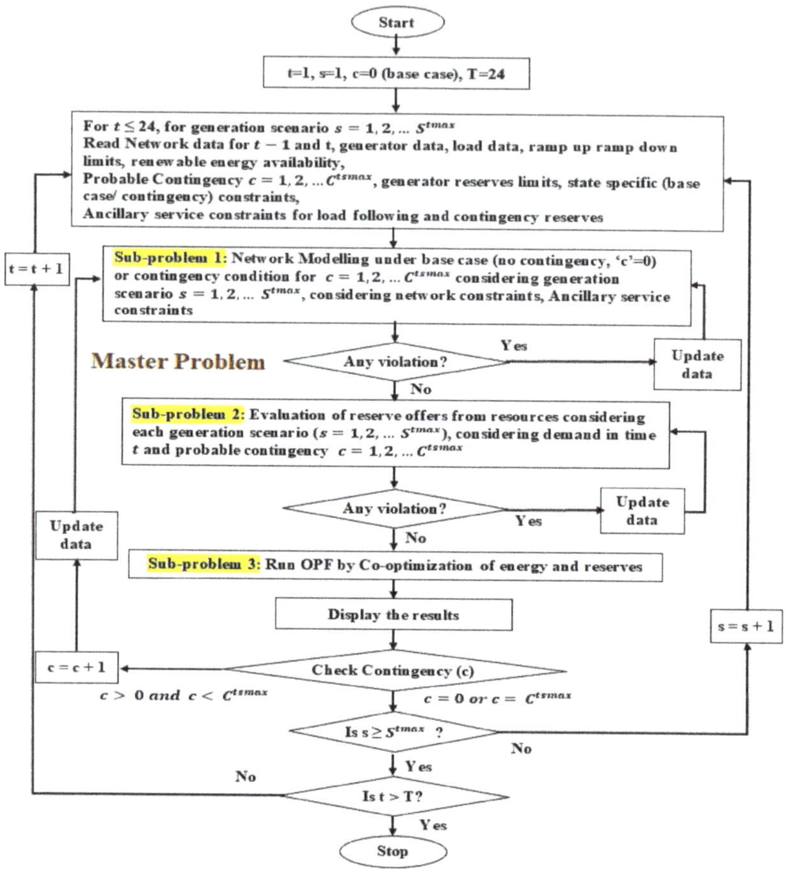

Figure 7. Flowchart for multiperiod stochastic security-constrained economic dispatch (SSCED).

The **Master problem** consists of three sub-problems.

The three subproblems are as follows:

- **Sub-problem 1:** Network modeling.
- **Sub-problem 2:** The evaluation of the reserve offer from resources considering the uncertainty of generation scenarios and probable contingency.
- **Sub-problem 3:** Objective function cost minimization by OPF Co-optimization of energy and reserves.

Initially, the energy and reserve dispatch are executed using the "master problem" for base cases (c = 0) in each generation scenario in time period t until $s \geq S^{tmax}$. The results are displayed for time period t, and the contingency condition is checked. If there is no contingency, c = 0, and generation scenarios are $s > 1$, then the generation scenario counter is incremented as $s = s + 1$ and the process repeats until $s \geq S^{tmax}$, i.e., the generation scenario number s becomes greater than, or equal to S^{tmax}, the maximum generation scenario S^{tmax} in time period t.

In this work, for the stochastic approach using more than one generation profile, i.e., $s > 1$ is considered. In stochastic security-constrained OPF, for the co-optimization of energy and reserves, the probabilistic contingencies are also considered. The contingency counter is greater than 0 ($C > 0$), and with multiple generation scenarios, the scenario counter s is also greater than 1 (i.e., $s > 1$). The maximum number of contingencies in a single time period t and generation scenario s is of analysis C^{tsmax}.

In subproblem 1, network modeling is carried out as described in Section 3.2. In this subproblem, the cost model for active power from conventional sources is carried out using Equation (1). Equation (2) gives the output from a WECS and its active power cost model is given in Equations (3) and (4). The uncertainty of renewable energy generation scenarios from one period to the next period is modeled by "probability transition scenarios" given by the "probability transition matrix" [16] in Equation (5). The approximation of uncertainty of the state of the system, i.e., the base case or state of contingency, is defined using Equations (6)–(9). If there are any violations, the data are updated so that the state of the system can be decided. The subproblem is executed with the updated data until there are no violations. The program then proceeds further to subproblem 2, for ramping evaluation. In this subproblem, the ramping reserve considers generation scenario s and contingency c, $c = 0$, for the base case (normal case) and $c > 0$ for a contingency case. For a high-probability central path, transitions from base scenarios in one period to base scenarios in the next period are given in Equation (10) to Equation (14) for load-following reserves. If the state of the system in the subproblem is a contingency state, then the maximum upward and downward contingency reserve is as explained in Equations (15) and (16), and the amount of the contingency reserve constraints are given in Equations (17) and (18).

After modeling the system in subproblem 1, subproblem 2 is executed as explained in detail in Section 3.3. The ramping evaluation is executed considering the central path and various generation scenarios present in the problem formulation data. In subproblem 2, the ramping reserve requirement in the base case or contingency case is calculated. At each time period t, the ramping reserve requirement corresponds to the reserve quantity, which is calculated based on the demand to be satisfied and the available generation from resources. In a normal state (base case), the load-following reserves are subject to constraints given by Equations (10) and (11). the load-following ramp-up and ramp-down reserves are estimated using Equations (13) and (14).

The contingency reserves ramping constraints given in (15) and (16) are followed while injecting the contingency condition ramping reserves. The ramping limits on transitions from base to contingency cases are given in Equations (17) and (18). If there are any violations in the constraints, then the system data are updated and the loop is again executed until the ramping evaluation is properly executed.

Then, subproblem 3, is executed. This subproblem is discussed in detail in Section 3.4. In this subproblem, the execution of OPF by the co-optimization of energy and reserves is carried out. The objective function is stochastic security-constrained economic dispatch, with a minimization of the total cost of energy and reserves, as given in Equation (19). In the equation for normal operating conditions, i.e., the base case having no contingency, the objective function is as given in Equation (20), and in contingency conditions, the objective function is as given in Equation (21). The objective function in Equation (19) consists of five terms. Each of the terms is executed, as given in Equations (22)–(26), and are

1. The optimization of the expected cost of active power dispatch and redispatch.
2. The optimization of the expected cost of load-following ramping (wear and tear).
3. The optimization of the cost of load-following ramp reserves.
4. The cost of endogenous contingency reserves.
5. The optimization of no-load, startup, and shutdown costs.

These terms in the objective function are subject to network constraints given in Equations (27) and (28), the unit commitment schedule in Equation (29), startup and shutdown events, as explained in Equations (30)–(32), and minimum uptime and minimum downtime constraints, which are explained in Equations (33)–(35). The OPF results are displayed giving the economic dispatch results for energy and reserves, considering the uncertainty of the generation profile and the contingencies.

When the three subproblems have been executed, the contingency for scenario 1 is checked; if the contingency number is less than the maximum contingency, i.e., $c < C^{tsmax}$, then the contingency number c is incremented by 1, i.e., $c = c + 1$, and then the data are updated. This updated data are utilized for executing the three subproblems and again the results are displayed considering contingency no. 2. This process continues until all of the contingencies for scenario 1 are addressed and the contingency counter c is equal to the maximum number of contingencies C^{tsmax} in scenario s, i.e., $c = C^{tsmax}$. If there is more than 1 generation profile, the scenario counter is incremented by 1, i.e., $s = s + 1$. The data are read considering generation profile no. 2 and the three subproblems are then executed. The program remains in the loop until all contingencies and dispatch decisions are displayed for scenario 2. The flowchart is generalized for the maximum number of scenarios S^{tmax}. Thus, the master problem is executed considering all contingencies for each generation scenario until $s \geq S^{tmax}$, for the time period t. Once $s \geq S^{tmax}$, the time period t is then incremented by 1, i.e., $t = t + 1$, and again, the data are read for the next time period and the loop continues to be executed until $t = 24$.

3.2. *Subproblem 1: Network Modeling*

In subproblem 1, network modeling is carried out. At each time step, as the system data are read, the available generator, its energy and reserve output, load as well as network data are updated. At each time step t, the generation and demand balance are evaluated considering the limitations on the power flow lines and DC power flow constraints. The modeling of active power output from CHP-based generators and active power from WECS output is carried out, as discussed in Sections 3.2.1 and 3.2.2. The cost of active power from these generators is taken into consideration. Modeling of the uncertainty of active power generation from WECS is carried out by probability-weighted scenarios, as discussed in Section 3.2.3, and the mathematical modeling of the probabilistic transition of states from the base case to the contingency state is carried out, as explained and discussed further in Section 3.2.4.

3.2.1. Cost of Active Power from CHP-Based Generators

The function of the total fuel cost for CHP-based generators (\$/h), such as natural gas-based and biogas-based generators in the system, can be presented as a second-order quadratic polynomial, as follows [17]:

$$C\left(P^{ij}\right) = \sum_{i=1}^{I}\left(a + bP^{i} + cP^{i^2}\right)\cdot(\$/\text{h}) \tag{1}$$

where $C(P^{ij})$ is the cost of active power production from the generator, i is the maximum number of generators, a (\$), b (MW), and c (\$/MW^2) are cost coefficients of the active power production function, and P^{ij} is the active power produced in MW by generator i at bus j.

3.2.2. Cost of Wind Energy Conversion System (WECS)

The wind turbine's output power and wind speed are related to the wind energy conversion system, as given in Equation (2). The wind speed is monitored every 10 m in height.

The following equation [18,19] can be used to express how the output power of a wind energy conversion system (WECS) relates to other factors:

$$P^{tw} = \begin{cases} 0, & W_w < W_{cin} \\ \frac{W_w - W_{cin}}{W_{Nor} - W_{cin}} * P^{twrated}, & W_{cin} \leq W_w \leq W_{Nor} \\ P^{twrated}, & W_{Nor} \leq W_w \leq W_{CO} \\ 0, & W_w \geq W_{CO} \end{cases} \tag{2}$$

where P^{tw} is the wind turbine output in MW at an instant (t), actual wind speed is W_w in m/s, $P^{twrated}$ is the rated power of the WECS, W_{cin}, W_{Nor}, and W_{CO} are the cut-in speed, rated speed, and cut-out speed, respectively.

The active power cost from the WECS consists of the following cost components

$$\sum_w C_w \left(P^{tw}\right) = \sum_w \left(C_{w,cc} + C_{w,op}\right) P^{tw} + C_{w,dp} \quad (\$/\text{h}) \tag{3}$$

where $C_{w,P}\left(P^{tw}\right)$ = the cost function of active power from the WECS in (\$/h);

$C_{w,cc}$ = the capital cost (\$/MWh); $C_{w,op}$ = the operation and maintenance cost (\$/MWh); $C_{w,dp}$ = the depreciation cost of the wind turbine w (\$/h).

For the wind turbine used in this analysis, the resource $i = w$ (4)

For uncertainty modeling, two wind profiles are considered for the system under study, as shown in Figure 8. Uncertainty modeling for wind power output scenarios by probability-weighted scenarios is discussed further in Section 3.2.3.

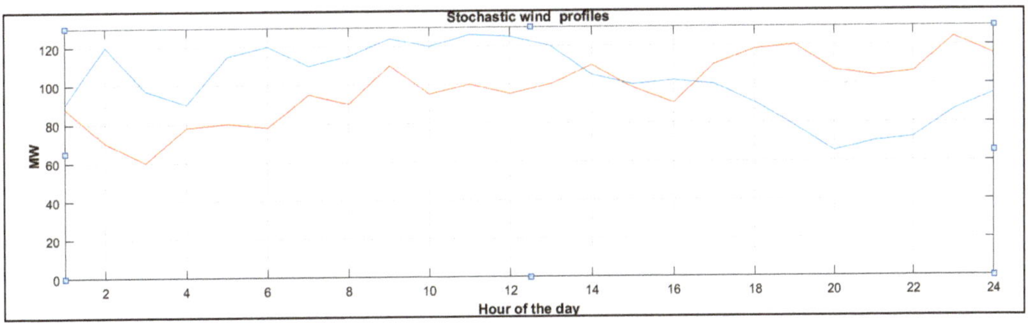

Figure 8. Stochastic wind power generation: two wind scenarios considered for analysis.

3.2.3. Modeling the Uncertainty of Wind Power Output Scenarios Using Probability-Weighted Scenarios

In this study, the joint distribution of uncertain parameters is used to generate scenarios that are then added to the problem structure as "multiple probability-weighted" [15] base cases to describe the uncertainty of wind power generation. Utilizing "probability matrices", as shown in Equation (5), with the full transition of generating scenarios, when periods of transition between low, average, and high-probability wind scenarios are permitted, the more general instance of SSCED can be accomplished.

$$\Phi^t = \begin{bmatrix} \phi^{t11} & \phi^{t12} & \cdots & \phi^{t1n_{St-1}} \\ \phi^{t21} & \phi^{t22} & \cdots & \phi^{t2n_{St-1}} \\ \vdots & \vdots & \ddots & \vdots \\ \phi^{tn_{St}1} & \phi^{tn_{St}2} & \cdots & \phi^{tn_{St}n_{St-1}} \end{bmatrix} \tag{5}$$

Here, in the stochastic model with multiple base scenarios per period, the transition matrix is a cell array of length n_t containing the transition probability matrices Φ^t of (5). That is, transmat{t} contains the $n_{S^t} \times n_{S(t-1)}$ matrix of transition probabilities from period $t-1$ to period t. The first element of the transmat matrix is a column vector of transition probabilities from period 0, (n_{S^t} = 1) to period 1. The columns of Φ^t sum to 1 and its coefficients are used to weight the wear and tear costs of ramping for load-following reserves and contingency reserves [13–15].

3.2.4. Mathematical Modeling of Probabilistic Transition of States from Base Case to Contingency State [16,20]

The individual state, i.e., scenario-based state and contingency state, is defined by state-specific probabilities ψ^{tsc} for period t, and can be derived from those in period $(t-1)$ in two steps [16].

Step 1: Determine the probability γ^{ts} that scenario s for s = 1, s = 2, etc., or any of its aforementioned contingencies will take place at a time t by using the formula

$$\begin{bmatrix} \gamma^{t1} \\ \gamma^{t2} \\ \vdots \\ \gamma^{tn_{S^t}} \end{bmatrix} = \Phi^t \begin{bmatrix} \psi^{(t-1)10} \\ \psi^{(t-1)20} \\ \vdots \\ \psi^{(t-1)n_{S^{t-1}}0} \end{bmatrix} \tag{6}$$

where

$$\gamma^{ts} = \sum_{c \in C^{ts}} \psi^{tsc} \tag{7}$$

Step 2: Since the sum of conditional probabilities ψ_0^{tsc} of contingencies c is 1, we simply scale each by the corresponding γ^{ts} to obtain the correct state-specific probabilities

$$\psi^{tsc} = \gamma^{ts}\, \psi_0^{tsc} \tag{8}$$

$$\gamma^t \equiv \sum_{s \in S^{t-1}} \psi^{(t-1)s0} = \sum_{s \in S^t, c \in C^{ts}} \psi^{tsc} < 1,\ t > 1 \tag{9}$$

The probability of transitioning to scenario $s = 2$ in period t given that scenario $s = 1$ was realized in period $t-1$ is assumed to be a known value $\phi^{ts_1s_2}$.

In this work, the transition can be explained, as shown in Figure 9. For multiperiod problem structures with contingencies, it is assumed that each period starts from the initial state of t = 0. The initial phase has a probability equal to 1. For period t, the base case according to renewable generation scenarios s = 1, s = 2 with probability $\Phi^{ts_2s_1}$ and the probability of being in a base case state for each generation scenario without a contingency state is the probability of ψ^{ts0}. Scenario 1 and scenario 2, in the system, might transit to any of a number of states in period t + 1 with probability γ^t. There are two types of uncertainty considered in this proposed problem structure—contingencies and parameter uncertainty (stochastic renewable energy generation). For this a multistage decision approach with scenario recombination and scenario trimming is considered to avoid the exploding number of scenarios.

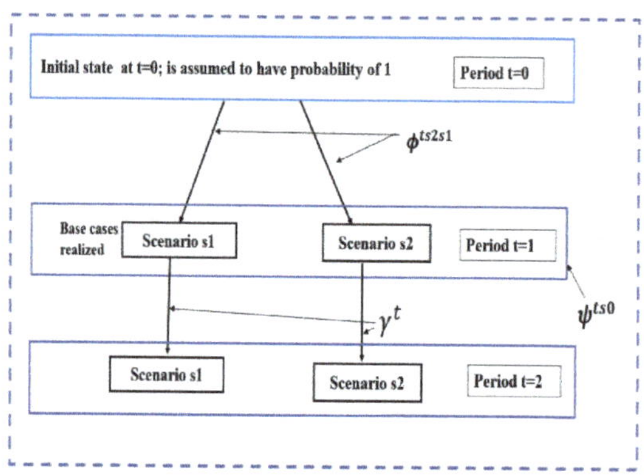

Figure 9. Probability transition states in the multiperiod stochastic economic dispatch problem.

As seen in Figure 9, the base case scenario to contingency states in period (t = 2) depends on transition probabilities. Additionally, the transition from base to contingency states, i.e., state-specific probabilities are used to model the probabilistic contingency occurrence.

3.3. Subproblem 2: Ramping Evaluation

In this subproblem 2 of the proposed master problem of SSCED, the reserves are defined by the maximum redispatch deviations across all scenarios and contingencies and the physical ramp rates limit. At each time period t, the ramping reserve requirement corresponds to reserve quantity, which is calculated based on the demand to be satisfied and available generation from CHP generators, its ramping capacity, and the ramping flexibility offered [21,22] by WECS considering its uncertainty.

3.3.1. Mathematical Modeling of Load-Following Ramping Reserves in Normal (Base Case-No Contingency) State

To minimize the mismatch between generation and demand in normal (base case) conditions, the ramp-up (δ_{+}^{ti}) and ramp-down (δ_{-}^{ti}) of the ramping units is subjected to respective ramp-up and ramp-down constraints, as given in Equations (10) and (11) [16].

$$0 \leq \delta_{+}^{ti} \tag{10}$$

$$0 \leq \delta_{-}^{ti} \tag{11}$$

In normal conditions, the load-following up (δ_{+}^{ti}) /down (δ_{-}^{ti}) ramping reserves are injected. The load-following ramp reserve definition is given in Equations (13) and (14) [16].

Load-following ramp reserve definition:

$$\forall \left\{ t \in T,\ i \in I^{tsc},\ s_1 \in S^{t-1},\ s_2 \in S^{t} \left| \zeta^{ts_2s_1} = 1 \right. \right\} \tag{12}$$

$$p^{(t-1)is_10} - p^{tis_20} \leq \delta_{-}^{(t-1)i} \tag{13}$$

$$p^{tis_20} - p^{(t-1)is_10} \leq \delta_{+}^{(t-1)i} \tag{14}$$

The ramping reserves wear and tear costs are defined as a simple quadratic "wear and tear" cost from the difference in a dispatch from one period to another, applied as a probability-weighted cost to each possible transition, as given in Equation (5), as well as up and down load-following ramping reserve costs. These reserve costs apply to maximum upward and downward transitions included in the central path scenarios.

3.3.2. Mathematical Modeling of Contingency Reserves Ramping in Contingency

There are now two types of uncertainty, contingencies and parameter. The reference power dispatch and reserve dispatch decisions from each generator are made before the uncertainty is realized, but the base case or contingency state specific dispatch decisions are based on the type of uncertainty. The contingency reserves ramping constraints are given in Equations (15) and (16) and are followed for injecting the ramping reserves in contingency conditions. The total ramping service provided by all generators is equal to the ramping required by the demand. The ramp-up and ramp-down will be such that it will be the maximum ramping that can take place according to the full transition probabilities-enabled central path, as described in Equations (17) and (18) [16,20].

- Reserve, redispatch, and contract variables:

$$0 \leq p_{+}^{tisc} \leq cr_{+}^{ti} \leq cr_{max+}^{ti} \tag{15}$$

$$0 \leq p_{-}^{tisc} \leq cr_{-}^{ti} \leq cr_{max-}^{ti} \tag{16}$$

- Ramping limits on transitions from base to contingency cases:

$$P^{tisc} - P_r^{ti} = P_+^{tisc} - P_-^{tisc} \tag{17}$$

$$-PR_-^i \leq P^{tisc} - P^{tis0} \leq PR_+^i, \quad c \neq 0 \tag{18}$$

3.4. Subproblem 3: Objective Function Cost Minimization by OPF Co-Optimization of Energy and Reserves

OPF by Co-optimization of energy and reserves is executed in this subproblem. The following objective function is considered in this subproblem.

Objective function: multiperiod stochastic security-constrained economic dispatch of energy and reserves has the following objective function $f(x)$ for the minimization of the total cost of energy and reserves comprising five components, as given in Equation (19) [16]

$$\min f(x) = f_P(P, P_+, P_-) + f_\delta(P) + f_{lf}(\delta_+, \delta_-) + f_{cr}(cr_+, cr_-) + f_s(u, v, w) \tag{19}$$

where $f(x)$ is comprised five components, and consideration of each component depends upon the normal state or contingency state of the system. The first term is generators active power dispatch and redispatch cost, the second term is load-following ramping wear and tear cost, the third term is expected load-following ramp reserve cost, the fourth term is endogenous contingency reserve cost (optional), and the fifth term (last term) is a startup and shutdown cost.

- In the normal state, the objective function becomes:

$$\min f(x) = f_P(P, P_+, P_-) + f_\delta(P) + f_{lf}(\delta_+, \delta_-) + f_s(u, v, w) \tag{20}$$

- In the contingency state, the objective function considered is as follows:

$$\min f(x) = f_P(P, P_+, P_-) + f_{cr}(cr_+, cr_-) + f_s(u, v, w) \tag{21}$$

Each of the five components is expressed in terms of the individual optimization variables as follows.

1. Optimization of the expected cost of active power dispatch and redispatch ($/h)

$$f_P(P, P_+, P_-) = \sum_{t\in T}\sum_{s\in S^t}\sum_{c\in C^{ts}} \psi_\alpha^{tsc} \sum_{i\in I^{tsc}} \left[C_P^{ti}\left(P^{tisc}\right) + C_{P+}^{ti}\left(P_+^{tisc}\right) + C_{P-}^{ti}\left(P_-^{tisc}\right)\right] \tag{22}$$

   where $C_P^{ti}\left(P^{tisc}\right)$ is the cost function for active power injected by unit i at time t. $C_{P+}^{ti}\left(P_+^{tisc}\right)$ is the cost function for upward deviation from the reference active power. $C_{P-}^{ti}\left(P_-^{tisc}\right)$ is the cost function for downward deviation from the reference active power quantity for unit i at time t for scenario s and contingency c.
2. Optimization of the expected cost of load-following ramping (wear and tear) ($/h)

$$f_\delta(P) = \sum_{t\in T} \gamma^t \sum_{\substack{s_1 \in S^{t-1} \\ s_2 \in S^t}} \phi^{ts_2s_1} \sum_{i\in I^{ts_20}} C_\delta^i\left(P^{tis_20} - P^{(t-1)is_10}\right) \text{ (\$/h)} \tag{23}$$

In the above equation, C_δ^i gives the ramping cost for a time period (t), of unit i, when the system is in the normal state, i.e., contingency $c = 0$, and there is a transition of the system from scenario 1 (s_1) in time period ($t - 1$) to scenario 2 (s_2) in time period t.

Ramping reserves are procured on the difference between the dispatches for adjacent periods ($t - 1$) having scenario 1, to a period (t) having scenario 2.

3. Optimization of cost of load-following ramp reserves ($/h)

$$f_{lf}(\delta_+, \delta_-) = \sum_{t\in T} \gamma^t \sum_{i\in I^t} \left[C_{\delta+}^{ti}\left(\delta_+^{ti}\right) + C_{\delta-}^{ti}\left(\delta_-^{ti}\right)\right] \text{ (\$/h)} \tag{24}$$

where,
$C^{ti}_{\delta+}$ is the cost of upward load-following ramp reserve for unit i at time t for transition to time $t+1$ in \$/MWh.
(δ^{ti}_{+}) is the upward load-following ramp reserve in MW.
$C^{ti}_{\delta-}$ is the cost of downward load-following ramp reserve for unit i at time t for transition to time $t+1$ in \$/MWh.
(δ^{ti}_{-}) is the downward load-following ramp reserve in MW.

4. Cost of endogenous contingency reserves in (\$/h)

$$f_{cr}(cr_{+},\ cr_{-}) = \sum_{t\in T} \gamma^{t} \sum_{i\in I^{t}} \left[C^{ti}_{R+}\left(cr^{ti}_{+}\right) + C^{ti}_{R-}\left(cr^{ti}_{-}\right)\right]\ (\$/\text{h}) \tag{25}$$

where,
$C^{ti}_{ac+}\left(cr^{ti}_{+}\right)$ = the cost function for the upward contingency reserves purchased from unit i at time t in \$/h.
$C^{ti}_{ac-}\left(cr^{ti}_{-}\right)$ = the cost function of the downward contingency reserve purchased from unit i at time t.

5. Optimization of no-load, startup, and shutdown costs in (\$/h)

$$f_{s}(u,\ v,\ w) = \sum_{t\in T} \gamma^{t} \sum_{i\in I^{t}} \left[C^{ti}_{P}\ (0)\ u^{ti} + C^{ti}_{v} v^{ti} + C^{ti}_{w} w^{ti}\right]\ (\$/\text{h}) \tag{26}$$

where,
$C^{ti}_{P}\ (0)\ u^{ti}$ = the no-load cost function for unit i time t in \$ per startup/shutdown.
$C^{ti}_{v}\ v^{ti}$ = the shutdown cost function for unit i time t in \$ per startup/shutdown.
$C^{ti}_{w}\ w^{ti}$ = the shutdown cost function for unit i time t in \$ per startup/shutdown.

This objective function is to be satisfied at each time step for multiple periods under consideration for analysis, modeled as discussed.

While running the OPF, the objective function is exposed to multiple generating units and network constraints. Uncertainty modeling is carried out in this subproblem to model various generation scenarios created by probabilistic-weighted functions, as given in Equation (5), and probabilistic endogenous contingency events, as given in Equation (6).

- Network operations Constraints:

The objective function in Equations (19) and (20) is subject to the following network constraints for all $t \in T$, all $s \in S^{t}$, all $c \in C^{ts}$, and all $i \in I^{tsc}$, beginning with the constraints that are separable by period.

Power balance equations:

$$\left(\theta^{tsc},\ V^{tsc},\ P^{tsc}\right) = 0 \tag{27}$$

Power flow limits, voltage limits, and any other OPF inequality constraints:

$$\left(\theta^{tsc},\ V^{tsc},\ P^{tsc}\right)\ \leq 0 \tag{28}$$

- Unit Commitment:

For stochastic problems modeling multiple scenarios and/or contingency states, there is a single commitment schedule shared by all states. That is, in the current formulation, a single binary variable is used to model the commitment for a given unit across all scenarios and contingencies in a given period.

- Generator injection limits and commitments:

$$u^{ti} P^{tisc}_{min}\ \leq\ P^{tisc}\ \leq\ u^{ti} P^{tisc}_{max} \tag{29}$$

- Startup and shutdown events:

$$u^{ti} - u^{(t-1)i} = v^{ti} - w^{ti} \tag{30}$$

$$0 \leq v^{ti} \leq 1 \tag{31}$$

$$0 \leq w^{ti} \leq 1 \tag{32}$$

- Minimum up and down times:

$$\sum_{y=t-T_i^-+1}^{t} v^{yi} \leq u^{ti} \tag{33}$$

$$\sum_{y=t-T_i^++1}^{t} w^{yi} \leq 1 - u^{ti} \tag{34}$$

$$u^{ti} = \{0, 1\} \tag{35}$$

4. Case Study and Results

4.1. System Configuration

The modified "IEEE 30 Bus system" [23,24] is considered for the implementation of the proposed SSCED. The grid-connected microgrid is assumed to have a natural gas-based plant at the slack bus, a local CHP-based biogas generation system, and 1-WECS. The data for the generators is given in Table 1. The schematic of the system is shown in Figure 10. For creating the uncertainty of renewable generation, two wind energy profiles [25] are considered for analysis, as shown in Figure 9. These profiles are obtained using predicted wind profiles in Pune, India. The W_{cin} is the cut-in velocity = 12 km/h, W_{CO} is the cut-out velocity = 90 km/h, W_{Nor} is the normal velocity at that instant, and W_w is the wind speed. The conversion efficiency is considered as 90%.

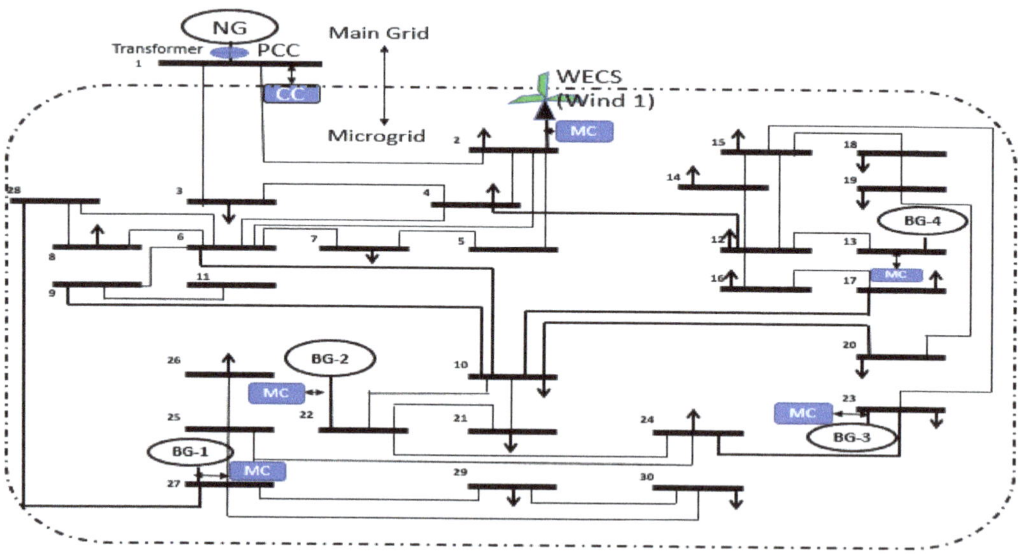

Figure 10. Modified IEEE 30 Bus system with a grid-connected microgrid.

The typical load profile is shown in Figure 11.

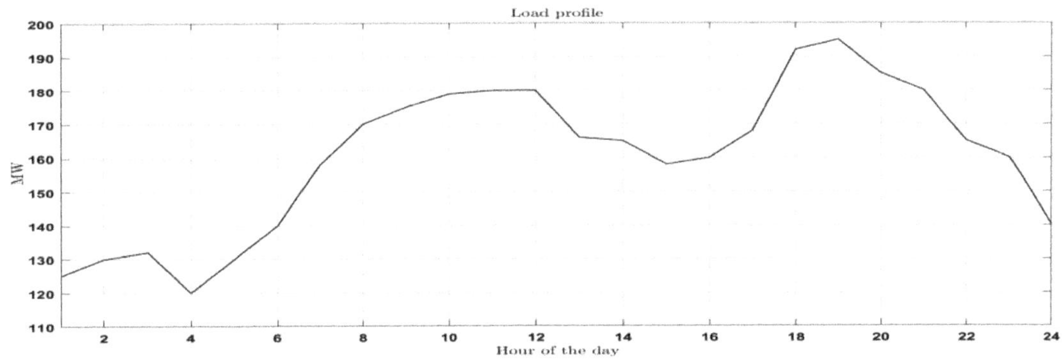

Figure 11. Typical load profile for Pune, India [26].

Table 1. Data for generators.

Bus Number	Generator Type (Fuel)	Pmax Pmin	$/MW	Ramping Capability [27]	Contingency Reserve Positive and Negative Active Reserve in MW
1	Natural gas	80 10	90	±5 to 8%	±10
2	WECS (Wind 1)	100 20	30	±30–60%	±30
27	Biogas Gen 1	55 10	80	±8 to 10%	±10
22	Biogas Gen 2	50 10	80	±8 to 10%	±10
23	Biogas Gen 3	30 10	80	±8 to 10%	±10
13	Biogas Gen 4	40 10	80	±8 to 10%	±10

4.2. Case Studies

1. **Case 1:** Base case—Normal State (load-following ramping reserve ancillary service)
2. **Case 2:** Contingency condition—Generator outage occurrence in Case 1, contingency reserve ancillary service
3. **Case 3:** Contingency condition—Underloading for 24 h (load decreased by 20% for all 24 h) of Case 1, contingency reserve ancillary service. These case studies are executed using Matpower Optimal Scheduling Tool (MOST) [16].

Assumptions:

- The voltages of DG units and PCC change.
- In the range of (5%), i.e., 95% to 105% of the rated value.
- All conventional and distributed energy sources have P/f droop control capability.
- Well-equipped communication infrastructure is present in grid-connected microgrids.
- The timescale considered for load-following ramp-up and ramp-down services reserves is 1 h for the simplicity of analysis.
- The frequency of the system is maintained at 50 Hz with ±0.5 Hz tolerance.

4.3. Results and Discussion

4.3.1. Case 1: Base Case—Normal State: Load-Following Ramping Reserve Ancillary Service

Power Generation in Case 1

After the execution of the master problem, the power generation obtained in the normal state of the system from all six generators including the WECS (Wind1) is shown in Figure 12. The uncertainty of the renewable energy output, i.e., stochastic wind energy generation, has two stochastic wind profiles and is modeled using probability-weighted scenarios. The scenarios are actuated such that the probability of transition of the output from WECS for scenario 1 of the generation profile is 0.1 and for scenario 2 of the generation profile is 0.8. The power generated considering the two generation profiles with the transition as per the transition probability values is shown in Figure 12.

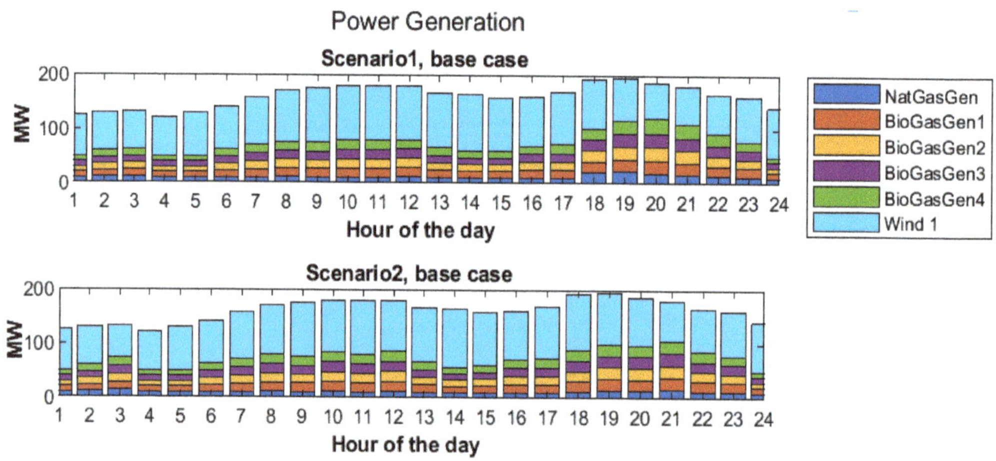

Figure 12. Power generation for scenario 1 and scenario 2: two base scenarios created using the transition matrices.

In normal conditions, for each ramping process in load-following, the wear and tear cost is considered from Equation (23), the cost of load-following ramping reserves is taken into consideration, as given in Equation (24), the constraint followed for load-following reserve is as per Equations (10) and (11), and the amount of load-following reserves is estimated by Equations (13) and (14).

The results obtained by the execution of proposed subproblem 2 of the master problem of SSCED are plotted in Figure 13a, showing the ramp-up reserves, and Figure 13b, showing the ramp-up reserve price. Similarly, Figure 13c shows the ramp-down reserves and Figure 13d shows the ramp-down reserve price, which is mainly due to the wear and tear costs associated with each load-following ramping.

As seen in Figure 13a, the ramping reserves follow the full transitions central path, which injects the high-probability ramping reserves from the WECS and other sources. The maximum ramp-up and ramp-down occur as per Equations (13) and (14).

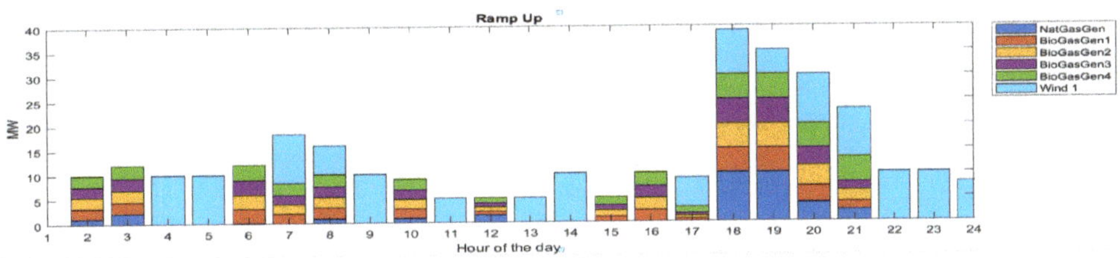

(a)

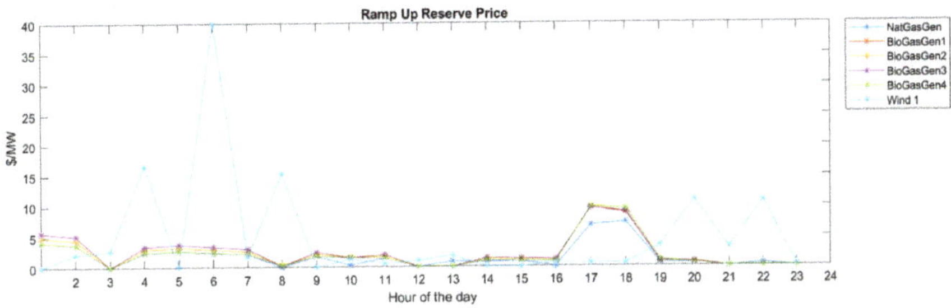

(b)

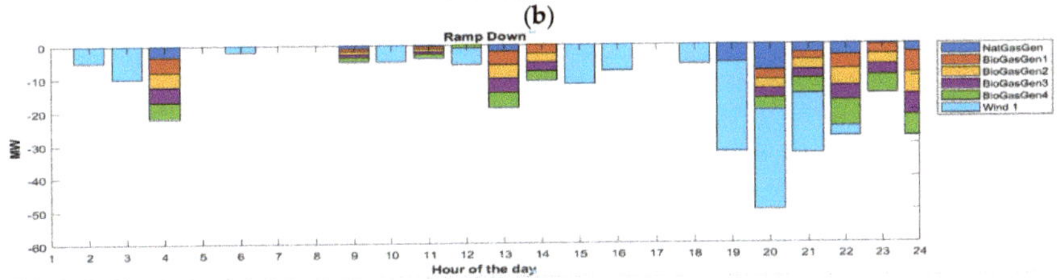

(c)

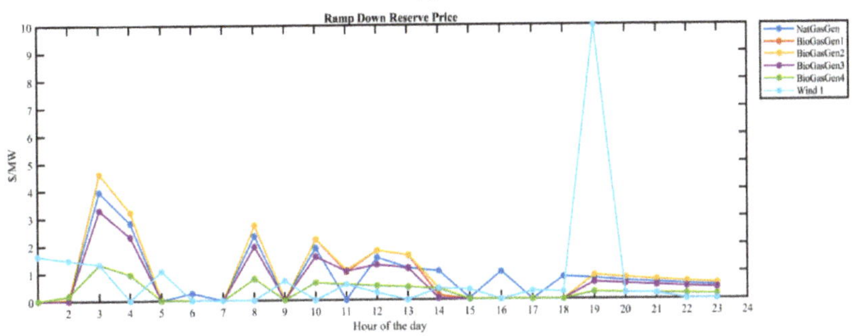

(d)

Figure 13. (**a**) Ramp-up reserves offered during load-following, (**b**) Ramp-up reserve price, (**c**) Ramp-down reserves offered during load-following by each resource, and (**d**) ramp-down reserve price.

Ramping Process in Case 1: Normal Condition (Base Case)

In normal conditions (base case), the load-following ramping can be explained in detail for each time period. For example, the load-following ramping process is explained considering the ramping period t = 3rd hour, and observed from Figure 13a,b, at period $(t-1)$ = 2nd hour and period $t+1$ = 4th hour. As discussed earlier, the ramping up/down of generators in time period t for unit i for transition to the time period $t+1$ from period t depends upon the power generated in time period $(t-1)$. Here, $t = 3$ if natural gas generator $i = 1$, the power generated in the base case in $(t-1)$ period = $P^{(3-1)110}$ = 11.18 MW. The power produced in scenario 2 in time period $t = 3$, i.e., P^{3120} is 13.38 MW. This natural gas generator ramps up by 2.19 MW in period t. A similar dispatch takes place for biogas generators no. 1, 2, and 3 and biogas generator no.4. For the wind generators (Wind 1), the generated power in period $t-1$ is 70 MW in scenario 1, and for scenario 2, it is 60 MW; therefore, it will ramp down by 10 MW, as shown in Figure 13c, and its price is shown in Figure 13d, according to Equations (13) and (14). This procedure is followed for each time step while solving the proposed SSCED master problem according to the updated data and system condition.

Operating Cost Minimization in Case 1: Normal Condition (Base Case)

The proposed SSCED gives the optimized objective function cost of \$9819/day. If the flexibility of wind energy is increased from 30% to 40%, the WECS injects power with more ramping capacity, and the overall objective function cost reduces to \$9057/day. The operation cost is minimized by almost 8.1%. It is, therefore, observed that the load-following ancillary service is efficiently and economically provided by the WECS, even with having an uncertain nature. By the use of the weighted-probability scenario method, its uncertainty is modeled properly, which is then converted into an advantage.

4.3.2. Case 2: Contingency Condition: Generator Outage Occurrence in Case 1, Contingency Reserve Ancillary Service

In this case, the contingency occurrence is simulated as a generator outage. Biogas generator no. 4, is the outage generator.

Power Generation in Case 2

SSCED is carried out in case 2 for probabilistic generator outage contingency conditions. This probabilistic contingency is modeled using a probability-weighted scenario and a state-specific probability function, as discussed earlier in subproblem 1.

As seen in Figure 14a, the power generation is shown for scenario 1 for the base case and contingency condition (generator outage-Gen 4), and the power generation for scenario 2, for the base case and contingency condition. Figure 14b shows the ramp-up reserves in MW, the WECS injects maximum ramp-up reserves along with the biogas-based Gen 1, Gen 2, and Gen 3, and helps to cope with the loss in generated power that was offered by biogas generator no.4, which is the outage generator.

Ramping Process in Case 2: Contingency Condition (Generator Outage)

For understanding the SSCED in the contingency condition ramping process in subproblem 2 of the master problem, the ramping at time period $t = 3$, $t-1 = 3-1 = 2$, i.e., time period 2 is considered. The two base case scenarios (scenario 1 and scenario 2) are actuated with a probability of transition of 0.1 and 0.8. For contingency cases, the fraction of the time slice is α, that is, the spent in the base case before the contingency occurs ($\alpha = 0$ means the entire period is spent in the contingency). The biogas-based generator 4 at bus 13 is in outage from $t = 1$, for the complete time period, as seen in Figure 14a. The contingency condition is actuated with $\alpha = 0$. The network constraints and contingency reserve constraints are followed while offering the contingency reserves, as seen in Figure 14b, with its price in Figure 14c, and maintaining the power supply to the load. Ramping limits on transitions from base to contingency cases follow Equations (15)–(18). As there are two renewable energy generation scenarios considered to model the uncertainty of renewable wind power generation, each generator will ramp-up/down to its maximum to achieve the central path of ramping according to the maximum difference between the power dispatched by that generator in contingency P^{tisc}, with the number of contingencies $c = 1$ and base case scenario P^{tis_0}, for scenario 1, and the same procedure will occur for scenario 2 power dispatch during contingency. For time period $t = 3$, $i = 1$ for the natural gas generator at bus no.1, in the base case, for generation scenario $s = 1$, the power produced by natural gas generator 1 = P^{3110} = 13.03 MW, and for the contingency condition $c = 1$, it produces P^{3111} = 13.03 MW. For generation scenario $s = 2$, for same time period $t = 3$, the power produced by generator number $i = 1$ in the base case = P^{3120} = 13.50 MW, and for the contingency condition, $c = 1$, in the same time period, $t = 3$, the generated power is $P^{3121} = 13.50$ MW. In this case, in order to get the maximum contingency reserve ramping, the reference dispatch for generator number 'i = 1' for natural gas generator, P_r^{31} is 13.50 MW. Therefore, according to Equation (17), $P^{3111} - P_r^{31} = -0.47$ MW; therefore, the generator will not ramp-up but will ramp down by 0.47 MW. The ramp-down reserve prices are shown in Figure 14e.

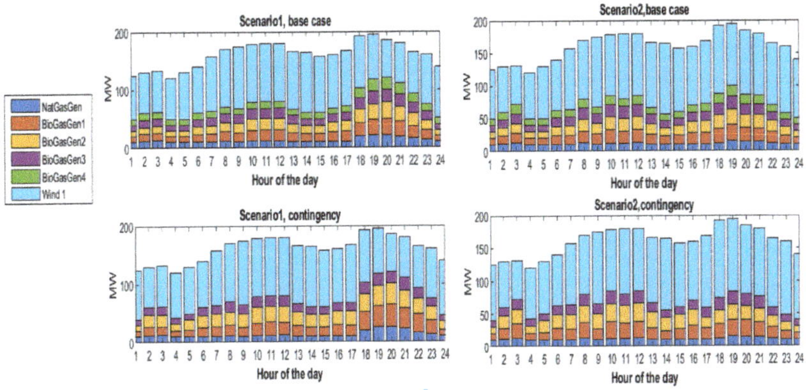

(a)

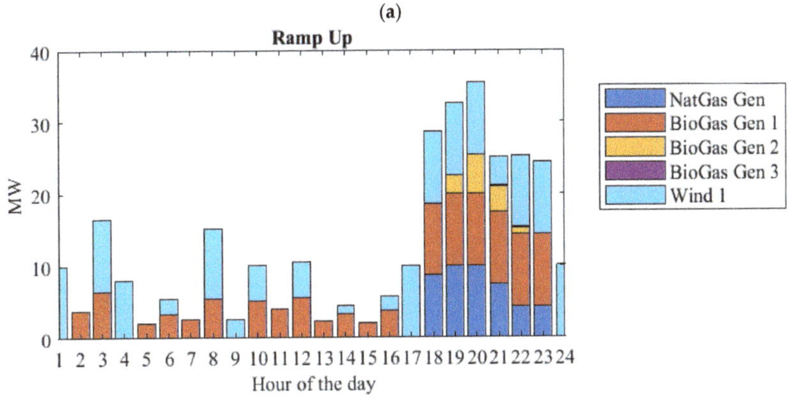

(b)

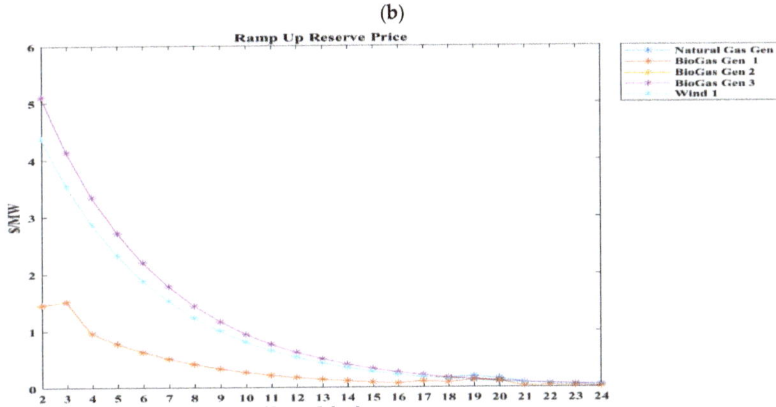

(c)

Figure 14. *Cont.*

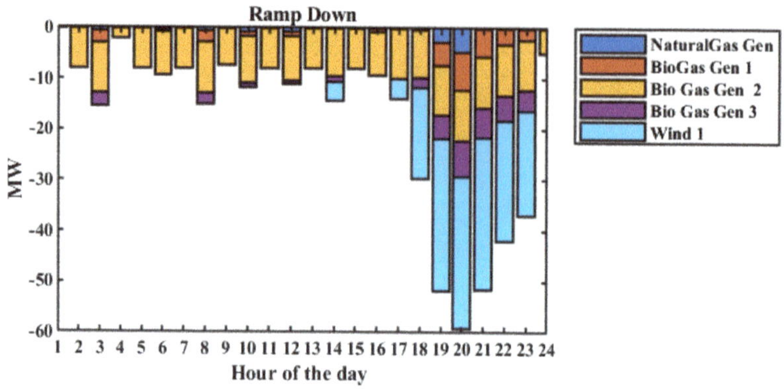

(d)

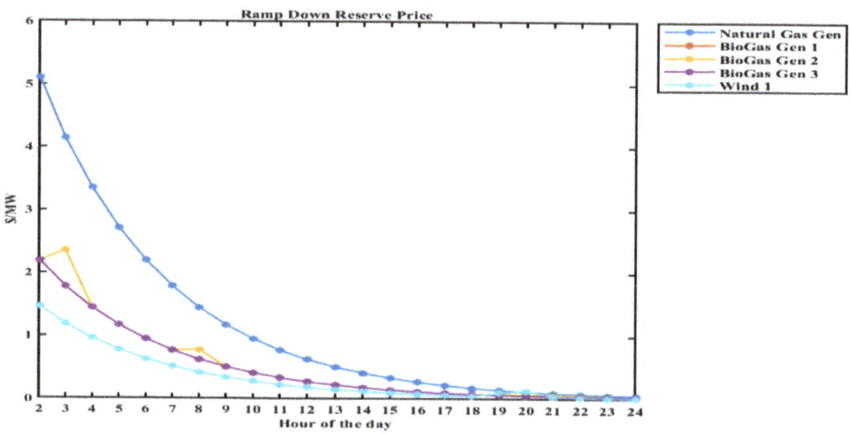

(e)

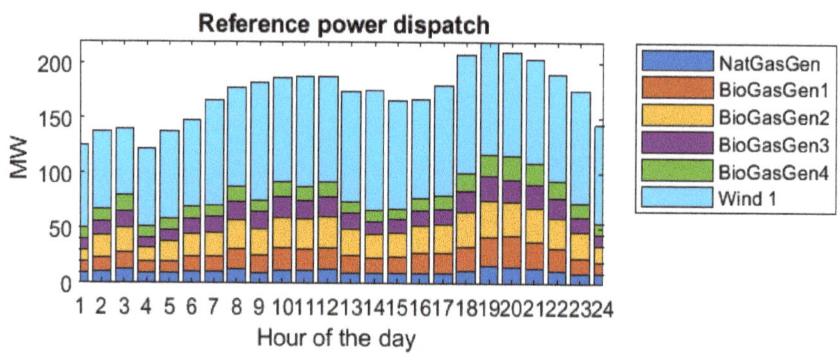

(f)

Figure 14. (**a**) Power generation for two base cases and respective probabilistic contingency cases, (**b**) ramp-up reserves injected during contingency, (**c**) ramp-up reserve price, (**d**) ramp-down reserve injected during contingency, (**e**) ramp-down reserve price, and (**f**) reference power dispatch during contingency.

It is to be noted that as per the reserve requirement, the generators provide both ramp-up and ramp-down contingency reserve quantities for the same time period $t = 3$. This can be explained as follows:

For same time period $t = 3$, for biogas-based generator 2, $i = 2$, in base case for scenario 1, $c = 0$, the power produced = P^{3210} is 12.3 MW, and for same time period $t = 3$, for biogas-based generator 2, $i = 2$ for contingency condition $c = 1$, P^{3211} = 14.72 MW.

For base case scenario 2, for $c = 0$, the power produced by the same generator is given by P^{3220} = 14.72 MW, and for the same scenario, contingency condition $c = 1$, the power generated is P^{3221} = 21.17 MW. Therefore, the maximum difference is observed for scenario $s = 2$ between the contingency and base case ramping up, i.e., 6.44 MW. The reference active power dispatch value of P_r^{32} is 14.72 MW, as seen in Figure 14f. This generator, therefore, ramps down at maximum for scenario 1 considering the power difference between probabilistic the contingency and the base case as 2.42 MW. As full trajectories in the transition probabilities are considered, the maximum ramping of reserves takes place. For all other generators, the same ramping process is followed to achieve the generation load balance in event of a loss of one of the generators. The WECS provides the maximum flexible ramping and, hence, is an ideal source for providing contingency spinning reserves. PR_+^i, PR_-^i are the upward/downward physical ramping limits for unit i for transitions from the base case ($c = 0$) to the contingency cases ($c = 1$). The upward/downward deviation from the active power (reference) quantity P^{tri} for unit i in the post-contingency state c of scenario s at time t is P_+^{tisc}, P_-^{tisc}. The reference dispatch is shown in Figure 14d.

Operating Cost Minimization in Case 2

After running the proposed SSCED successfully, the objective function cost is minimized to $27,622/day. The WECS flexibility to offer reserves is set to 30%. The expected energy cost is predominantly governed by the wind energy system. When the flexibility of the WECS is increased to 40% ramping capability, the minimized objective function cost then becomes $27,605/day. There is a reduction in objective function cost and the expected energy price also reduces.

4.3.3. Case 3: Contingency Condition: Underloading for 24 h (Load Decreased by 20% for All 24 h) of Case 1, Contingency Reserve Ancillary Service

Power Generation in Case 3

During probabilistic underloading contingency conditions, the SSCED is carried out using the transition probabilities and state-specific probability function. Figure 15a shows the power generated in the base case and contingency case for generation scenarios 1 and 2.

Ramping Process in Case 3: Contingency Condition

Figure 15b shows Ramp-up/down reserves injected and their prices during contingency. It shows that the WECS ramps down at its maximum capacity and maintains the system's stability. Figure 15c shows the reference dispatch utilized as a contingency reserve limit. For understanding security-constrained economic dispatch, in underloading contingency conditions, consider the time period $t = 3$ h. The load is reduced from 172 MW to 132 MW and the generators ramp-up/down accordingly to fulfill the reduced load. The active power generated by generator 1, i.e., the natural gas generator in the base case for scenario $s = 1$, $P^{3110} = 11.42$ MW, and the active power generated in the contingency scenario 1 = P^{3111} = 10 MW. At the same time period for generator 1 in the base case scenario $s = 2$, the power generated, $P^{3120} = 13.25$ MW, and for scenario $s = 2$, the contingency condition, generator 1 will produce P^{3121} = 10 MW, so the maximum ramping reserve obtained is in this condition; therefore, the generator will ramp down by 3.25 MW. This procedure is repeated for all generators including the WECS. The reference power dispatch is shown in Figure 15c.

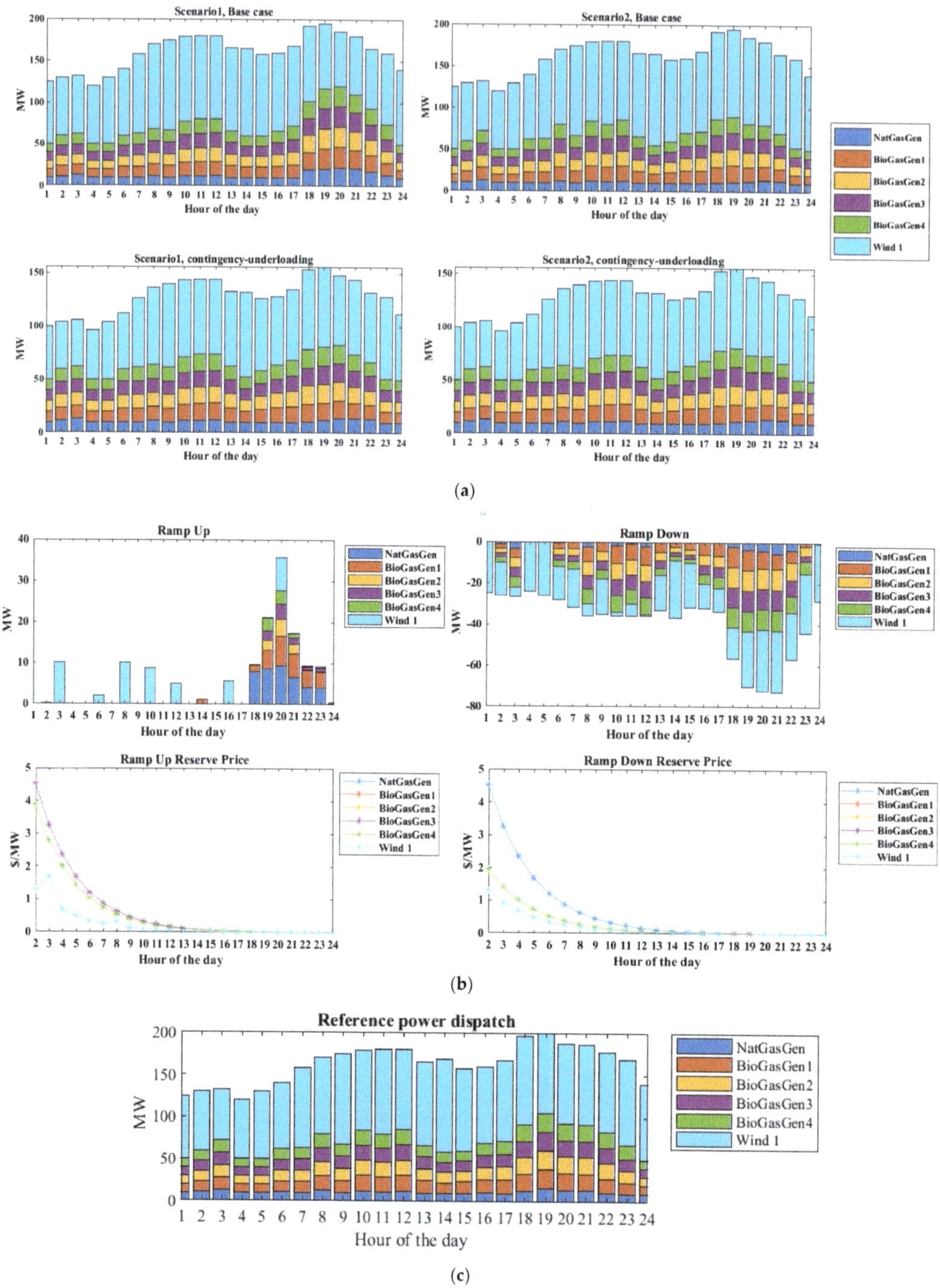

Figure 15. (**a**) Power generation for two base cases and respective probabilistic contingency cases created using transition probabilities. (**b**) Ramp-up/down reserves injected and their prices during contingency. (**c**) Reference power dispatch during contingency.

The maximum ramping estimated from (15)–(17), comparing the power generated in base case scenario 1, $s = 1$, and scenario $s = 2$ with their respective contingency states, is finally considered for ramping. Furthermore, the constraint of ramping limit in (18) is followed to obtain the final contingency ramp-up ramp-down value.

Operation Cost Minimization in Case 3

Operation costs are $18,623/day with the ramping capability of the WECS. The impact of resource flexibility on objective cost reduction is clearly observed from the results. If the flexibility of the WECS is increased to 40%, the overall objective function cost decreases to $18,598/day. As WECS is a renewable energy- based source, it is cheaper than other sources. Even if the WECS provides greater flexibility, the per-unit ramping reserve cost is less than the ramping reserve cost of other generators.

As observed from all three cases, the proposed SSCED is ideal for OPF with the co-optimization of energy and reserves in normal and contingency conditions considering uncertainty. It is also observed from Table 2 that if the flexibility of the WECS is increased to 40%, then the objective function cost, i.e., the cost of energy and reserves, reduces. Additionally, the system performance is improved in the case of contingencies, as the increased flexible ramping improves the generation load balance considering the low-cost generator, i.e., the WECS.

Table 2. Reduction in objective function cost.

Case	WECS Flexibility	Objective Function Cost ($/Day)	Increase in WECS Flexibility	New Objective Function Cost ($/Day)
Case 1	30%	9819	40%	9017
Case 2	30%	27,622	40%	27,605
Case 3	30%	18,623	40%	18,598

5. Conclusions

This work attempted to solve the stochastic security-constrained economic dispatch of load-following and contingency reserves ancillary service during uncertainty through the use of a grid-connected microgrid. An attempt was made to model the uncertainty of wind power output and the uncertainty of the probabilistic contingency conditions using "probability-weighted scenarios and transition matrices" [13]. The wind energy source is purposely included in this analysis to demonstrate its capability in providing a flexible ramping reserve.

Its flexibility is advantageous in normal and probable contingency conditions. The results show that there is a reduction in operating costs by increasing flexibility from 30% to 40% with the WECS providing the reserves.

The methodology implemented in each case under study demonstrates the potential technical, economical, and reliability benefits of load-following ramping reserve ancillary service and contingency reserves ancillary service in major contingency, which comes under frequency control ancillary services.

This work can be extended considering energy storage and its ramping capability providing frequency control reserves, enhancing the system performance, and reducing the operation cost drastically, in normal and contingency conditions. It can be incorporated for stochastic modeling in SSCED, considering its losses, state of charge, storage cycle, and all of its constraints.

Author Contributions: Conceptualization, K.M.K. and G.A.V.; methodology, K.M.K. and G.A.V.; software, K.M.K. and G.A.V.; formal analysis, K.M.K. and G.A.V.; investigation, K.M.K.; resources, K.M.K. and G.A.V. data curation, K.M.K. writing—original draft preparation, K.M.K. and G.A.V.;

writing—review and editing, K.M.K. and G.A.V.; visualization, K.M.K. All authors have read and agreed to the published version of the manuscript.

Funding: This research received no external funding.

Data Availability Statement: Data available in a publicly accessible repository. The data presented in this study are openly available in [16,20,23–27].

Conflicts of Interest: The authors declare no conflict of interest.

Nomenclature

t	Index of time periods (1 h).
T	Set of indices of periods in the planning horizon, typically $\{1 \ldots n_t\} = 24$ h.
s	Index of scenarios.
S^t	Indexes of every possible scenario taken into consideration at time t.
S^{tmax}	Maximum number of generation profiles (scenarios) in time period t.
c	Index of post-contingency cases (c = 0 for the base case, meaning that there was no contingency).
C^{tsmax}	Maximum number of contingency indices taken into account in scenario s at time period t.
i	Injections index (generation units).
I^t	Indices of all the units (generators, dispatchable, or curtailable loads) that are available for dispatch in any situation at time t.
p^{rti}	Quantity/reference of the active power dispatching active power to unit I at time t.
cr_+^{ti}, cr_-^{ti}	Unit i at time t provides an active contingency reserve quantity that is moving upward or downward.
cr_{max+}^{ti}, cr_{max-}^{ti}	Maximum capacity restrictions for the unit i at time t can go up or down.
P_{min}^{tisc}, P_{max}^{tisc}	Limits on active injection for unit i at time t in a post-contingency state c of scenario s.
δ_+^{ti}, δ_-^{ti}	Unit i is required to provide ramping reserves for loads moving up or down for the changeover to time t + 1 at time t.
θ^{tsc}, V^{tsc}, p^{tsc}	Angles and magnitudes of the voltage as well as active injections of power for the post-contingency condition (state c) of scenario (s) at time (t).
δ_{max+}^{ti}, δ_{max-}^{ti}	Limits for unit i upward/downward load-following ramping reserves at time t for the transition to time t + 1.
ψ_α^{tsc}	Probability Contingency k in the scenario s at the time t, adjusted for α.
α	For contingency cases, the fraction of the time slice that is spent in the base case before the contingency occurs.
Δ	Length of the time slice for scheduling, usually 1 h.
uc^{ti}	Binary commitment state: 1 if a unit is online, otherwise 0, for unit i in period t.
ψ_0^{tsc}	Probability conditional on reaching period t without deviating from the main path in a contingency in periods 1 … $t-1$ and on scenario s being realized in some way of the contingency c in scenario s at period t. (base or contingency). The basic case, or conditional probability of no contingency, is ψ_0^{ts0}.
$\phi^{ts_2s_1}$	Probability of moving from scenario s_1 to scenario s_2 in period t, given that s_1 was completed in period $t-1$.
$\zeta^{ts_2s_1}$	Whether the transition to scenario s_2 in Period t, assuming that scenario s_1 from period $t-1$ should be included in the load-following ramp requirements, is indicated by a binary-valued mask.
ψ^{tsc}	Estimated likelihood of contingency c in scenario s at time t using transition probabilities $\phi^{ts_2s_1}$. Conditional probabilities of contingencies ψ_0^{tsc}.
PR_+^i, PR_-^i	Physical ramping upper/lower limits. for unit i when moving from base (c = 0) to contingency scenarios.
u^{ti}	For unit i in period t, the commitment state is binary: 1 if the unit is online, otherwise 0.
v^{ti}, w^{ti}	For unit I in period t, there are binary startup and shutdown states: 1 if the unit experiences a startup or shutdown event in period t, otherwise 0.
$C_P^{ti}(\cdot)$	Cost formula for i at time t for active injection.
$C_{P+}^{ri}(\cdot)$, $C_{P-}^{ri}(\cdot)$	Cost for a deviation from the active power reference quantity for unit i at time t in either an upward or downward direction.
$C_{ac+}^{ti}(\cdot)$, $C_{ac-}^{ti}(\cdot)$	Cost formula for a contingency reserve that was bought from unit i at time t.
$C_{\delta+}^{ti}(\cdot)$, $C_{\delta-}^{ti}(\cdot)$	Cost of the ramp reserve for an upward or downward load for unit i at time t for the transition to time t + 1.
$C_\delta^i(\cdot)$	On the difference between the dispatches for unit i in neighboring periods, there is a quadratic, symmetric ramping cost.
$\widetilde{C}_P^{ti}(\cdot)$	With the no-load cost eliminated, the cost function for active power injection has been modified to read as $\widetilde{C}_P^{ti}(P) \equiv C_P^{ti}(P) - C_P^{ti}(0)$.

References

1. Chowdhury, S.P.; Chowdhury; Crossley, P. *Microgrids and Active Distribution Networks*; The Institution of Engineering and Technology: London, UK, 2009.
2. Braun, M. Technological Control Capabilities of Der To Provide Future Ancillary Services. *Int. J. Distrib. Energy Resour.* **2007**, *3*, 191–206.
3. Oureilidis, K.; Malamaki, K.N.; Gallos, K.; Tsitsimelis, A.; Dikaiakos, C.; Gkavanoudis, S.; Cvetkovic, M.; Mauricio, J.M.; Maza Ortega, J.M.; Ramos, J.L.M.; et al. Ancillary Services Market Design in Distribution Networks: Review and Identification of Barriers. *Energies* **2020**, *13*, 917. [CrossRef]
4. Smeers, Y.; Martin, S.; Aguado, J.A. Co-optimization of Energy and Reserve with Incentives to Wind Generation: Case Study. *IEEE Trans. Power Syst.* **2022**, *37*, 2063–2074. [CrossRef]
5. Available online: http://www.forumofregulators.gov.in/data/Reports/SANTULAN-FOR-Report-April2020.pdf (accessed on 2 December 2022).
6. Shah, C.; Wies, R. Three-stage Power Flow & Flexibility Reserve Co-Optimization for Converter Dominated Distribution Network Lookahead Model using Blockchain & S-ADMM—I: Method. *TechRxiv* **2022**, TechRxiv.11-08-2022. [CrossRef]
7. Ozay, C.; Celiktas, M.S. Stochastic optimization energy and reserve scheduling model application for alaçatı, Turkey. *Smart Energy* **2021**, *3*, 100045. [CrossRef]
8. Jin, Y.; Wang, Z.; Jiang, C.; Zhang, Y. Dispatch and bidding strategy of active distribution network in energy and ancillary services market. *J. Mod. Power Syst. Clean Energy* **2015**, *3*, 565–572. [CrossRef]
9. Cong, P.; Tang, W.; Zhang, L.; Zhang, B.; Cai, Y. Day-ahead active power scheduling in active distribution network considering renewable energy generation forecast errors. *Energies* **2017**, *10*, 1291. [CrossRef]
10. Kurundkar, M.K.; Karve, M.G.; Vaidya, M.G. Comparative performance analysis of Firefly algorithm and Particle swarm Optimization for Profit Maximization of Grid connected Microgrid providing energy and ancillary service. *Solid State Technol.* **2021**, *64*, 4610–4626.
11. Ye, H.; Li, Z. Pricing the Ramping Reserve and Capacity Reserve in Real Time Markets. *arXiv* **2015**, arXiv:1512.06050.
12. Fang, X.; Sedzro, K.S.A.; Hodge, B.S.; Zhang, J.; Li, B.; Cui, M. *Providing Ramping Service with Wind to Enhance Power System Operational Flexibility*; National Renewable Energy Laboratory, University of Texas at Dallas: Golden, CO, USA, 2020.
13. Murillo-Sánchez, C.E.; Zimmerman, R.D.; Anderson, C.L.; Thomas, R.J. Secure Planning and Operations of Systems with Stochastic Sources, Energy Storage and Active Demand. *IEEE Trans. Smart Grid* **2013**, *4*, 2220–2229. [CrossRef]
14. Makarov, Y.V.; Loutan, C.; Ma, J.; de Mello, P. Operational Impacts of Wind Generation on California Power Systems. *IEEE Trans. Power Syst.* **2009**, *24*, 1039–1050. [CrossRef]
15. Ela, E.; O'Malley, M. Probability-Weighted LMP and RCP for Day-Ahead Energy Markets using Stochastic Security-Constrained Unit Commitment. In Proceedings of the 12th International Conference on Probabilistic Methods Applied to Power Systems, Istanbul, Turkey, 10–14 June 2012; National Renewable Energy Laboratory, University College Dublin: Golden, CO, USA, 2012.
16. Zimmerman, R.D.; Murillo-S, C.E.; Thomas, R.J. MATPOWER: Steady- State Operations, Planning and Analysis Tools for Power Systems Research and Education. *IEEE Trans. Power Syst.* **2011**, *26*, 12–19. [CrossRef]
17. Bär, K.; Wageneder, S.; Solka, F.; Saidi, A.; Zörner, W. Flexibility Potential of Photovoltaic Power Plant and Biogas Plant Hybrid Systems in the Distribution Grid, Katharina Bar. *Chem. Eng. Technol.* **2020**, *43*, 1571–1577. [CrossRef]
18. Akrami, A.; Doostizadeh, M.; Aminifar, F. Power system flexibility: An overview of emergence to evolution. *J. Mod. Power Syst. Clean Energy* **2019**, *7*, 987–1007. [CrossRef]
19. Kaushik, E.; Prakash, V.; Mahela, O.P.; Khan, B.; El-Shahat, A.; Abdelaziz, A.Y. Comprehensive Overview of Power System Flexibility, during the Scenario of High Penetration of Renewable Energy in Utility Grid. *Energies* **2022**, *15*, 516. [CrossRef]
20. Wang, H.; Murillo-Sanchez, C.E.; Zimmerman, R.D.; Thomas, R.J. On Computational Issues of Market-Based Optimal Power Flow. *IEEE Trans. Power Syst.* **2007**, *22*, 1185–1193. [CrossRef]
21. Dvorkin, Y.; Kirschen, D.S.; Ortega-Vazquez, M.A. Assessing flexibility requirements in power systems. *IET Gener. Transm. Distrib.* **2014**, *8*, 1820–1830. [CrossRef]
22. Ethan, D.; Avallone. *Market Design Concepts to Prepare for Significant Renewable Generation Flexible Ramping Product: Market Design Concept Proposal*; Market Issues Working Group, NYISO: Rensselaer, NY, USA, 26 April 2018.
23. Alsac, O.; Stott, B. Optimal Load Flow with Steady State Security. *IEEE Trans. Power Appar. Syst.* **1974**, *PAS 93*, 745–751. [CrossRef]
24. Ferrero, R.; Shahidehpour, S.; Ramesh, V. Transaction Analysis In Deregulated Power Systems Using Game Theory. *IEEE Trans. Power Syst.* **1997**, *12*, 1340–1347. [CrossRef]
25. Available online: https://weather.com/en-IN/weather/today/l/18.52,73.86?par=google (accessed on 30 November 2022).
26. Available online: https://prayaspune.org/peg/electricity-load-patterns (accessed on 2 December 2022).
27. Joshi, M.; Palchak, J.D.; Rehman, S.; Soonee, S.K.; Saxena, S.C.; Narasimhan, S.R. *Ramping Up the Ramping Capability, India's Power System Transition*; National Renewable Energy Laboratory: Golden, CO, USA, 2020.

energies

MDPI

Article

Predictive Controller Based on Paraconsistent Annotated Logic for Synchronous Generator Excitation Control

João Inácio Da Silva Filho [1,*], Raphael Adamelk Bispo de Oliveira [1], Marcos Carneiro Rodrigues [1], Hyghor Miranda Côrtes [1], Alexandre Rocco [1], Mauricio Conceição Mario [1], Dorotéa Vilanova Garcia [1], Jair Minoro Abe [2], Claudio Rodrigo Torres [3], Viviane B. Duarte Ricciotti [4], Antonio Carlos Duarte Ricciotti [4], Arnaldo de Carvalho, Jr. [5] and Germano Lambert-Torres [1,6,*]

1 Laboratory of Applied Paraconsistent Logic, Santa Cecilia University–UNISANTA, Oswaldo Cruz Street 288, Santos 11045-907, SP, Brazil
2 Graduate Program in Production Engineering, Paulista University, José Maria Whitaker Avenue, 320, São Paulo 04057-000, SP, Brazil
3 Post Graduation Program in Management and Technology in Productive Systems-Paula Souza State Center for Technological Education (CEETEPS), Bandeirantes Street, 169, São Paulo 01124-010, SP, Brazil
4 Academic Department of Electrical Engineering, Federal University of Rondônia, Porto Velho 76801-058, RO, Brazil
5 Federal Institute of Education, Science and Technology of São Paulo (IFSP), Cubatão 11533-160, SP, Brazil
6 Gnarus Institute, Itajuba 37500-052, MG, Brazil
* Correspondence: inacio@unisanta.br (J.I.D.S.F.); germanoltorres@gmail.com (G.L.-T.)

Abstract: This study presents a new Model Predictive Controller (MPC), built with algorithms based on Paraconsistent Annotated Logic (PAL), with application examples in the excitation control of a synchronous generator. PAL is a non-classical evidential and propositional logic that is associated with a Hasse lattice, and which presents the property of accepting the contradiction in its foundations. In this research, the algorithm was constructed with a version of the PAL that works with two information signals in the degrees of evidence format and, therefore, is called Paraconsistent Annotated Logic with annotation of two values (PAL2v). For the validation of the algorithmic structure, the computational tool MATLAB® Release 2012b, The MathWorks, Inc., Natick, MA, United States was used. Simulations were performed which compared the results obtained with PPC-PAL2v to those obtained in essays with the AVR (Automatic Voltage Regulator) controls in conjunction with the PSS (Power System Stabilizer) and the conventional MPC of fixed weights. The comparative results showed the PPC-PAL2v to display superior performance in the action of the excitation control of the synchronous generator, with a great efficiency in response to small signals.

Keywords: paraconsistent logic; paraconsistent annotated logic; power grids; stability control; predictive models; energy efficiency; power system stability

Citation: Da Silva Filho, J.I.; de Oliveira, R.A.B.; Rodrigues, M.C.; Côrtes, H.M.; Rocco, A.; Mario, M.C.; Garcia, D.V.; Abe, J.M.; Torres, C.R.; Ricciotti, V.B.D.; et al. Predictive Controller Based on Paraconsistent Annotated Logic for Synchronous Generator Excitation Control. *Energies* **2023**, *16*, 1934. https://doi.org/10.3390/en16041934

Academic Editor: Abu-Siada Ahmed

Received: 24 January 2023
Revised: 10 February 2023
Accepted: 12 February 2023
Published: 15 February 2023

1. Introduction

An Electrical Power System (EPS) is eventually subject to disturbances that can cause stability problems in the generation and distribution of energy [1]. When an EPS changes from one operating point to another, the voltage and speed of the synchronous generator can consequently enter oscillatory states at low frequency. For small-scale disturbances, such as load and voltage variations in the electrical network, effective excitation circuits and speed controllers are required to keep the system within the proper conditions for its operation [1].

The excitation control, in addition to providing voltage control at the generator terminals and reactive power control, plays an important role in maintaining stability, and response to small signals [1–3]. In order to mitigate these problems of instability under these conditions, network analyzers and special devices have been developed to aid in the analysis of system behavior. For example, one of the most important circuits of this type is the AVR, or Automatic Voltage Regulator, which allows the analysis of the interconnection in order to minimize the oscillations. However, with the expansions of the industrial parks and consequent demand for electricity, the electricity transmission networks have expanded, and thus, there is a great variety of parameters of power systems with extreme operating conditions [3–5]. To avoid the risk of power failure, the EPSs had to be adapted to these new situations in which. Despite the good performance of the AVR—Automatic Voltage Regulator—in the SEPs, it was verified that, in some situations, when switching for example large blocks of load, this type of regulator does not provide a sufficient response in magnitude and speed so as to dampen the oscillations in the system effectively. Therefore, in order to alleviate these problems, a supplementary excitation controller called the Power System Stabilizer (PSS) has been added to reduce the electromechanical oscillations of the generator [3–10].

As described in [1–4], the stabilizing action of the PSS in the control mesh is in the feedback receiving the addition of the effects of the inertia of the generator, with the final result of a faster damping of the frequency and output voltage. In this way, with the addition of the PSS (Power System Stabilizer) to the AVR (Automatic Voltage Regulator), any negative effect can be eliminated to damping the oscillations after a disturbance. However, it has been found that the combination of these two controls can generate inconvenient control situations [8–10]. On these aspects, it can be verified in their applications that AVR and the PSS are normally projected separately, since both have different objectives in the process, to soften the excitation failures [5].

The difference between the AVR and PSS is that the AVR is developed to meet requirements for voltage regulation and the PSS is designed to damp electromechanical oscillations [8,9]. Therefore, the coordination of the two objectives in joint actions can present problems when there is a change in the operating conditions of the system [1,3,4]. Recently, studies have been developed to discover new techniques with which to improve the performance of these types of control, and thus reduce the risk of failure in the EPS. For instance, a linear optimum control used for excitation control is highlighted in [3]. This control was developed to minimize the variations in system state. In order to work with voltage and velocity deviations, a cost function was formulated [4,6–12]. In Ref. [4] a controller with actions was presented, minimizing the cost function, and an optimized control was obtained. In this approach, the results were compared to those obtained by the PSS, and a wide frequency band could be obtained with a good response. However, because it is based on a specific operating point, this controller did not perform well when subjected to large variations in the system operating point. In order to achieve this, several methods of analysis and stability control using artificial intelligence (AI) and techniques with non-classical logic have been developed for these power systems, such as: Backstepping, Fuzzy Control, Direct Feedback Linearization (DFL), Artificial Neural Network (ANN), LgV, among others [5,13–16].

The adaptive controllers use Integrated Circuits (IC microcontrollers) with special algorithms based on nonlinear models to improve control performance [17,18]. In some cases, the algorithms use the precision of the predictions of the model making a common approach to consider the uncertainty in the parameters. For this, it adds an extra term in the cost function of a Minimum Variance controller, which penalizes the uncertainty in the parameters of the nonlinear approximation. Another similar approach has been proposed in [19], which is based on minimizing two separate cost functions. The first minimization is used to improve the parameter estimation and the second to generate the system output in accordance with a given reference signal [20]. Thus, adaptive controls can change their parameters according to the new operating point through preprogrammed rules by using neural networks or other non-classical techniques. However, depending on the number of instructions and the hardware of the microcontroller, the processing may not meet the speed required by the controlled system [21].

With the help of microcontrollers, special algorithms that consider both past states and future states, estimates can be used together with the adaptive control. In the interval, or step of actuation, a finite horizon of time is given in which these estimates present the minimum of possible uncertainties. This gives an excitation controller capable of defining a voltage-optimized value in a one-step prediction horizon. This type of Predictive Controller, despite the good performance in control, has a high computational cost since the algorithm needs an adjustment strategy for the cost matrix variables of the quadratic cost function [18,19].

In order to extract a better performance of this control method, the techniques involving predictive control can be improved using algorithms based on non-classical logics that are capable of responding well to uncertainties and with a greater speed. In this case, theories based on paraconsistent logics, which do not ignore contradictions and, instead, extract information from them, can present researchers with a good choice in adaptive and predictive control [17,19–21].

The Paraconsistent Annotated Logic (PAL) used in this work is one of the non-classical logics [22,23] suitable for reasoning with data that can bring inconsistent information. Therefore, it is possible to extract from its foundation efficient algorithms of predictive control. These can be applied in the control of excitation of a generator installed in an EPS [24–28]. With these considerations, this paper demonstrates the development and analysis of an excitation control for a synchronous generator, integrating the foundations of the Paraconsistent Logic Annotated in its special form of two values, PAL2v [24], into a Model Predictive Control (MPC). The application of the PAL2v will be performed in order to obtain a mechanism of intelligent self-adjustment to the classic MPC, one that can offer greater efficiency in the maintenance of the stability of the EPS when submitted to variations in load and in tension.

In addition to this introduction, the text of this work is presented as follows: in Section 2, the main foundations of Paraconsistent Logic and Paraconsistent Annotated Logic are presented. At the end of Section 2, the algorithms of PAL2v and the Paraconsistent Analysis Network (PANnet), used in the proposed Model Predictive Control, are presented. In Section 3, we present initial information about Predictive Control, based on MPC (Model Predictive Control), and the cost function equations used in the mathematical method. In Section 4 (Materials and Methods), we present the configuration of the Predictive Control, based on MPC, with the algorithms of PAL2v and its mathematical logic configuration in control of the excitation of a synchronous generator installed in an EPS. Additionally, in Section 4, the details of the computational tool MATLAB® Release 2012b implementation of the Predictive Controller, based on MPC built with the algorithms of PAL2v (PPC-PAL2v), are presented. In Section 5, the results obtained by the simulation of the PPC-PAL2v Controller and the comparative graphical results with the AVR and PSS are presented. In Section 6, discussions about the results, obtained in the tests performed, are presented. In Section 7, the final considerations on the application of the PPC-PAL2v controller which was applied to the excitation control of an EPS synchronous generator are presented.

2. Paraconsistent Logic

Paraconsistent Logic (PL) belongs to the class of non-classical logics whose fundamental structures differ from classical binary logic, opposing the law of non-contradiction. Due to these fundamentals, PL can be adapted acting as theoretical support for the algorithms that constitute computational systems for the treatment of uncertainties [22–24].

PL can be studied through an annotation concept in which its representation is considered a lattice. This can be the one of 4 vertices (Lattice FOUR), according to Figure 1a. Each annotation assigns to the proposition, P, a Paraconsistent Logical State, ε_τ, that is represented at the vertex of the lattice. With an annotated logic structuring, PAL allows its algorithms to manipulate inconsistent information, where logical states are represented at the vertices of its associated lattice. Thus, the values of information signals obtained by measurements are represented by normalized degrees of evidence and considered in the annotation that gives logical connotation to the given proposition, P [24–28]. The four vertices lattice, τ, may be associated via the Paraconsistent Annotated Logic with the annotation composed of two degrees of evidence (μ, λ). Through performing analysis with the Paraconsistent Annotated Logic of annotation with two values (PAL2v), one can establish an improved representation of how much the annotations, or evidence, express the knowledge about a proposition, P [24–28]. In this case, a value, μ, represents the evidence favorable to proposition, P, and the other value, λ, represents the evidence unfavorable to proposition P. In the PAL2v, an associated lattice, τ, formed by pairs of independent degrees of evidence is used in the formalization of PAL2v, where μ is the degree of evidence favorable to P and λ is the degree of evidence unfavorable to P, such that [24–28]: $\tau = \{(\mu, \lambda) \mid \mu, \lambda \in [0, 1] \subset \mathcal{R}\}$.

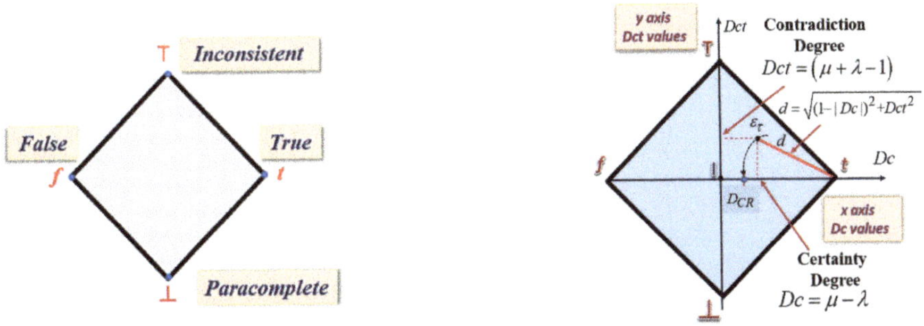

(a) (b)

Figure 1. Paraconsistent Annotated Logic Representation—PAL. (**a**) Lattice FOUR associated at PAL. (**b**) Real Certainty (D_{CR}) PAL2vlattice.

For better representation of an annotation, and also for the practical use of the PAL2 lattice in the treatment of uncertainties, some algebraic interpretations involving a Unitary Square in the Cartesian Plane (USCP) and the representative lattice of PAL2v are made. With the representation of the degrees of evidence in an USCP, one can be applied geometric and linear transformations to obtain points of intersection represented in the lattice τ associated with PAL2v and obtain equations resulting in algorithms for practice applications. The transformations between USPC and PAL2v–lattice are defined through three steps: scale change (T1), rotation (T2) and translation (T3) [24–28]. By making the composition of the three phases that generated the transformations T3, T2, T1, we obtain the final transformation represented by Equation (1) [24]:

$$T(X,Y) = (x - y, x + y - 1) \tag{1}$$

Relating the components of the transformation according to the usual nomenclature of PAL2v: $x = \mu \rightarrow$ degree of evidence favorable to proposition P and $y = \lambda \rightarrow$ degree of evidence unfavorable to proposition P. The first term obtained in the pair of values of the transformation equation is: $X = x - y = \mu - \lambda$, which we call degree of certainty—Dc. Therefore, the degree of certainty [24–26] is obtained by Equation (2):

$$Dc = \mu - \lambda \tag{2}$$

Its values, which belong to the set $\mathcal{R}$, vary in the closed interval +1 and -1, and are on the horizontal axis of the lattice, which is called the "Degrees of Certainty Axis". When Dc results in +1, it means that the logical state resulting from the paraconsistent analysis is True, and when Dc results in -1 means that the logical state resulting from the analysis is False [24–28]. The second term obtained in the pair of values of the transformation equation has: $Y = x + y - 1 = \mu + \lambda - 1$, which is called the degree of contradiction—Dct. Therefore, the degree of contradiction Dct [24–28] is obtained Equation (3):

$$Dct = \mu + \lambda - 1 \quad (3)$$

Their values, which belong to the set $\mathcal{R}$, vary in the closed interval +1 and −1, and are on the vertical axis of the lattice, which is called the "Axis of degrees of Contradiction". When *Dct* results in +1, it means that the logical state resulting from the paraconsistent analysis is Inconsistent, and when *Dct* is −1, the logical state resulting is either Paracomplete, or Undetermined [24].

The paraconsistent analysis that resulted in the calculations of certainty (*Dc*) and degree of contradiction (*Dct*) (Equations (2) and (3), respectively) produced values that are interpolated in the lattice at an internal point (*Dc*, *Dct*), according to Figure 1b. The distance *d* of the line from the point of maximum degree of certainty, *t*, represented at the right vertex of the lattice, to the point of interpolation, is calculated by Equation (4) [24]:

$$d = \sqrt{(1-|Dc|)^2 + Dct^2} \quad (4)$$

The projection of the distance, *d*, in the axis of certainty values gives the point whose value will be considered the degree of real certainty—D_{CR}. In Figure 1b, this condition is shown where the value of the real certainty degree—D_{CR}—is obtained [24].

If the certainty degree (*Dc*) calculated by Equation (2) results in a negative value, the distance, *d*, will be obtained from the point of certainty False, *f*, represented at the left vertex of the lattice, to the point of internal interpolation (*-Dc*, *Dct*). It is verified that, at any point in the lattice of values, it is possible to obtain the degree of real certainty, D_{CR}. The values of negative *Dct* do not modify the means of obtaining D_{CR}. Therefore, the value of the true degree of D_{CR} is obtained from the determination of distance, *d*, according to the conditions shown by Equation (5) and (6) [24,25]:

For $Dc > 0$

$$D_{CR} = 1 - \sqrt{(1-|D_C|)^2 + D_{ct}{}^2} \quad (5)$$

For $Dc < 0$

$$D_{CR} = \sqrt{(1-|D_C|)^2 + D_{ct}{}^2} - 1 \quad (6)$$

These D_{CR} values are normalized to become the actual resulting degree of evidence, as follows in Equation (7) [24,29,30]:

$$\mu_{ER} = \frac{D_{CR} + 1}{2} \quad (7)$$

where: μ_{ER} = resulting degree of evidence and D_{CR} = degree of real certainty.

With these fundamental equations, it is possible to construct Paraconsistent Annotated Logic (PAL2v) algorithms for applications in analysis and logical signal processing. These algorithms are capable of being interconnected to form Decision Analysis networks with different topologies [24].

2.1. *Algorithms of PAL2v Used in the Paraconsistent Model*

Paraconsistent Systems or Analysis Nodes are algorithms extracted from the Paraconsistent Logic, capable of forming Paraconsistent Analysis Networks for the treatment of information signals [24,26,31–34]. With their inputs being fed by the evidence degrees taken from external measurements or from the Uncertain Knowledge database, the Paraconsistent algorithms use the equations obtained from the PAL2v methodology and present the results in the form of the actual resulting evidence degrees, μ_{ER}. The PAL2v equations and their interpretation allow for the creation of algorithms for direct applications. In this work, three types of PAL2v algorithms are used in Paraconsistent Predictive Control; the algorithm for extracting the degree of evidence; the Paraconsistent Analysis Node (PAN) algorithm; and the Paraconsistent Logic Maximization Algorithm.

These three algorithms will be described below [24,29,30,32]:

2.1.1. Algorithm for Extracting the Degree of Evidence

The algorithm, named the Extractor of Evidence Degree, calculates the value of the degree of evidence of the quantity when measured through a function which has been considered in a Discourse Universe or Interval of Interest. Depending on the application, a straight-line equation can be used to obtain the value of the degree of evidence. The Extractor of Evidence Degree is exposed in Algorithm 1 below.

Algorithm 1: Extractor of Evidence Degree

---

1. Enter the Maximum Limit Value of the magnitude in your unit of measure to form the Discourse Universe.
   Maximum Limit Value (Max_{value}) = ...
2. Enter the Minimum Limit Value of the magnitude in your unit of measure to form the Discourse Universe.
   Minimum Limit Value (Min_{value}) = ...
3. Display the Measured Value of Greatness in your unit of measurement.
   Value Greatness X (X value) = ...
4. Determine the Degree of Evidence favorable, μ, through the equations, considering the conditions:

$$\mu_1 = \begin{cases} 1 & \text{If } Xvalue \geq Max_{value} \\ \frac{Xvalue - Max_{value}}{Max_{value} - Min_{value}} & \text{If } Xvalue \in [Min_{value}, Max_{value}] \\ 0 & \text{If } Xvalue \leq Min_{value} \end{cases}$$

5. Determine the unfavorable Degree of Evidence, λ, by complementing the favorable degree of Evidence:
$$\lambda = 1 - \mu$$
6. End.

The symbolic representation of extractor of Degree of Evidence is seen in Figure 2a.

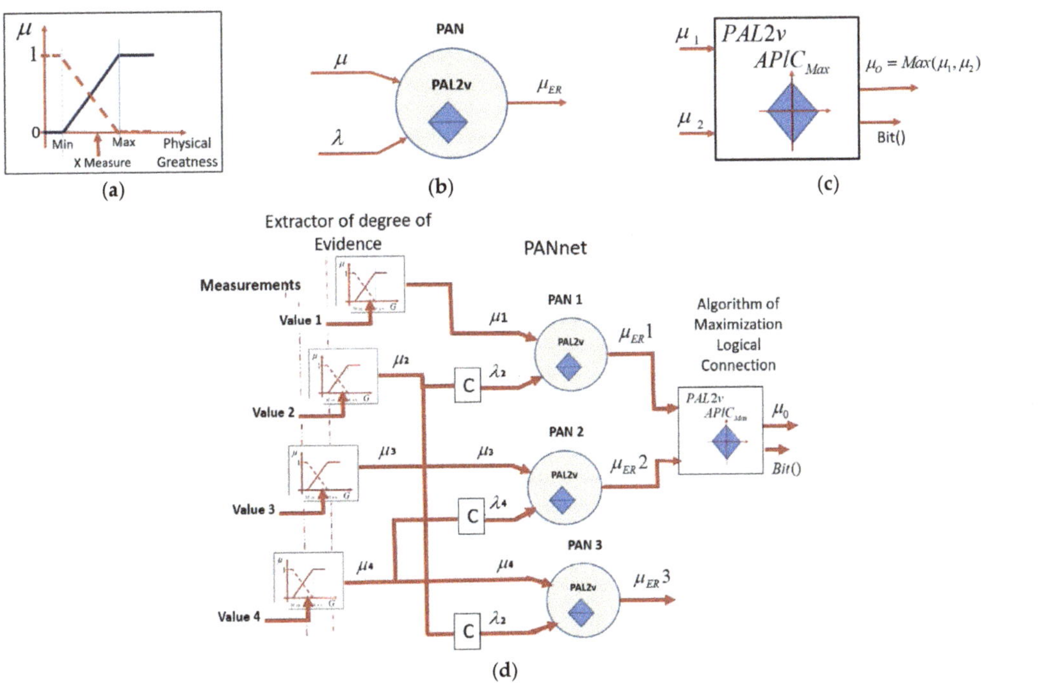

Figure 2. Symbolic representation of PAL2v algorithms and Paraconsistent Analysis Network (PANnet). (**a**) Extractor of degree of evidence. (**b**) PAN—Paraconsistent Analysis Node. (**c**) Maximization Logical Connection Algorithm—$APIC_{Max}$. (**d**) Structure of a PANnet with 3 PANs, two outputs and one signaling bit.

2.1.2. PAN Paraconsistent Analysis Node Algorithm

The Algorithm Paraconsistent Analysis Node (PAN) receives two information signals, represented by degrees of evidence, and presents as a final result a single value of the resulting degree of evidence. The degree of evidence resulting from the output is a value that expresses a representation of the analysis, where the effect of the contradiction between the two values applied in its inputs is null [24]. In the construction of PAN, the equations of PAL2v are used, and its structure is the minimum cell of analysis of a Paraconsistent System of Treatment of Uncertainties [24,29,31,32].

The description of the PAN is shown in Algorithm 2.

Algorithm 2: Paraconsistent Analysis Node—PAN

1. Present two input values:
   μ */ favorable Evidence degree $0 \leq \mu \leq 1$ */
   λ */ unfavorable Evidence degree $0 \leq \lambda \leq 1$ */
2. Calculate the Degree of Certainty: $Dc = \mu - \lambda$
3. Calculate the Degree of Contradiction: $Dct = (\mu + \lambda) - 1$
4. Calculate the Normalized Degree of Contradiction: $\mu ct = \frac{\mu+\lambda}{2}$
5. Calculate the distance d (projection on the axis (horizontal) of the degrees of certainty on the PAL2v lattice):
   $d = \sqrt{(1-|Dc|)^2 + Dct^2}$
6. Determine the output signal.
   If $d > 1$, then do: $S1 = 0.5 \rightarrow$ Consider Undefinition and go to End.
   Otherwise go to the next item
7. Determine the Real Certainty Degree
   If $Dc > 0$ Calculate: $D_{CR} = (1 - d)$
   If $Dc < 0$ Calculate: $D_{CR} = (d - 1)$
8. Calculate the resultant Real Evidence Degree: $\mu_{ER} = \frac{D_{CR}+1}{2}$
9. Present the results in the output: Do $S1 = \mu_{ER}$
10. End

In the PAN the projection of the distance value d, Equation (4) on the axis of the degrees of certainty establishes the extraction of the effect of the contradiction between the degrees of evidence of the inputs. Thus, the value of the output, evidence degree, represents the value of the resulting evidence regarding the proposition analyzed without the effect of the contradiction [26,27]. In Ref. [27], details about the PAN algorithm are shown, including its data flowchart with the steps for the PAL2v analysis.

2.1.3. Paraconsistent Algorithm of Maximization Logical Connection—$APIC_{Max}$

The Paraconsistent Algorithm of Logical Maximization Connection has the function of establishing logical connectives between signals, representing degrees of evidence. In this work, we use the algorithm that makes a logical connection of maximization OR [24]. For maximization, a simple analysis is initially made by determining the normalized resulting degree of evidence, the result of which will inform which of the two input signals is the highest value [26]. With this information, the algorithm establishes the output signal as being the maximum value.

The equation used in the algorithm and the conditions that determine the outputs for a maximization process are presented in the description of Algorithm 3 below:

The symbolic representation of $APIC_{Max}$ is seen in Figure 2c.

Algorithm 3: Maximization Logical Connection—$APlC_{Max}$

Since the input variables are:
μ_{1A}, such that: $0 \leq \mu_{1A} \leq 1$, and μ_{1B}, such that: $0 \leq \mu_{1B} \leq 1$,
then
1. Do: $\mu_{1A} = \mu_1$ and $\lambda_2 = 1 - \mu_{1B}$
2. Calculate the Degree of Certainty by equation (2): $Dc = \mu_1 - \lambda_2$
3. Calculate the resulting normalized Evidence Degree by: $\mu_{ER} = \frac{(\mu_1 - \lambda_2) + 1}{2}$
4. Select the highest value in the output S1 by making the conditionals:
   If: $\mu_{ER} \geq 0,5 \rightarrow \mu_{1A} \geq \mu_{1B} \rightarrow$ Present in Output S1 the value of μ_{1A}
   Indicate in the output $S2 \rightarrow$ *Bit*() = 1
   If not: $\rightarrow$ Present in Output S1 the value of μ_{1B}
   Indicate in the output $S2 \rightarrow$ *Bit*() = 0
5. End

2.2. *Paraconsistent Analysis Network (PANnet)*

The paraconsistent system of treatment of uncertainties defined by the PAN can be used in several fields of knowledge. With its application, incomplete and contradictory information will receive adequate treatment through the equations of PAL2v [24–26].

As the output of a PAN has normalized value, this result value can be used as a degree of evidence for another PAN, and thus different configurations of analysis networks can be created. These configurations are called Paraconsistent Analysis Networks (PANnet). In this way, the computational structures of PANnet are used for the logical treatment of signals according to the purpose for which they are proposed.

Figure 2d shows a PANnet, composed of 3 PANs configured in the way it will be used in this work.

Initially, the modeling algorithms, which are called the Evidence Degree Detection Extractors, create the normalized degrees of evidence for the analyses in the PANnet [24–27] from the values extracted from measures of physical quantities. The PAN1 and PAN2 algorithms have their outputs, interconnected through the Paraconsistent Maximization Logic Algorithm which selects the highest value between the two outputs μ_{Er1} and μ_{Er2}. A signaling *Bit*() indicates which of the two output values of the PANs are in the μ_0 output. This PANnet configuration still has the PAN3 that uses the degrees of evidence from information sources. To compose this annotation, it uses the favorable degree of evidence, μ_4, and the degree of unfavorable evidence, $\lambda_2 = 1 - \mu_2$. The PAN3 performs the paraconsistent analyses and displays the result at its separate output, $S_{\mu 3}$. This PANNet configuration will be used in the treatment of the signals to control the excitation of the synchronous generator, and its operation will be explained in detail below.

Recently, PAL2v algorithms, forming paraconsistent analysis networks (PANnet), have been used successfully in several applications which cover different fields of knowledge. As an example, we can mention in [26,27], where PAL2v algorithms were used to detect skin cancer by analyzing Raman spectroscopy signals. In Ref [28], the PAL2v algorithms were used to support the operation of electricity transmission networks. In Ref [31], the authors used the PAL2v algorithms to support the control of a flow loop, forming a Hybrid Proportional Integral (PI) Controller.

3. Model-Based Predictive Control—MBPC

The characteristic of Model-Based Predictive Control—MBPC—is its ability to predict the future responses of the controlled system. This feature distinguishes it from other types of controllers [1–4]. The Model Predictive Control MPC uses information signals, represented by discrete steps (k + 1) (or pulses) through a sample frame. If the prediction of the future output path of the process {$\hat{y}$ (k + i), i = 1, … , P}, then the controller calculates the necessary control action {Δu (k + i), i = 0, 1, … , $m - 1$} so that the difference between the predicted path and the user-specified reference (Set Point) is minimized. The MPC can be described as a state space model and is calculated with the state-space equations, which are generally used for the mathematical modeling of a time-varying physical system [14,15].

3.1. State-Space Equations in Discrete Time

The state-space equations can be described as:

$$\dot{x}(t) = A_t x(t) + B_t u(t)$$
$$y(t) = Cx(t)$$

where $x(t)$ is the state vector of the system, A_t is the state matrix, B_t is the matrix related to the input, $u(t)$ is the input variable (manipulated variable), $y(t)$ is the output vector, and $Cx(t)$ is the output matrix.

Considering a sampling interval Δ_t, it is possible to define a value $k = 1,2,3,\ldots N$, such that $k = \frac{t}{\Delta_t}$, therefore, making the system discrete in time. Through Equation of Differences, we obtain the set of simplified discrete time state equations, represented by Equation (8) [1–5]:

$$\dot{x}(k) = \frac{x(k+1) - x(k)}{\Delta t} \rightarrow x(k+1) = Ax(k) + Bu(k+1) \text{ and } y(k) = Cx(k)$$

$$y(k+1) = Cx(k+1) \tag{8}$$

where: $A = A_t\Delta t - I$, $B = B_t\Delta t$.

In the set of equations, it is possible to observe that the variables correspond to the future state given the conditions of the system in the present. The MBPC algorithm, at each instant of time, attempts to optimize the behavior of the controlled system by calculating the future values of the manipulated variables to achieve this goal [13]. An illustrative block diagram of a predictive controller is shown in Figure 3, where the plant is driven by a control generated in a system termed Optimizer. In this work, an Optimizer Block, with algorithms based on PAL2v, will be constructed [14,15].

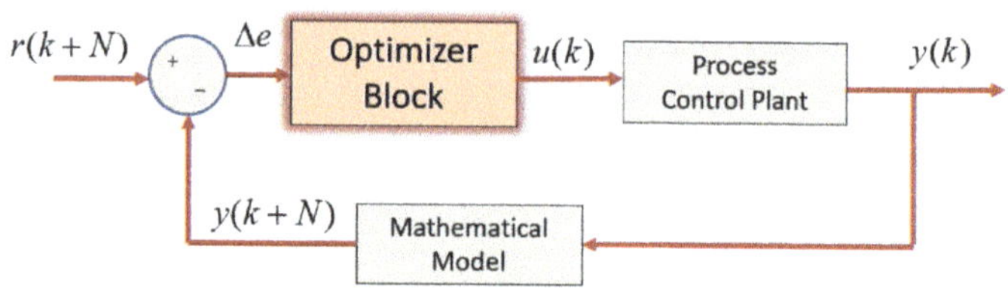

Figure 3. Simplified Model-Based Predictive Control-MBPC Blocks Diagram algorithms.

3.2. Quadratic Cost Function

The MBPC makes an instantaneous time, t, measurement of the output variable of the plant and, through the mathematical model of the controlled system, makes a prediction. Based on this measurement and the reference r, a prediction horizon, N, is stipulated and future values for the manipulated variable are calculated by minimizing of an equation called the quadratic cost function [13–15].

The quadratic cost function is represented by Equation (9) [15]:

$$J = [r - y(k+1)]^T Q[r - y(k+1)] + \Delta u^T R \Delta u \tag{9}$$

where $\Delta u = u(k+1) - u(k)$.

Since matrices Q and R are called weights or cost matrices, they are used to determine degrees of importance to certain inputs and outputs, and can be represented in Equation (10):

$$y(k+1) = CAx(k) + CBu(k) + CB\Delta u \tag{10}$$

Putting Equation (10) into (9) and applying the minimization of the function J, $\frac{\delta J}{\Delta u} = 0$ results in Equation (11):

$$\Delta u = \frac{(CB)^T QE}{[(CB)^T QCB + R]} \tag{11}$$

where:

The matrix Q is a positive definite matrix that is related to the states.

The matrix R, is a positive semi definite matrix that is related to the input.

The matrices Q and R are usually diagonal, with positive elements on their diagonal.

4. Materials and Methods

In this work, a Paraconsistent Predictive Controller-PPC-PAL2v was developed. This was composed of a computational architecture, constructed from algorithms based on PAL2v. The PPC-PAL2v was applied to the model of a synchronous machine connected to an Electrical Power System (EPS) where for its performance, we established:

1. Voltage regulation at the generator terminals and;
2. Damping of the electromechanical oscillation that is directly related to the rotor speed variations.

For the study of the control model developed in this work, the EPS used in the modeling has a synchronous generator connected to the infinite bus, as represented in the circuit of Figure 4a.

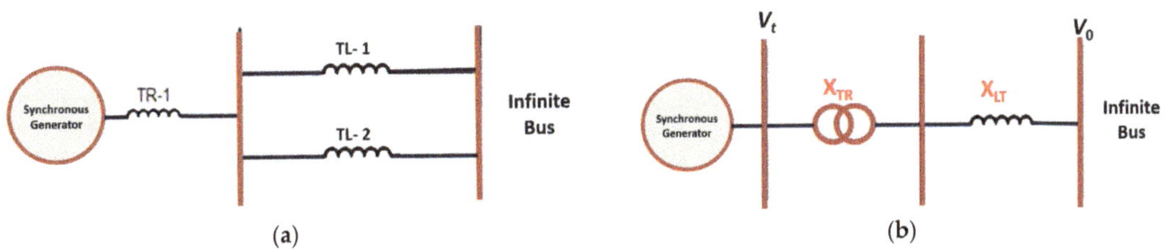

Figure 4. EPS model with synchronous generator connected to the infinite bus. (**a**) The Electrical Power System. (**b**) Electrical Power System Modeling is a figure. Schemes follow the same formatting.

The synchronous generator is connected via a voltage transformer to two parallel transmission lines which, at the other end, are connected to the infinite bus. Figure 4b shows the model of the EPS, where v_t is the voltage at the terminals of the generator and v_o is the voltage in the infinite bus. The reactance, X_{TR}, is the transformer and the X_{LT} transmission lines connected in parallel.

From a mathematical model created, a simulation was performed in the MATLAB® software R2012b to analyze the excitation control actions for the electric field of the synchronous generator, integrating the fundamentals of the Paraconsistent Annotated Logic in its special two-valued form (PAL2v) into a Model Predictive Control (MPC).

The dynamic variations of the synchronous generator will be expressed in the simulation in the face of the variations of voltage and speed caused by the small perturbations applied to the EPS.

All the mathematical logical modeling of the PPC-PAL2v was performed based on the dynamic conditions of the EPS, where it was then applied to act on the excitation control of the synchronous generator.

Table 1 shows the data concerning the reactance of the EPS transformers used in this work.

Table 2 shows the data of the synchronous generator used.

Table 1. Electrical Power System data.

Electrical Power System Parameters	
X_{TR}	0.15
X_{LT1}	0.5
X_{LT2}	0.93

Table 2. Synchronous Generator Data.

Nominal Values		Parameters				Saturation	
Nominal power	555 MVA	X_d	1.81 p.u.	X_l	0.15 p.u.	A_{sat}	0.031
Nominal voltage	24 kV	X_q	1.76 p.u.	R_a	0.003 p.u.	B_{sat}	6.93
Frequency	60 Hz	X'_d	0.30 p.u.	T'_{d0}	8.00 s	Ψ_{T1}	0.80
Number of poles	2	X'_q	0.65 p.u.	T'_{q0}	1.00 s		
Cte of Inertia	3.5 MWs/MVA	X''_d	0.23 p.u.	T''_{d0}	0.03 s		
Cte of Damping	0	X''_q	0.25 p.u.	T''_{q0}	0.07 s		

The variables selected to achieve optimum performance in voltage regulation at the generator terminals and the damping of electromechanical oscillations are electromagnetic torque, T_e, voltage at the generator terminals, V_t and rotor speed, ω_r. These variables, in the MBPC representation, are generated through the outputs, $y(k+1)$, predicted by the synchronous generator model, and the reference values to be applied in the equations of the reference trajectories that are processed in the optimization algorithm PAL2v. In the control of the voltage variations at the generator terminals, the variable is used Δe_{vt}, which is expressed by the difference between the voltage deviation in the step, k, and the deviation predicted in the step $k+1$, according to the following discrete time state Equation (12):

$$\Delta e_{vt}(k+1) = \Delta_{vt}(k) - \Delta_{vt}(k+1) \tag{12}$$

Being the voltage reference $r_{vt} = \Delta_{vt}(k)$.

The voltage deviation in the step k is determined from the set-point of the voltage V_0 and the voltage in the step k, called $V_t(k)$, at the terminals of the generator. As for the damping characteristic of electromechanical oscillations, there are two variables that participate in this function: $\Delta e_{\omega r}$, which is related to the rotor speed, and Δe_{Te}, related to the electromagnetic torque. These two variables are expressed, respectively, by Equation (13) and Equation (14).

$$\Delta e_{\omega r}(k+1) = \Delta\omega_r(k+1) - \frac{1}{2H}\Delta_t \Delta Te(k+1) \tag{13}$$

Being the reference of the speed of the rotor $r_{\omega r} = \frac{1}{2H}\Delta_t \Delta Te(k+1)$

$$\Delta e_{Te}(k+1) = T_m - T_e(k) - \Delta T_e(k+1) \tag{14}$$

Being the torque reference $r_{Te} = T_m - T_e(k)$, where: $\Delta e_{\omega r}(k+1)$ is the prediction value of the model of the speed deviation of the generator and $\Delta e_{Te}(k+1)$ is the prediction of the model of the deviation of the electromagnetic torque.

4.1. General Aspects of the Paraconsistent Predictive Model

In the optimization of the control, the values of the current speed $\omega_r(k)$, the values of the speed in the previous instant $\omega_r(k-1)$, and the reference voltage (Set Point) V_0 are used. The variables involved in the PPC-PAL2v model applied to the excitation control of the synchronous generator are shown in Figure 5.

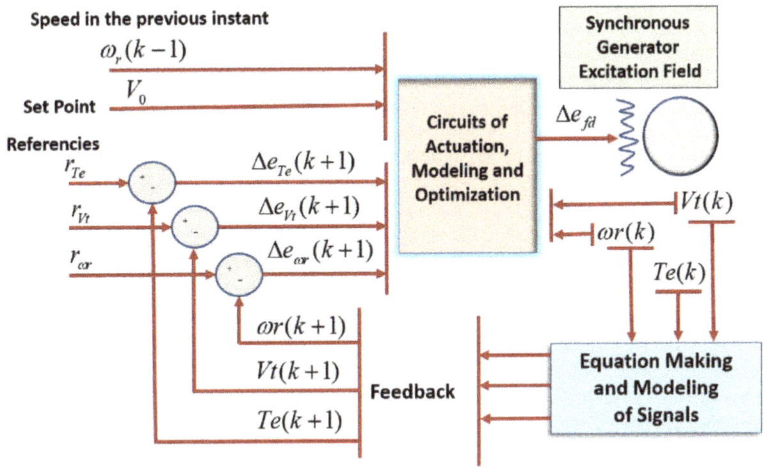

Figure 5. Variables involved in the PPC-PAL2v model applied to the excitation control of the synchronous generator.

4.2. *Paraconsistent Predictive Controller-PPC-PAL2v*

In order to meet the voltage regulation and damp effects of electromechanical oscillations, the PPC-PAL2v was structured in functional blocks, where each block has specific actions to compose joint actions in the predictive control in the excitation of the synchronous generator. For better understanding, the details of the control configuration used in the PPC-PAL2v are described below.

- *Structural configuration of PPC-PAL2v*

The structural configuration of the model-based Paraconsistent Predictive Controller—PPC-PAL2v—will be presented in two main blocks which are interconnected by the signal flow of the variables involved. The general configuration of the PPC-PAL2v is shown in Figure 6, with its signal flows in its two main blocks: OT-PAL2v—Optimizer— and MM—mathematical model of the synchronous generator.

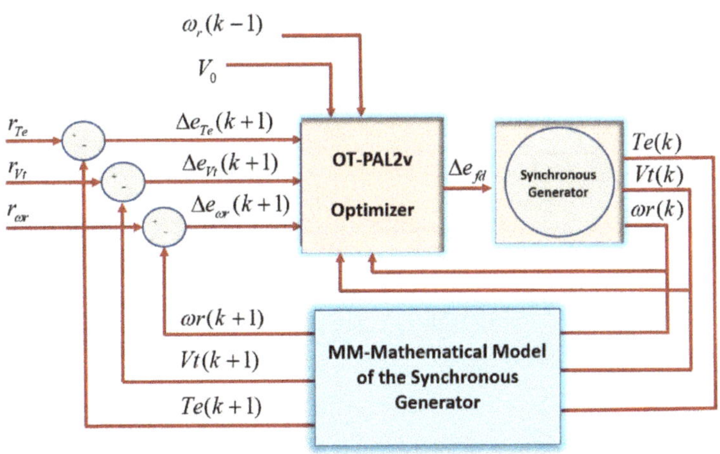

Figure 6. General PPC-PAL2v configuration with the two master blocks that make up the PAL2v Predictive Controller.

- *OT-PAL2v—Optimizer Block*

In the Predictive Controller model, the PAL2v algorithms were used in the construction of the Optimizer block (OT-PAL2v). In Figure 7, the composition of Optimizer PAL2v with three secondary blocks is shown, as follows: Paraconsistent Analysis Network (PANnet), Weight Adjustment Mechanisms (WAM) and Field Voltage Setting Determination Block (FVSD).

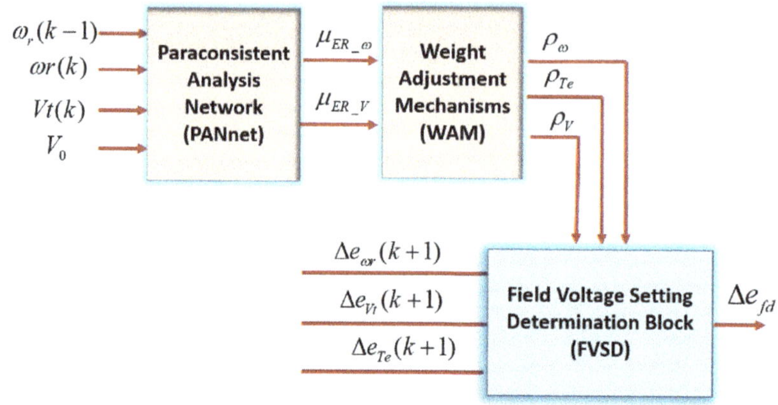

Figure 7. Optimizer block (OT-PAL2v) flowchart.

As shown in Figure 8, the Paraconsistent Analysis Network (PANnet), inserted into the Optimizer-PAL2v block, has the objective of performing the logical treatment of the normalized values. These values are obtained by measuring the generator speed, ω_r and the voltage, V_t.

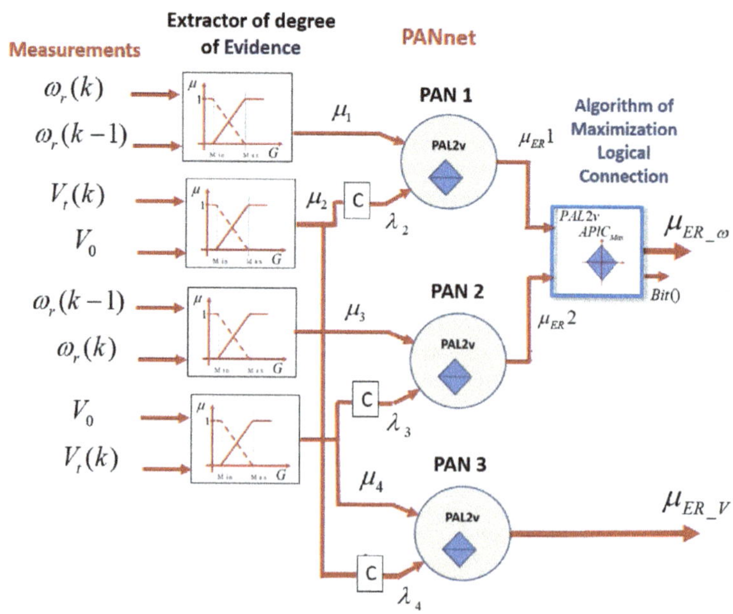

Figure 8. Paraconsistent Analysis Network (PANnet), inserted into the Optimizer-PAL2v block to determine the weight factors.

The result of the paraconsistent logical treatment in the signals will serve to determine the weight factors used in the block of Adjustment Mechanisms.

The PANnet configuration was developed by analyzing the conditions of the quantities involved in the control, and the optimization of the signals was performed through the configurations of PANs.

Table 3 shows the analysis of the field voltage variation according to the speed and voltage variation conditions at the generator terminals. Two points of contradiction are observed:

Table 3. Analysis of the Variations of Voltage and Speed.

	Signal	Condition	Value of e_{fd}
$\Delta\omega_r$	Positive	Acceleration	Increases
ΔV_t	Positive	Overvoltage	Decreases
$\Delta\omega_r$	Positive	Acceleration	Increases
ΔV_t	Negative	Undervoltage	Increases
$\Delta\omega_r$	Negative	Deceleration	Decreases
ΔV_t	Negative	Undervoltage	Increases
$\Delta\omega_r$	Negative	Deceleration	Decreases
ΔV_t	Positive	Overvoltage	Decreases

1. When there is acceleration of the rotor and the voltage is above the desired voltage V_0;
2. When there is deceleration of the rotor and the voltage is below the desired value V_0.

From the conditions indicated in Table 3, we can define the equations of the algorithms that extract the degrees of evidence that represent the values for the input signals of PPC-PAL2v.

- *PPC-PAL2v Input Variables with Normalized and Selected Values*

As seen in Figure 8, the PANnet, inserted into the Optimizer-PAL2v block, has algorithms for extracting the degree of evidence that normalizes and selects the signals of the input variables. These are: the speed at the present instant $\omega_r(k)$, speed at the previous instant $\omega_r(k-1)$, voltage at the generator terminals at the present instant $V_t(k)$ and the reference voltage (Set Point) V_0.

The first select/normalization that results in μ_1 for PAN 1 is performed by considering the acceleration condition above the desired value, so the mathematical function that extracts the degree of evidence and selects the output for this condition is:

$$\mu_1 = \begin{cases} 0 & if\ [\omega_r(k) - \omega_r(k-1)] \leq 0 \\ \frac{[\omega_r(k)-\omega_r(k-1)]}{20.41\times10^{-9}} & if\ 0 < [\omega_r(k) - \omega_r(k-1)] \leq 20.41 \times 10^{-9} \\ 1 & if\ [\omega_r(k) - \omega_r(k-1)] > 20.41 \times 10^{-9} \end{cases} \quad (15)$$

The second select/normalization that results in λ_2 = 1-μ_2 for PAN 1and λ_4 = 1-μ_2 for PAN 3 is made considering the maximum deviation of 5% of the voltage at the generator terminals above the desired voltage, and so the mathematical function that extracts the degree of evidence for this condition is:

$$\mu_2 = \begin{cases} 1 & if\ [V_t - V_0] \leq 0 \\ \frac{0.05-[V_t-V_0]}{0.05} & if\ 0 <[V_t - V_0] \leq 0.05 \\ 0 & if\ [V_t - V_0] > 0.05 \end{cases} \quad (16)$$

The third select/normalization that results in μ_3 for PAN 2 is made considering a deceleration condition with a maximum tolerance of 20.41×10^{-9} below the desired value. Therefore, the mathematical function that extracts the degree of evidence for this condition is:

$$\mu_3 = \begin{cases} 0 & if\ [\omega_r(k-1) - \omega_r(k)] \leq 0 \\ \frac{[\omega_r(k-1)-\omega_r(k)]}{20.41\times10^{-9}} & if\ 0 < [\omega_r(k-1) - \omega_r(k)] \leq 20.41 \times 10^{-9} \\ 1 & if\ [\omega_r(k-1) - \omega_r(k)] > 20.41 \times 10^{-9} \end{cases} \quad (17)$$

The fourth select/normalization that results in λ_3 = 1-μ_4 for PAN 2 and μ_4 for PAN 3 is made considering the maximum deviation of 5% of the voltage at the generator terminals above the desired voltage. Therefore, the mathematical function that extracts the degree of evidence for this condition is:

$$\mu_4 = \begin{cases} 1 & if\ [V_0 - V_t] \leq 0 \\ \frac{0.05-[V_0-V_t]}{0.05} & if\ 0 < [V_0 - V_t] \leq 0.05 \\ 0 & if\ [V_0 - V_t] > 0.05 \end{cases} \quad (18)$$

The inputs to PAN 3 are the same normalized values μ_2, obtained by Equation (16), and μ_4, obtained by Equation (18).

4.3. Detail of the Paraconsistent Analysis Network (PANnet)

The PANnet is the first block of the OT-PAL2v-Optimizer and is where the values of the speed $\omega_r(k)$ at actual instant, the speed at the previous instant $\omega_r(k-1)$, the voltage at the terminals of the generator at the actual instant $V_t(k)$ and the reference voltage V_0 are applied. In the PANnet inputs, all these values were transformed into degrees of evidence, and from these paraconsistent variables PANnet defines the degree of real evidence for speed $\mu_{ER\_\omega_r}$. This represents the evidence (intensity) e_{fd} of the need to increase or decrease to satisfy both the voltage control and the damping of electromechanical oscillations.

The result, in terms of the degree of real evidence for the voltage μ_{ER_V} that is simultaneously obtained in PANnet, represents evidence (intensity) of the need for tension control. Based on these control conditions, the function of each PAN in the PANnet configuration shown in Figure 8 is described below:

- PAN 1 $\Rightarrow$ Analysis of the increase condition of e_{fd};
- PAN 2 $\Rightarrow$ Analysis of the decrease condition of e_{fd};
- PAN 3 $\Rightarrow$ Analysis of the variation of K_v.

After the analysis performed by PANs 1 and 2, the resultant evidence degrees, μ_{ER1} and μ_{ER2}, corresponding to the need to increase e_{fd} and the need to decrease e_{fd}, are passed through a Paraconsistent Logic Maximization Algorithm $APIC_{Max}$. The $APIC_{Max}$ acts on only the greater value between the two, and sends the resulting signal directly to the Weighting Mechanism Block. Thus, dominant degree of evidence that is represented by μ_{ER_ω} will provide the increase or decrease in e_{fd} to satisfy both the voltage control and the damping of electromechanical oscillations. Simultaneously, after the analysis performed by the PAN3, the degree of evidence resultant of its output is represented by μ_{ER_V}. This will act in the Weights Adjustment Mechanism (WAM) Block for the control of the tension.

The WAM Block is described below.

4.4. Weights Adjustment Mechanism Block (WAM)

In the Mechanism of Adjustment of Weights, an analysis of the degrees of evidence, received from the PANs that compose the PANnet, is made to decide how to modify the weights in order to achieve optimal control. In the outputs of this block are:

$K_\omega(p_\omega)$ which represents the weight relative to the speed adjustment.

$K_V(P_V)$ which is the weight relative to the voltage.

$K_{Te}(p_{Te})$ which is the weight relative to the electromagnetic torque.

Thus, in the PAL2v-Optimizer block, the resulting signals from the Mechanism of Adjustment of Weights block are obtained through the equations of the multiplier factors for the Q-Matrix, as shown below.

- *Equations of the multiplier factors for the Q-Matrix*

From the resultant degrees of evidence provided by PANnet, we calculate the multiplier factors ρ_ω, ρ_V and ρ_{Te} which, respectively, change the weights, $K_\omega = 8$, $K_V = 1$, $K_{Te} = 0.05$. The dominant degree of evidence μ_{ER_ω} varies in the interval $[0,1]$, and when the output of Paraconsistent Logic Maximization Algorithm-$APIC_{Max}$ provides zero (or one) value bit, the multiplier factors vary according to Equation (19):

$$\rho_\omega = \rho_{Te} = 2\mu_{ER\_\omega} + 1 \tag{19}$$

Likewise, since the value of the resultant evidence degree μ_{ER_V} referring to the voltage is within the range, then the multiplier of the weight relative to the voltage is expressed by Equation (20):

$$\rho_V = -\mu_{ER-V} + 2 \tag{20}$$

Once the weights and factors have been defined, it is possible to express the Q matrix in its complete form through Equation (21):

$$Q = \begin{bmatrix} \rho_\omega K_\omega & 0 & 0 \\ 0 & \rho_V K_\omega & 0 \\ 0 & 0 & \rho_{Te} K_{Te} \end{bmatrix} \tag{21}$$

- *Field Voltage Setting Determination block (FVSD)*

The Field Voltage Setting Determination Block (FVSD) operates using a matrix equation in which the field voltage variation is effectively obtained to achieve the required values of the controlled variables.

The selected variables, which are required to reach the optimum performance in voltage regulation at the generator terminals and the damping of electromechanical oscillations, are: (a) electromagnetic torque T_e; (b) voltage at the generator terminals V_t and; (c) rotor speed ω_r. In the representation of the PPC-PAL2v, these variables generate, through the outputs, $y(k+1)$, predicted by the synchronous generator model and the reference values, the reference trajectories that are processed in the optimization algorithm PAL2v. The constants, represented in the equations of the trajectories, are dependent on the initial conditions and the inductances of the system and are obtained in the small signal stability study.

4.5. MM-Mathematical Model of the Synchronous Generator

The PPC-PAL2v model must act to regulate the voltage at the generator terminals and damp the electromechanical oscillation directly related to the rotor speed variations. These two premises can be described mathematically through the output vector of the state space formulation, as shown Equation (22) [18]:

$$y(k) = \begin{bmatrix} w_r(k) \\ V_t(k) \end{bmatrix} \tag{22}$$

where: w_r is the rotor speed and V_t is the voltage at the terminals of the synchronous generator.

To obtain the equations of the model in the MM (Mathematical Model of the synchronous Generator), the simplification of a synchronous machine with the inclusion of the dynamic effect of the field circuit is performed [10,12,18]. The state space matrix of the corresponding model is represented by Equation (23):

$$\begin{bmatrix} \Delta\dot{\omega}_r \\ \Delta\dot{\delta} \\ \Delta\dot{\Psi}_{fd} \end{bmatrix} = \begin{bmatrix} a_{11} & a_{12} & a_{13} \\ a_{21} & 0 & 0 \\ 0 & a_{32} & a_{33} \end{bmatrix} \begin{bmatrix} \Delta\omega \\ \Delta\delta \\ \Delta\Psi \end{bmatrix} + \begin{bmatrix} b_{11} & 0 \\ 0 & 0 \\ 0 & \mathrm{b}_{33} \end{bmatrix} \begin{bmatrix} \Delta T_m \\ \Delta E_{fd} \end{bmatrix} \tag{23}$$

where: ω_r is the rotor speed, δ is the power angle and Ψ_{fd} is the field flow.

The elements of the state matrix are function of the system parameters and the following equations show how they are obtained.

$$\begin{aligned} a_{11} &= \frac{K_D}{2H} \quad a_{12} = \frac{K_1}{2H} \quad a_{13} = \frac{-K_3}{2H} \\ a_{21} &= \omega_0 \quad a_{32} = -\frac{\omega_0 \, R_{fd}}{L_{fd}} m_1 L_{ads} \\ a_{33} &= -\frac{\omega_0 \, R_{fd}}{L_{fd}} \left[1 - \frac{L'_{ads}}{L_{fd}} + m_2 L'_{ads} \right] \\ b_{11} &= \frac{1}{2H} \quad b_{32} = \frac{\omega_0 \, R_{fd}}{L_{adu}} \end{aligned} \tag{24}$$

where: K_1, K_2, m_1 and m_2 are constants that depend on the parameters of the system and its initial conditions of operation. K_D is the damping coefficient and H is the inertia constant.

- *Dynamic analysis of the synchronous generator*

In the aspect of the dynamic analysis of the synchronous generator, the state variables are represented by the state space matrix. Therefore, the dynamics of the synchronous generator through the differential equations from these state variables is described as:

$$\text{For rotor speed } \rightarrow \omega_r \rightarrow \dot{\omega}_r(k) = \frac{1}{2H}(T_m(k) - T_e(k))$$

$$\text{For power angle } \rightarrow \delta \rightarrow \dot{\delta}(k) = \omega_0 \Delta \omega_r(k)$$

$$\text{For field flow } \rightarrow \Psi_{fd} \rightarrow \dot{\Psi}_{fd}(k) = \omega_0 \Delta e_{fd}(k)$$

where T_m is the mechanical torque, T_e is the electromagnetic torque and H is the inertia constant. Therefore, the state of the system can be described at any instant t, or step k, by the matrix represented by Equation (25):

$$x(k) = \begin{bmatrix} \delta(k) \\ \omega_r(k) \\ \Psi_{fd}(k) \end{bmatrix} \tag{25}$$

This modeling is performed through the equations of the reference trajectories, as explained below.

- *Equations of the reference trajectories*

The control signal, that is, the input $u(k)$ of the state space equations, is represented in the dynamic analysis of the synchronous generator by the voltage applied to the field circuit e_{fd}.

The equations of the reference trajectories are expressed by measurable quantities of the synchronous generator. For the reference trajectory related to the voltage at the terminals, we have: $\Delta e_{Vt}(k+1) = \Delta V_t(k) - K_3 \Delta \delta(k) - K_4 \Delta \Psi_{fd}(k)$.

Where the final analysis is performed by Equation (26):

$$\Delta e_{Vt}(k+1) = (-K_3 \omega_0 \Delta_t) \Delta \omega_r(k) + \Delta V_t(k) + (-K_4 \omega_0 \Delta_t) \Delta e_{fd}(k) \tag{26}$$

For the reference trajectory related to the speed of the synchronous generator, we have: $\Delta e_{\omega r}(k+1) = 2\Delta\omega(k) - \Delta\omega(k-1) - \frac{\Delta_t}{2H}(K_1 \Delta \delta(k) + K_2 \Delta \Psi_{fd}(k))$.

Where the final analysis is performed by Equation (27):

$$\Delta e_{\omega r}(k+1) = (-1)\Delta\omega(k-1) + \left(2 - \frac{K_1 \omega_0 {\Delta_t}^2}{2H}\right) \Delta \omega_r(k) + \left(-\frac{K_2 \omega_0 {\Delta_t}^2}{2H}\right) \Delta e_{fd}(k) \tag{27}$$

Additionally, for the reference trajectory related to the electromagnetic torque of the generator, we have: $\Delta e_{Te}(k+1) = -\frac{2H}{\Delta_t}\Delta\omega(k-1) + \frac{2H}{\Delta_t}\Delta\omega(k) - K_1\Delta\delta(k) - K_2\Delta\Psi_{fd}(k)$.

Where the final analysis is performed by Equation (28):

$$\Delta e_{Te}(k+1) = \left(-\frac{2H}{\Delta_t}\right)\Delta\omega(k-1) + \left(\frac{2H}{\Delta_t} - K_1\omega_0\Delta_t\right)\Delta\omega(k) + (-K_2\omega_0\Delta_t)\Delta e_{fd}(k) \quad (28)$$

Therefore, the reference trajectory in its matrix form is represented by Equation (29):

$$\Delta e_{Te}(k+1) = \mathrm{A}_1\Delta\omega(k-1) + \mathrm{A}_2\Delta\omega(k) + \mathrm{A}_3\Delta V_t(k) + \mathrm{E}_2\Delta e_{fd}(k) \quad (29)$$

where the multiplier matrices used in the simplification are:

$$E_2 = \begin{bmatrix} -\mathrm{K}_4\omega_0\Delta t \\ \dfrac{-\mathrm{K}_2\omega_0\Delta t^2}{2H} \\ -\mathrm{K}_2\omega_0\Delta t \end{bmatrix} \quad \mathrm{A}_1 = \begin{bmatrix} 0 \\ -1 \\ \dfrac{-2H}{\Delta t} \end{bmatrix}$$
$$\mathrm{A}_2 = \begin{bmatrix} -\mathrm{K}_3\omega_0\Delta t \\ 2 - \dfrac{\mathrm{K}_1\omega_0\Delta t^2}{2H} \\ \frac{2H}{\Delta t} - \mathrm{K}_1\omega_0\Delta t \end{bmatrix} \quad \mathrm{A}_3 = \begin{bmatrix} 1 \\ 0 \\ 0 \end{bmatrix} \quad (30)$$

Reducing the matrix equation even more by joining the first three terms, results in Equation (31):

$$\mathrm{E}_1 = \mathrm{A}_1\Delta\omega(k-1) + \mathrm{A}_2\Delta\omega(k) + \mathrm{A}_3\Delta V_t(k) \quad (31)$$

Equation (29) can be rewritten as Equation (32):

$$\Delta e_{Te}(k+1) = \mathrm{E}_1 + \mathrm{E}_2\Delta e_{fd}(k) \quad (32)$$

The quadratic cost function can be written as Equation (33):

$$\mathrm{J} = \Delta e^T \mathrm{Q}\Delta e + \Delta e_{fd}{}^T \mathrm{R}\Delta e_{fd} \quad (33)$$

By doing (32) in (33), we get the quadratic cost function represented as Equation (34):

$$\mathrm{J} = \left(\mathrm{E}_1 + \mathrm{E}_2\Delta e_{fd}(k)\right)^T \mathrm{Q}\left(\mathrm{E}_1 + \mathrm{E}_2\Delta e_{fd}(k)\right) + \Delta e_{fd}{}^T \mathrm{R}\Delta e_{fd} \quad (34)$$

Rearranging the terms, it remains:

$$\mathrm{J} = \mathrm{E}_1{}^T\mathrm{QE}_1 + (\mathrm{E}_1{}^T\mathrm{QE}_2 + \mathrm{E}_2{}^T\mathrm{QE}_1)\Delta e_{fd} + (\mathrm{E}_2{}^T\mathrm{QE}_2 + \mathrm{R})\Delta e_{fd}{}^2 \quad (35)$$

Applying the minimization of the function J, that is: $\frac{\delta \mathrm{J}}{\Delta e} = 0$
$(\mathrm{E}_1{}^T\mathrm{QE}_2 + \mathrm{E}_2{}^T\mathrm{QE}_1) + 2(\mathrm{E}_2{}^T\mathrm{QE}_2 + \mathrm{R})\Delta e_{fd} = 0$
The final equation is represented as Equation (36):

$$\Delta e_{fd} = \frac{-(\mathrm{E}_1{}^T\mathrm{QE}_2 + \mathrm{E}_2{}^T\mathrm{QE}_1)}{2(\mathrm{E}_2{}^T\mathrm{QE}_2 + \mathrm{R})} \quad (36)$$

The Q matrix from Equation (21) is included in Equation (36) and was developed with the weights extracted from the analysis and modeling of PAL2v. In the model of the PPC-PAL2v, the factors obtained from the characteristics of the EPS are applied in Equation (36) to optimize the value for the field excitation control of the synchronous generator according to the variables involved in the process.

5. Results

In this work, two particular cases are presented for the performance evaluation of the PPC-PAL2v:

Case 1—Variation of 5% of the mechanical power of the generator.

Case 2—Variation of 2% in the reference voltage at the generator terminals.

Comparisons of results are made in relation to those obtained with the AVR and PSS, acting together and using the conventional MPC (Model Predictive Controller).

The results which are obtained for these two conditions with a prediction horizon N = 1 are shown below.

Results in case 1: At the 10 s instant, the synchronous generator begins to suffer a decrease in the mechanical torque of 0.018 pu/min, which, in nominal values of the generator, is equivalent to 10 MW/min. This causes an initial imbalance between mechanical torque and electric torque, and the machine rotor begins to decelerate.

Figure 9a shows the comparisons of velocity deviation, resulting in the 3 control methods: AVR+PSS, conventional MPC and the PPC-PAL2v.

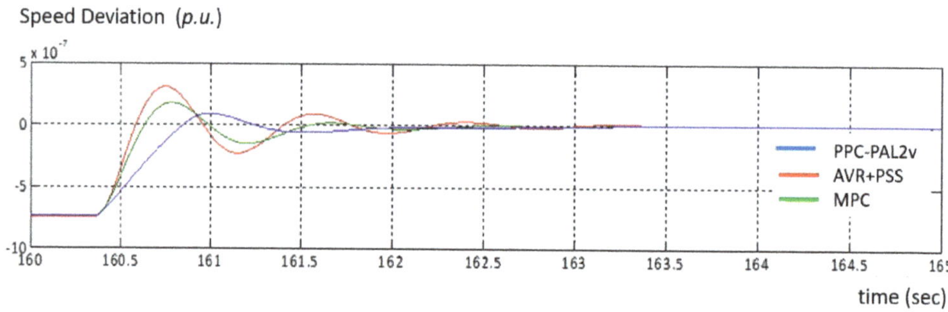

(a)

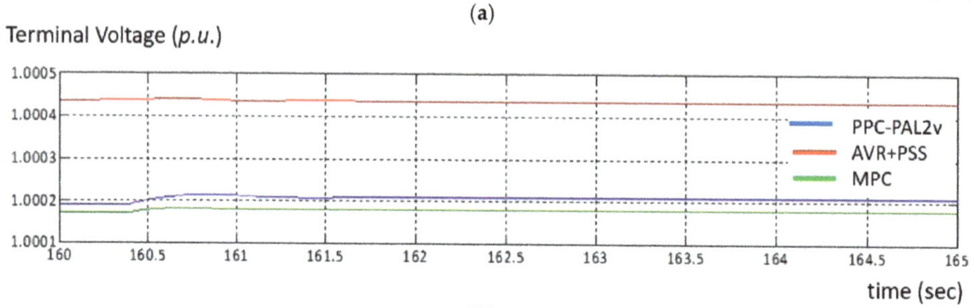

(b)

Figure 9. Case 1—Graphical results with deviations in speed and voltage values at the generator terminals using the three control methods: AVR+PSS, conventional MPC and the PPC-PAL2v. (**a**) Comparison of the deviation in Velocity—Case 1. (**b**) Comparison of the voltage generator values deviation—Case 1.

The graphs show the behavior of the speed at the moment when the mechanical torque reaches its final value of −5% of its initial value.

Figure 9b shows the comparisons of values resulting from voltage at the generator terminals in the 3 control methods: AVR+PSS, conventional MPC and the PPC-PAL2v.

The graphs show the behavior of voltage values at the generator terminals at the moment when the mechanical torque applied reaches the stabilization.

Figure 10a–c show the degrees of evidence of the outputs of the Paraconsistent Analysis Network algorithms.

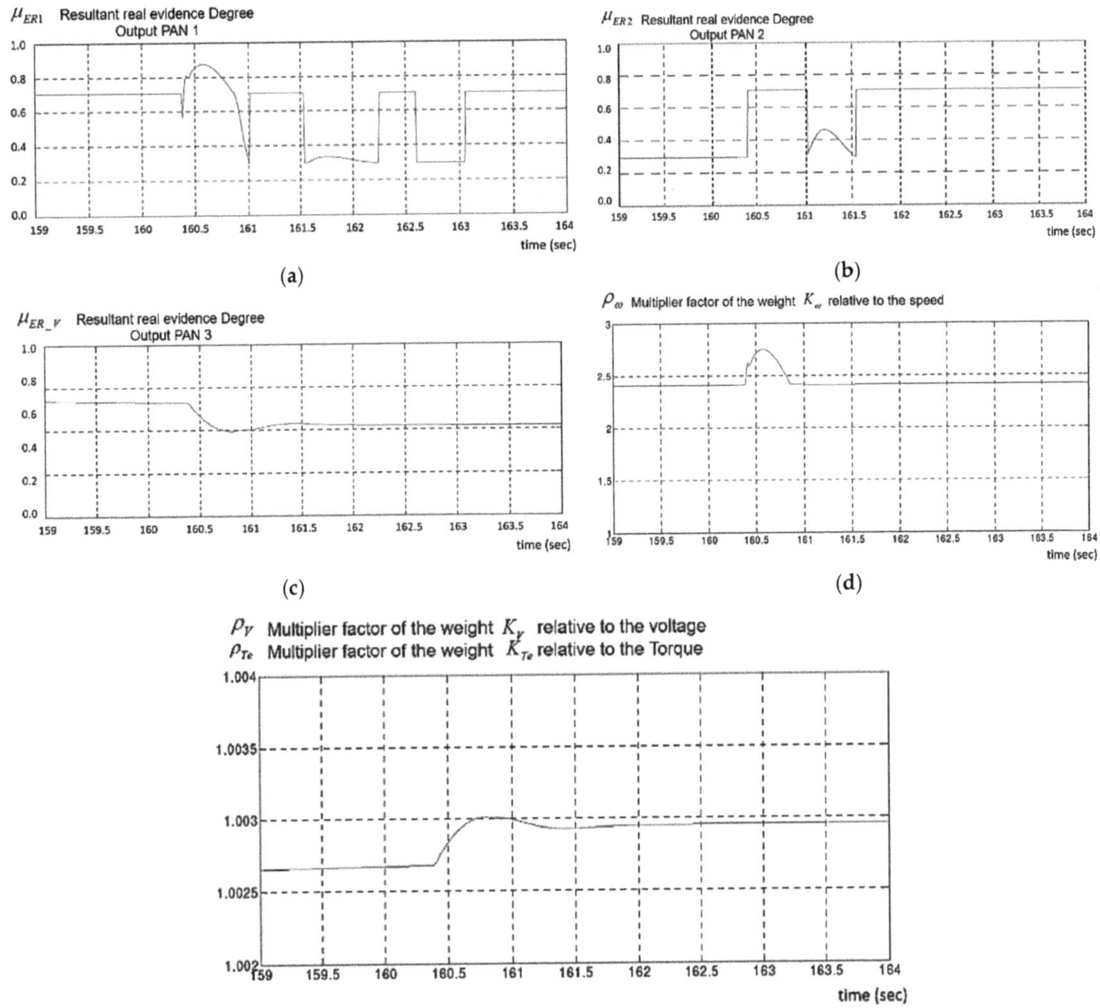

(e)

Figure 10. Case 1—Graphical results with evidence degrees at PANnet outputs and the Matrix Q weights using the PPC-PAL2v. (**a**) Resultant real evidence Degree-Output PAN 1. (**b**) Resultant real evidence Degree-Output PAN 2. (**c**) Resultant real evidence Degree-Output PAN 3. (**d**) Multiplier factor of the weight relative to the speed. (**e**) Multiplier factor of the weight relative to voltage at the generator terminals and multiplier factor of the weight relative to the torque.

Figure 10d,e show the behaviors of the multiplier factors of weights that act in the Q Matrix in this case 1.

Results in case 2: The performance of the voltage at the generator terminals, in case 2, when a decrease of 2% of the reference voltage in the generator bus is realized at the instant of time 10 s.

Figure 11a shows the comparisons of velocity deviation resulting in the 3 control methods: AVR+PSS, conventional MPC and the PPC-PAL2v.

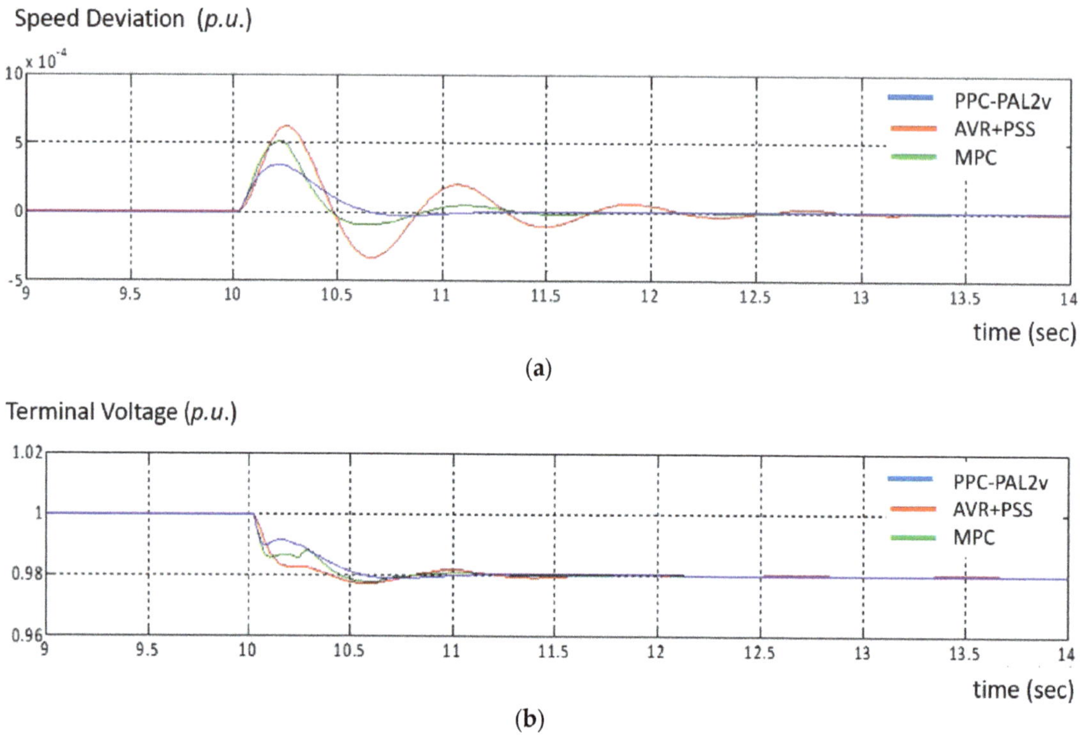

Figure 11. Case 2—Graphical results with deviations in speed and voltage values at the terminals using the three control methods: AVR+PSS, conventional MPC and the PPC-PAL2v. (**a**) Comparison of the deviation in Velocity—Case 2. (**b**) Comparison of the voltage generator values deviation—Case 2.

The graphs show the behavior of the speed in the moment when the mechanical torque reaches its final value of −2% of its initial value.

Figure 11b shows the comparisons of values resulting from voltage at the generator terminals in the 3 control methods: AVR+PSS, conventional MPC and the PPC-PAL2v.

The graphs show the behavior of voltage values at the generator terminals in the moment when the mechanical torque applied reaches the stabilization.

Figure 12a–c show the degrees of evidence of the outputs of the Paraconsistent Analysis Network algorithms.

Figure 12d,e show the behaviors of the multiplier factors of weights that act in the Q Matrix in this case 2.

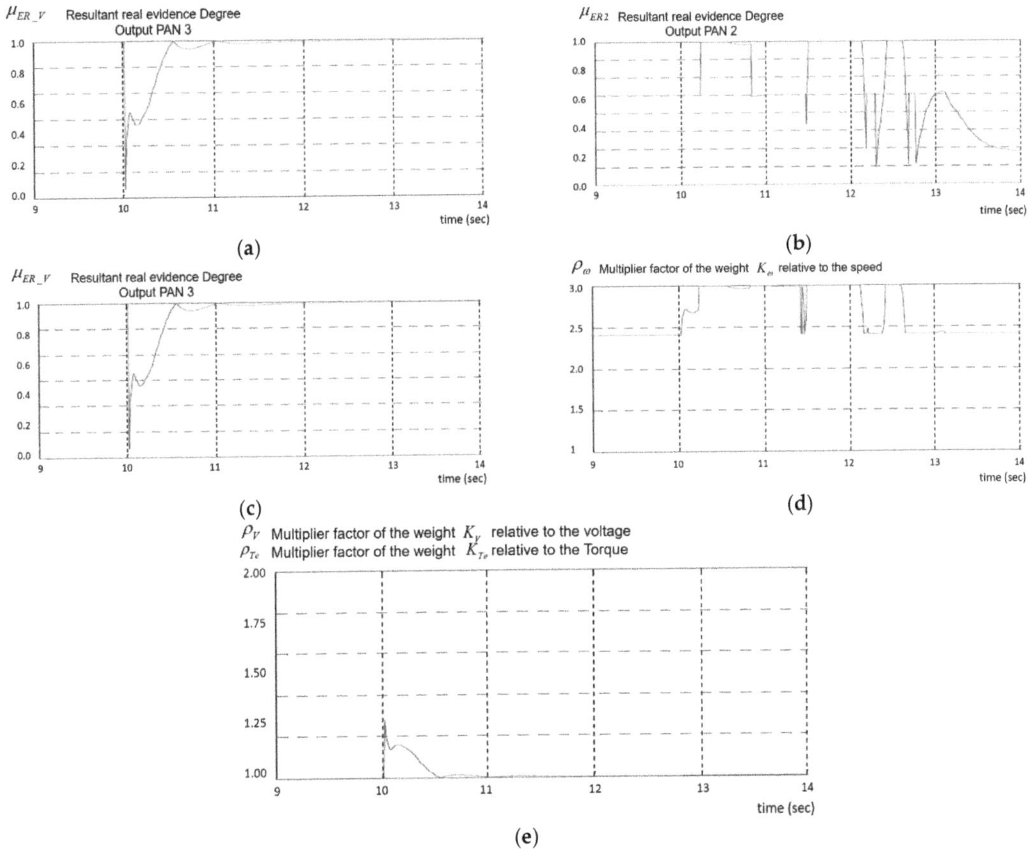

Figure 12. Case 2—Graphical results with evidence degrees at PANnet outputs and the Matrix Q weights using the PPC-PAL2v. (**a**) Resultant real evidence Degree-Output PAN 1. (**b**) Resultant real evidence Degree-Output PAN 2. (**c**) Resultant real evidence Degree-Output PAN 3. (**d**) Multiplier factor of the weight relative to the speed. (**e**) Multiplier factor of the weight relative to voltage at the generator terminals and multiplier factor of the weight relative to the torque.

6. Discussion

Case 1: In case 1, at the initial moment the velocity deviation, the PPC-PAL2v and conventional MPC controllers presented better performance than the AVR+PSS in terms of the peak values of the oscillations and the speed in the oscillation damping. As can be seen in Figure 9, the PPC-PAL2v and conventional MPC controllers reach a practically constant value after 12 s, where the speed is lower than the nominal value, while the AVR+PSS controller tends to have an accommodation with a lower speed than the other two.

The performance of the PPC-PAL2v cushions the electromechanical oscillation faster compared to the conventional MPC. In Figure 9a, in 160.4 s the mechanical torque reaches the final value, but electric torque continues to decrease in a short time, causing the acceleration of the rotor. Thus, its velocity increases, reaching stability in 162 s for PPC-PAL2v, 162.25 s for MPC and 163 s for AVR+PSS.

In Figure 9b, when the synchronous generator initiates the descent of the mechanical torque, the voltage at the terminals of the generator decreases and soon after an ascending ramp begins. The conventional MPC controller starts its rise by a value above the PPC-PAL2v, and the AVR+PSS controller has a higher slope than the other two controllers.

When the mechanical torque reaches its final value, with increasing rotor speed, the voltage tends to rise in a small step and stabilizes within the 5% variation band determined by the controllers, with the value for PPC-PAL2v a little above the accommodation value for the conventional MPC, and below the value for the AVR+PSS.

Figure 10 shows the values resulting from the actions of the Paraconsistent Algorithms to perform the control in the circuit under study. We can verify that, for the optimized control obtained by PPC-PAL2v in case 1, there was no saturation of any output values of the PANs nor of the multiplier factors of the weight.

Case 2: It is observed in Figure 11a that, for the deviation in velocity, the PPC-PAL2v and MPC controllers presented better performance than the AVR+PSS for damping the oscillation. It is found that the MPC achieves stability after 12 s, while the PPC-PAL2v controller approaches the desired value faster. Regarding the voltage performance at the generator terminals in case 2, shown in Figure 11b, until the instant 10.3 s, the AVR+PSS controller has a more efficient response in the voltage adjustment than the PPC-PAL2v and MPC controllers, while the MPC controller walks faster to 0.98 p.u. than the PPC-PAL2v at the initial time. After 10.6 s, the PPC-PAL2v crosses 0.98 p.u., and accommodates quickly compared to AVR+PSS and MPC controllers.

Figure 12 shows the values resulting from the actions of the Paraconsistent Algorithms to perform the control in the circuit under study. We can verify that, for the optimized control obtained by PPC-PAL2v in case 2, there was saturation in some moments of the output values of the PANs and in the multiplier factors of the weight. However, the system was able to carry out the control more efficiently than the other methods.

7. Conclusions

In this paper, we presented a predictive controller, implemented with the fundamentals of Paraconsistent Annotated Logic (PAL) in their extended form, which we denoted Paraconsistent Annotated Logic with annotation of two values—PAL2v. This Paraconsistent Predictive Controller (PPC-PAL2v) was applied in the control of a Synchronous Generator in the case of stability to small signals. For the validation tests of the PPC-PAL2v, a mathematical model of a Single-Machine Infinite Bus (SMIB) was developed, as well as the model of an Automatic Voltage Regulator (AVR) in conjunction with a Power System Stabilizer (PSS) and a classical Model Predictive Control (MPC). After this, the results of the performances of the three control methods were compared. From the computational simulation of the proposed model, it was verified that the PPC-PAL2v effectively presented a superior result to the classical control of the AVR in conjunction with the PSS in the two cases studied. Regarding the MPC, which presents constant weight adjustment, the PPC-PAL2v was also superior due to its characteristic of variable adjustment in a continuous space of values with predetermined limits. The application of the algorithms of PAL2v, forming the Paraconsistent Analysis Network (PANnet) which acts on the stability of the small signals of power systems, is innovative and has shown very promising properties. The results found in this research will serve as a basis for new developments and future applications of the PAL2v in the control and automation of electric power systems.

Author Contributions: Resources and methodology, J.I.D.S.F. and R.A.B.d.O.; Formal analysis, M.C.R., H.M.C. and A.R.; original draft preparation, M.C.M., D.V.G. and J.M.A.; writing, C.R.T. and V.B.D.R. writing—review and editing, A.C.D.R., A.d.C.J. and G.L.-T. All authors have read and agreed to the published version of the manuscript.

Funding: This research received no external funding.

Data Availability Statement: Not applicable.

Conflicts of Interest: The authors declare no conflict of interest.

References

1. Kundur, P.; Dandeno, P.L. Implementation of Advanced Generator Models into Power System Stability Programs. *IEEE Power Eng. Rev.* **1983**, *PER-3*, 31. [CrossRef]
2. Rogers, G. *Power System Oscillations*; Kluwer Academic Publishers: Alphen aan den Rijn, The Netherlands, 2000; pp. 28–50. [CrossRef]
3. Yu, Y.-N.; Moussa, H.A.M. Optimal Stabilization of a Multi-Machine System. *IEEE Trans. Power Appar. Syst.* **1972**, *PAS-91*, 1174–1182. [CrossRef]
4. Kazmierkowski, M.P. Power System Dynamics: Stability and Control, Third Edition [Book News]. *IEEE Ind. Electron. Mag.* **2020**, *14*, 94–95. [CrossRef]
5. Ali, M.A.S.; Mehmood, K.K.; Kim, C.-H. Power System Stability Improvement through the Coordination of TCPS-based Damping Controller and Power System Stabilizer. *Adv. Electr. Comput. Eng.* **2017**, *17*, 27–36. [CrossRef]
6. Kumar, K.; Prakash, A.; Parida, S.K.; Ghosh, S.; Kumar, C. Coordinated Tuning of AVRs and PSSs for Local and Inter-Area Modes of Oscillation in Eastern Regional Grid of India. In Proceedings of the 2021 IEEE 2nd International Conference on Smart Technologies for Power, Energy and Control (STPEC), Bilaspur, India, 19–22 December 2021. [CrossRef]
7. Izdebski, M.; Małkowski, R.; Miller, P. New Performance Indices for Power System Stabilizers. *Energies* **2022**, *15*, 9582. [CrossRef]
8. Law, K.; Hill, D.; Godfrey, N. Robust controller structure for coordinated power system voltage regulator and stabilizer design. *IEEE Trans. Control Syst. Technol.* **1994**, *2*, 220–232. [CrossRef]
9. Arrillaga, J.; Arnold, C.P.; Harker, B.J. *Computer Modeling of Electrical Power Systems*; John Wiley & Sons: Hoboken, NJ, USA, 1983. [CrossRef]
10. CIGRE. Practical use of expert systems in planning and operation of power systems. *Électra* **1993**, *2*, 30–67.
11. Eremia, M.; Shahidehpour, M. *Handbook of Electrical Power System Dynamics: Modeling, Stability, and Control*; John Wiley & Sons: Hoboken, NJ, USA, 2013; ISBN 9781118497173. [CrossRef]
12. Škultéty, J.; Miklovičová, E.; Bars, R. Predictive Synchronous Generator Excitation Control Based on Laguerre Model. *J. Electr. Eng.* **2013**, *64*, 173–179. [CrossRef]
13. Bolognani, S.; Peretti, L.; Zigliotto, M. Design and Implementation of Model Predictive Control for Electrical Motor Drives. *IEEE Trans. Ind. Electron.* **2008**, *56*, 1925–1936. [CrossRef]
14. Mao, C.; Fan, J.; Malik, O.; Hope, G. Studies of real-time adaptive optimal excitation controller and adaptive optimal power system stabilizer. *IEEE Trans. Energy Convers.* **1992**, *7*, 598–605. [CrossRef]
15. Jin, N.; Pan, C.; Li, Y.; Hu, S.; Fang, J. Model Predictive Control for Virtual Synchronous Generator with Improved Vector Selection and Reconstructed Current. *Energies* **2020**, *13*, 5435. [CrossRef]
16. Clarke, D.W.; Mothadi, C.; Tuffs, P.S. Generalized Predictive Control Part 1 & 2. *Automatica* **1987**, *23*, 137–160.
17. Landau, I.D. Evolution of Adaptive Control. *J. Dyn. Syst. Meas. Control.* **1993**, *115*, 381–391. [CrossRef]
18. Bittanti, S.; Piroddi, L. Neural implementation of GMV control schemes based on affine input/output models. *IEE Proc.-Control Theory Appl.* **1997**, *144*, 521–530. [CrossRef]
19. Fabri, S.; Kadirkamanathan, V. Dual adaptive control of nonlinear stochastic systems using neural networks. *Automatica* **1998**, *34*, 245–253. [CrossRef]
20. Ilina, I.-D.; Tudorache, T. Determination of Operation Characteristics of a Synchronous Generator by Static Experimental Tests. *Adv. Electr. Comput. Eng.* **2016**, *16*, 93–100. [CrossRef]
21. Zhang, Y.; Chen, G.P.; Malik, O.P.; Hope, G.S. A multi-input power system stabilizer based on artificial neural networks. In Proceedings of the IEEE WESCANEX 93 Communications, Computers and Power in the Modern Environment—Conference Proceedings, Saskatoon, SK, Canada, 17–18 May 1993; pp. 240–246. [CrossRef]
22. Blair, H.A.; Subrahmanian, V. Paraconsistent logic programming. *Theor. Comput. Sci.* **1989**, *68*, 135–154. [CrossRef]
23. Abe, J.M.; Nakamatsu, K.; Da Silva Filho, J.I. Three decades of paraconsistent annotated logics: A review paper on some applications. *Procedia Comput. Sci.* **2019**, *159*, 1175–1181. [CrossRef]
24. Da Silva Filho, J.I.; Lambert-Torres, G.; Abe, J.M. *Uncertainty Treatment Using Paraconsistent Logic: Introducing Paraconsistent Artificial Neural Networks*; Frontiers in Artificial Intelligence and Applications; IOS Press: Amsterdam, The Netherlands, 2010; Volume 211, p. 328. [CrossRef]
25. Da Silva Filho, J.I.; Rocco, A. Power systems outage possibilities analysis by Paraconsistent Logic. In Proceedings of the 2008 IEEE Proceedings of the Power and Energy Society General Meeting—Conversion and Delivery of Electrical Energy in the 21st Century, Pittsburgh, PA, USA, 20–24 July 2008. [CrossRef]
26. Da Silva Filho, J.I.; Nunes, C.V.; Garcia, D.V.; Mario, M.C.; Giordano, F.; Abe, J.M.; Pacheco, M.T.T.; Silveira, L.; Silveira, L., Jr. Paraconsistent analysis network applied in the treatment of Raman spectroscopy data to support medical diagnosis of skin cancer. *Med. Biol. Eng. Comput.* **2016**, *54*, 1453–1467. [CrossRef]
27. Ribeiro, J.C.; Cardoso, G.; Silva, V.B.; Oliveira, A.L.; Ricciotti, A.C.D.; Oliveira, P.T. Paraconsistent analysis network for uncertainties treatment in electric power system fault section estimation. *Int. J. Electr. Power Energy Syst.* **2021**, *134*, 107317. [CrossRef]

28. Ricciotti, A.C.D.; Da Silva Filho, J.I.; Oliveira, R.A.B.; Ricciotti, V.B.D.; Côrtes, H.M.; Nicolini, M.A. A new strategy of modulation based on Space Vector Modulation and Annotated Paraconsistent Logic for a three-phase converter. In Proceedings of the 2019 IEEE 15th Brazilian Power Electronics Conference and 5th IEEE Southern Power Electronics Conference (COBEP/SPEC), Santos, Brazil, 1–4 December 2019; pp. 1–6. [CrossRef]
29. Garcia, D.V.; Da Silva Filho, J.I.; Silveira, L.; Pacheco, M.T.T.; Abe, J.M.; Carvalho, A.; Blos, M.F.; Pasqualucci, C.A.G.; Mario, M.C. Analysis of Raman spectroscopy data with algorithms based on paraconsistent logic for characterization of skin cancer lesions. *Vib. Spectrosc.* **2019**, *103*, 102929. [CrossRef]
30. Da Silva Filho, J.I.; Camargo, J.M.; Santos, M.R.; Onuki, A.; Mario, M.; Ferrara, L.; Garcia, D.; Pereira, J.; Rocco, A. Support at Decision in Electrical Systems of subtransmission through selection of Topologies by a Paraconsistent Simulator. *IEEE Lat. Am. Trans.* **2016**, *14*, 1993–1999. [CrossRef]
31. Coelho, M.S.; Da Silva Filho, J.I.; Côrtes, H.M.; de Carvalho, A.; Blos, M.F.; Mario, M.C.; Rocco, A. Hybrid PI controller constructed with paraconsistent annotated logic. *Control Eng. Pract.* **2018**, *84*, 112–124. [CrossRef]
32. Da Silva Filho, J.I.; Abe, J.; Marreiro, A.; Martinez, A.; Torres, C.; Rocco, A.; Côrtes, H.; Mario, M.; Pacheco, M.; Garcia, D.; et al. Paraconsistent Annotated Logic Algorithms Applied in Management and Control of Communication Network Routes. *Sensors* **2021**, *21*, 4219. [CrossRef]
33. Côrtes, H.M.; Santos, P.E.; Da Silva Filho, J.I. Monitoring electrical systems data-network equipment by means of Fuzzy and Paraconsistent Annotated Logic. *Expert Syst. Appl.* **2021**, *187*, 115865. [CrossRef]
34. de Carvalho, A.; Justo, J.F.; Angelico, B.A.; de Oliveira, A.M.; Da Silva Filho, J.I. Rotary Inverted Pendulum Identification for Control by Paraconsistent Neural Network. *IEEE Access* **2021**, *9*, 74155–74167. [CrossRef]

energies

MDPI

Article

Cross-Correlated Scenario Generation for Renewable-Rich Power Systems Using Implicit Generative Models

Dhaval Dalal [1], Muhammad Bilal [1], Hritik Shah [1], Anwarul Islam Sifat [1], Anamitra Pal [1,*] and Philip Augustin [2]

1 School of Electrical, Computer, and Energy Engineering, Arizona State University, Tempe, AZ 85281, USA
2 Salt River Project (SRP), 6504 East Thomas Road, Scottsdale, AZ 85251, USA
* Correspondence: anamitra.pal@asu.edu

Abstract: Generation of realistic scenarios is an important prerequisite for analyzing the reliability of renewable-rich power systems. This paper satisfies this need by presenting an end-to-end model-free approach for creating representative power system scenarios on a seasonal basis. A conditional recurrent generative adversarial network serves as the main engine for scenario generation. Compared to prior scenario generation models that treated the variables independently or focused on short-term forecasting, the proposed implicit generative model effectively captures the cross-correlations that exist between the variables considering long-term planning. The validity of the scenarios generated using the proposed approach is demonstrated through extensive statistical evaluation and investigation of end-application results. It is shown that analysis of abnormal scenarios, which is more critical for power system resource planning, benefits the most from cross-correlated scenario generation.

Keywords: dynamic time warping; generative adversarial network; power system planning; renewable energy; scenario generation

check for updates

Citation: Dalal, D.; Bilal, M.; Shah, H.; Sifat, A.I.; Pal, A.; Augustin, P. Cross-Correlated Scenario Generation for Renewable-Rich Power Systems Using Implicit Generative Models. *Energies* **2023**, *16*, 1636. https://doi.org/10.3390/en16041636

Academic Editor: Germano Lambert-Torres

Received: 10 January 2023
Revised: 24 January 2023
Accepted: 2 February 2023
Published: 7 February 2023

1. Introduction

Reliability planning for the transmission system of the electric power system is essential in keeping the grid operational in times of high uncertainty/variability. Traditionally, this exercise involved managing only one variable—load. The scenarios used to evaluate the system operation and resilience were worst-case loading scenarios derived from historical data with some growth projections. However, globally coordinated initiatives for carbon emission reduction have led to increased emphasis on planning integrated energy systems that feature rapid growth in renewable generation (RG), energy efficiency, and high electrification rates [1–3]. Particularly, due to the increasing proliferation of large RG sources, power system resource planning studies must include additional variables, viz., solar and wind generation. The variability associated with these new variables, which depends on weather conditions, such as solar irradiation and/or wind speed, makes reliability evaluation of renewable-rich power systems a more complex and challenging problem [4–7]. To address this problem, power system planners create *synthetic scenarios* that are aimed at capturing actual system conditions [8–10]. Two strategies that have been extensively used for creating synthetic scenarios are: *classical techniques*, which try to fit a model onto the distribution and then attempt to generate scenarios from the fitted model, and *machine learning approaches*, which learn the distribution from large amounts of historical data and are then able to produce similar scenarios. A brief overview of these two strategies is provided below.

The classical techniques typically rely on probabilistic modeling to generate new scenarios. These include methods that employ Latin hypercube sampling (LHS) [11], generalized dynamic factor model (GDFM) [12], generalized auto-regressive score (GAS) models [13], vine copula methods [14], principal component analysis (PCA) [15], and generalized Gaussian mixture models (GGMM) [16], amongst others. However, despite their complexity, these models cannot fully capture all the correlations between the variables. Consequently, many *classical* studies still focus on one variable at a time [17]. As the power systems become more complex, it will become increasingly hard to extract models that capture all of the system's characteristics using probabilistic methods.

Machine learning models offer flexibility and versatility when generating new scenarios. Particularly, neural-network-based approaches eliminate the need for the extraction of relevant features from the available data. In [18,19], a simple single-layered neural network and radial basis function network (RBFN) were used to forecast wind power ramp-up events and distributions, respectively. However, complex tasks, such as multivariate scenario generation that is being considered in this paper, would require more complex (deeper) architectures.

In recent times, generative adversarial networks (GANs) [20] have emerged as a popular deep-learning algorithm for scenario generation. This implicit generative model is capable of transforming raw noise into meaningful information. Therefore, it can work on a variety of datasets, such as two-dimensional images and one-dimensional time-series data. Furthermore, it can generate samples that replicate the ones available in the data and other, more varied samples *not present* in the original dataset. The performance of a stand-alone GAN has been improved for scenario generation applications by using a hybrid model strategy, tweaking the error function, and/or adding appropriate conditions. For example, recurrent neural networks (RNN) with long short-term memory (LSTM) and reinforcement learning algorithms were added to the GAN model to produce wind power generation scenarios in [21]. This hybrid model strategy was tested on two case studies, and it created varied and believable scenarios in both of them. Similarly, a Wasserstein distance-based error function was embedded into convolutional GANs to improve performance in [22]. This approach was extended to condition-based solar and wind power scenario generation in [23,24]. These models accurately predicted wind ramp events and peak values but treated the variables (namely, solar and wind), independently.

Since the power system is a complex network of interconnected generation sites and electricity consumers, both residential and large-scale, the relationship between the different modes of power production and the nuances of load demand (i.e., their correlations) must be systematically considered. In line with this realization, a convolutional GAN with an LSTM-based sequence encoder was proposed in [25] to perform day-ahead forecasting of correlated photovoltaic (PV) and wind production sequences from meteorological data. In [26], correlated scenarios were generated to determine the most cost-effective generation procedure for optimizing a large-scale hydro–wind–solar hybrid system. Correlated GANs were also used in the cost-optimal scheduling of a battery energy storage system (BESS) to increase the BESS-PV system's incentive revenue [27]. Although GAN-based architectures have been applied to generate correlated scenarios for power systems, the *cross-correlation between RG and load* has not been well-explored. In addition, the scenario-generation techniques developed in [25–27] only focused on generating short-term forecasting scenarios.

To better facilitate long-term reliability planning, there is a genuine need to capture the cross-correlation present between the variables (RG and loads) while creating representative scenarios. The interdependence between these variables occurs naturally in the historical data. However, if the variables are treated independently during the scenario generation process, there is a risk of losing this interdependence and generating less meaningful scenarios. Particularly, under abnormal conditions, ignoring these cross-correlations and using the independent scenario generation approach can result in grossly misleading outcomes (see Section 4). At the same time, note that incorporating cross-correlation during multivariate scenario generation is a more challenging task. To accomplish this task, a sophisticated implicit generative model is proposed, as explained below.

1.1. Major Contributions

In this paper, a conditional recurrent GAN is proposed to generate cross-correlated scenarios on a seasonal basis. The labeling of the historical data for GAN training is an important aspect of the methodology. In the presence of multiple variables, the determination of normal/abnormal days is not straightforward. Instead of relying on normal and abnormal labels assigned based on visual inspection, a data-driven technique is developed to create seasonal labels. Then, based on the labels, normal and abnormal day assignments are made for each season. Along with cross-correlation, this approach also captures the temporal correlations present in the time-series data of each variable.

For the generation of statistically similar but distinct correlated scenarios, the conditional recurrent GAN is modeled with the use of RNN-LSTM. An RNN-LSTM incorporated GAN is able to better process and reproduce the long-term modalities and temporal aspects in time-series data compared to a conventional GAN which does not consider these properties. Exploiting the temporal modeling capabilities of RNN-LSTMs along with the latent conditional feature modeling power of label-incorporated GANs helps enhance the relevance of the generated scenarios for different end-applications.

An extensive validation of the proposed approach is also provided in which correlated scenarios are compared against uncorrelated scenarios for an actual power system application, namely, optimal power flow (OPF). For normal conditions, uncorrelated synthesis of scenarios has a performance similar to correlated scenarios. However, during abnormal conditions, the results obtained using correlated scenario generation are more realistic than those obtained using uncorrelated scenario generation.

In summary, the novel contributions of this paper are as follows:

- Creation of a fine-tuned cross-correlated conditional recurrent GAN (C^2RGAN) for multivariate scenario generation. This implicit generative model is scalable and yields relevant abnormal scenarios to augment limited historical data.
- Formulation of a data-driven labeling process for historical data to eliminate the subjectivity associated with manual labeling.
- Demonstration of the validity of correlated scenario generation for the power system OPF application in terms of cost and voltage angle distribution.

1.2. *Paper Organization and Key Terms*

Some of the salient terms used in this paper are explained here to provide the appropriate context.

- *Normal day* refers to a day that follows the typical seasonal pattern.
- *Abnormal day* refers to a day where one of the variables (RG and/or load) deviates significantly from the typical seasonal pattern. This is different from an abnormal operating condition/event that typically refers to line faults and sudden or unexpected load-shed/generator shut-down.
- *Scenario generation* refers to creation of *representative scenarios* for long-term resource planning. This is different from scenario forecasting, which is typically used for short-term day-ahead planning.
- *Cross-correlated scenarios*, one of the main contributions of this paper, refer to those *representative scenarios* that capture the inherent correlations between the variables. Implicit generative models are employed to extract these correlations.

The rest of the paper is structured as follows. Section 2 presents relevant insights drawn from the data-driven label assignment of the historical data used for the analysis conducted here. Section 3 provides a detailed look into the GAN architecture, selection, design, training, and implementation. Section 4 delves into an extensive analysis of the results obtained using the proposed method and their comparison with the uncorrelated scenario generation results. The conclusion is provided in Section 5.

2. Data-Driven Label Assignment

A two-variable (load and solar generation) dataset was employed in this research. The primary requirement is to generate cross-correlated labels for normal and abnormal days in the dataset so that the C^2RGAN can be trained conditionally. Additionally, it was determined that the seasonal variations in load and solar generation are significant enough to warrant the breakdown of the data by seasons first and then classify normal and abnormal days within each season. This methodology creates more homogeneity within each labeled dataset and allows the GAN to be trained better. For example, a normal winter day is sufficiently different from a normal summer day as both load and solar generation are significantly lower for the former in comparison to the latter.

2.1. Seasonal Classification

Rather than relying on a calendar-based approach to classify seasons, seasonal breakdowns are identified based on the available historical data. This enables us to capture spatial determinants of seasonal variations, such as geographic insolation and local weather patterns as well as geographic load patterns (e.g., heavy loads in winter for the colder climates and heavy loads in summer for the hotter climates). Moving daily averages of load and solar generation are plotted first, and the plot is then partitioned based on the pattern transition in both load and solar generation. Figure 1 depicts the resultant partitioning for the available historical data. The non-summer/non-winter days are classified as shoulder days. The characteristics of each season are captured in Table 1. Note that Shoulder A and Shoulder B are combined for training the GAN as they represent very similar (average) characteristics despite different slope polarities.

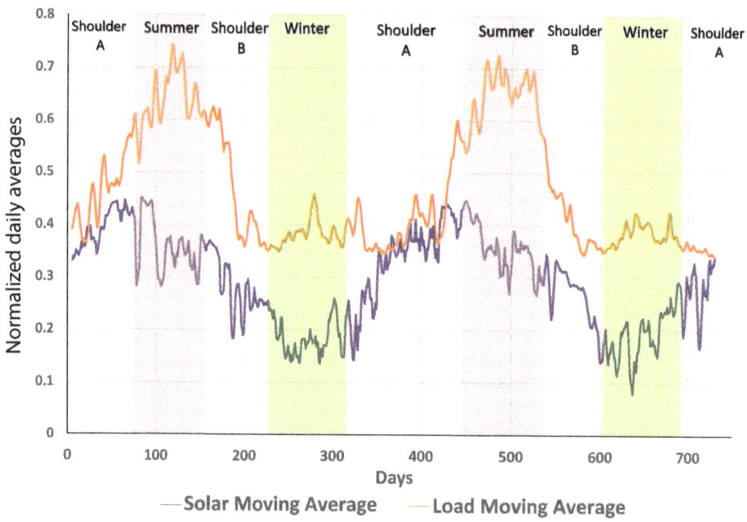

Figure 1. Seasonal classification of two years of historical data.

Table 1. Seasonal patterns.

Season	Load Pattern	Solar Pattern
Shoulder A	Low (increasing at the end)	Increasing to peak
Summer	High (peaking in the middle)	Decreasing
Shoulder B	Decreasing	Decreasing
Winter	Flat and low	Flat and low

2.2. Normal and Abnormal Days Classification

From a power system reliability planning perspective, understanding abnormal conditions is much more critical than normal conditions. Abnormal conditions require special attention, as any mismatch or sudden change in solar generation and load patterns could impact the net load significantly. If the generated abnormal scenarios have a good correlation to the actual historical abnormal scenarios, they can enhance the analysis and understanding of such scenarios. Additionally, since the historical data does not have many abnormal scenarios, this type of scenario generation helps evaluate the system under such conditions by generating additional scenarios that are *similar but distinct*.

Normal and abnormal classification must be based on a metric that can provide clear and consistent differentiation between normal and abnormal days. In many cases, the choices of normal and abnormal days are based on ad hoc decisions. Instead, the methodology employed in this work uses the distance of any given day from a reference normal day to decide on the labeling of that day. While this methodology is universally applicable, it was determined that the application to the seasonally partitioned data was more appropriate.

To implement this methodology, a *reference normal day* must be identified for each season. The representative scenario generation methodology developed in [28] generates median representations for all seasons. The *reference normal day* is selected from the seasonal cluster by identifying the day with the shortest distance to the median seasonal representation. Since the daily profiles are time-series representations, dynamic time warping (DTW) is employed as it provides a better measure of (dis)similarity between days than the Euclidean distance measure [29,30]. DTW computes the best alignment between two time series by identifying the path with the minimum time-normalized distance between them. This is given by (1).

$$P^* = \arg\min_{P} \left[\frac{\sum_{s=1}^{k} d(p_s)}{k} \right] \tag{1}$$

where $d(p_s)$ is the distance between time-series points i_s and j_s, k is the length of the warping path, and P is the warping function. For univariate time-series data (e.g., hourly solar profile per day), the time series $i_s, j_s \in \mathbb{R}^{24\times 1}$, and the DTW operation identifies the smallest distance by permuting through the different paths from hour 1 to hour 24. When considering multivariate time-series data, each day is represented by a matrix $d_i \in \mathbb{R}^{24\times m}$ (where m is the number of variables), and the DTW operation is performed for each variable. DTW is particularly important for the multivariate case as it can accurately capture the *cross-correlations* between variables (such as load and solar generation), and it identifies similarities between patterns even if they are time-displaced.

For n days in a given seasonal dataset, DTW creates a symmetrical $\mathbb{R}^{n\times n}$ matrix that has the DTW distances between each pair of days. Next, the distance of each day in a season to its *reference normal day* is computed, and the sorted distances are plotted. The slope change can be detected by taking the second derivative of the DTW distance plots. The data points beyond this change point represent the abnormal days as they have farther and faster-growing distances from the reference normal day. This is illustrated in Figure 2. Applying the criterion outlined above, the summer season is split into 143 normal days and 39 abnormal days (from 2 years of data). A similar classification was obtained for the winter (146 normal and 36 abnormal days) and shoulder (327 normal and 40 abnormal days) seasons.

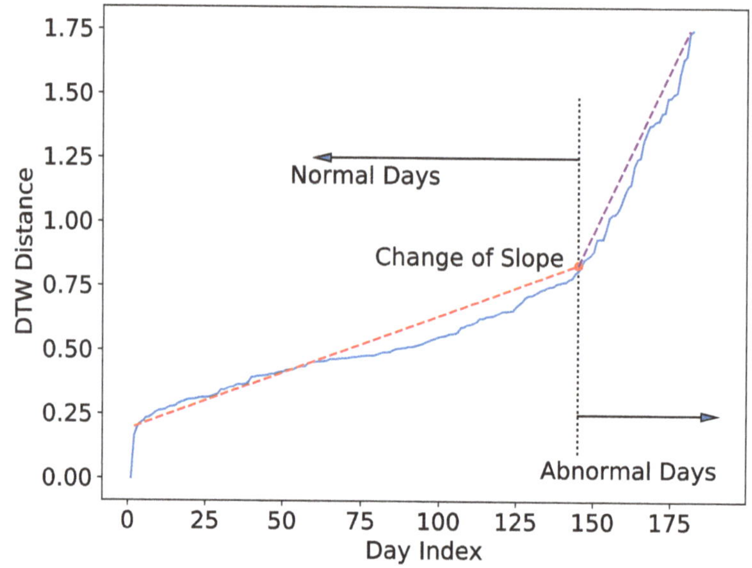

Figure 2. Summer season distances to reference normal day and normal/abnormal classification. Raw data (blue), fitted data for normal days (red), fitted data for abnormal days (purple).

3. Proposed Implicit Generative Model and Its Implementation

GANs are composed of two neural networks battling against one another (see Figure 3). The first neural network is called the generator, which aims to generate the synthesized samples. The second neural network is called the discriminator (or the critic). The discriminator's job is to differentiate between the real and the generated samples. The main objective of a GAN is to learn the distribution of a real dataset and map it to a separate latent space, from which more samples, similar to the original dataset, can be synthesized.

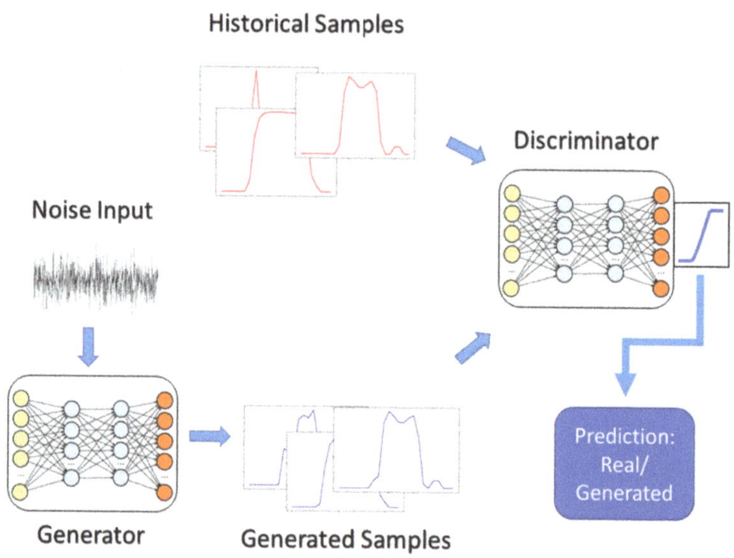

Figure 3. Architecture of generative adversarial networks (GANs).

Let us have a dataset, X, with samples x_i^t for time $t \in T$, and with dimensions i, whose distribution, $\mathbb{P}_x$, is to be learned by the generative model. Noise vector inputs, z, are sampled from a latent space, $\mathbb{P}_z$, and the multi-layer perceptrons within the generator, $G(z, \theta_g)$, are trained to map $\mathbb{P}_z$ to $\mathbb{P}_x$, without explicitly training on $\mathbb{P}_x$. This is accomplished by the generator producing samples as close to the real data's distribution as possible (denoted by $\mathbb{P}_x$). In contrast, the discriminator, $D(x, \theta_d)$, tries to distinguish the real samples from the generated ones and forces the generator to perform better. As the training progresses, the generator becomes better at producing realistic-looking samples, while the discriminator gets better at distinguishing generated samples from the real ones. The losses of the generator and the discriminator are expressed as,

$$L_G = -\mathbb{E}_Z[\log(D(G(z)))] \tag{2}$$

$$L_D = -\mathbb{E}_X[\log(D(x))] + \mathbb{E}_Z[\log(D(G(z)))] \tag{3}$$

The training of the generator and the discriminator can be summarised as a two-player mini–max game with the value function $V(G, D)$,

$$\min_G \max_D V(G, D) = \mathbb{E}_X[\log(D(x))] + \mathbb{E}_Z[\log(1 - D(G(z)))] \tag{4}$$

3.1. *Proposed Conditional Recurrent GAN*

GANs can be trained conditionally by incorporating labels in the training dataset, allowing the generator the ability to generate samples based on a certain event or condition. The label, y, can be any auxiliary information that can be appended to the real samples, x. The generator will then learn to associate a certain class of data with its associated label. After training has been finished, the generator can be forced to produce only a certain class of samples by appending the corresponding label, y, to all the noise vectors. The value function of the conditional GANs, conditioned on the label y, can be written as,

$$\min_G \max_D V(G, D) = \mathbb{E}_X[\log(D(x|y))] - \mathbb{E}_Z[\log(D(G(z|y)))] \tag{5}$$

Since the available historical data was a multivariate *time series*, it was necessary to include recurrent layers in both the generator and the discriminator. Recurrent layers in the generator model retain the time-series long-term modulations and help generate sequences that capture the fluctuations of the real data. In the discriminator, the recurrent layers help identify the sequential data better. The recurrent model of choice was RNN-LSTM, making the proposed machine learning model a conditional recurrent GAN. Note that the LSTM layer ensures that the recurrent GAN is properly trained to capture both short-term (daily) and long-term (seasonal) patterns in the time-series data. Furthermore, it leads to the generation of more homogeneous and valid training data for the GANs, which eventually leads to more consistent generated scenarios as the output of the GAN. The generator and the discriminator models consist of three stacked LSTM layers, along with a linear output layer. The hyperparameters of the models were tuned by comparing the observed outputs to the expected results. To optimize the discriminator output, it was trained thrice as much as the generator to maintain the best estimation ratio between the data density and the model density [31]. The model details are given in Table 2.

Table 2. Design details of cross-correlated conditional recurrent GAN (C^2RGAN) model.

	Layer Type	Input Size	Output Size
Generator	LSTM	259	128
	LSTM	128	128
	LSTM	128	128
	Fully connected	128	24 × 2
Discriminator	LSTM	27 × 2	128
	LSTM	128	128
	LSTM	128	128
	Fully connected	128	1

3.2. Overall Implementation

The proposed approach of systematic model-free data segmentation and scenario generation using implicit generative models has been summarized in Figure 4. First, the historical data is preprocessed by normalizing the different variables to their peak values and creating daily profiles. Next, the data is segmented in preparation for GAN training. The available data is classified by season, and a representative day is selected for each season. Finally, the normal/abnormal classification is performed for each season, leading to the generation of six datasets (three seasons and normal/abnormal for each season). The next phase involves training the generator and discriminator using the labeled correlated datasets. The hyperparameters are tuned, and the loss functions are monitored to achieve an equilibrium that indicates a fully trained GAN model. In the next phase, the GAN model is fed labeled noise to generate similar but distinct scenarios for each of the six datasets. Finally, statistical validation of the generated scenarios is performed before moving onto OPF-based validation concerning the historical data and against the existing methodology of uncorrelated scenario generation.

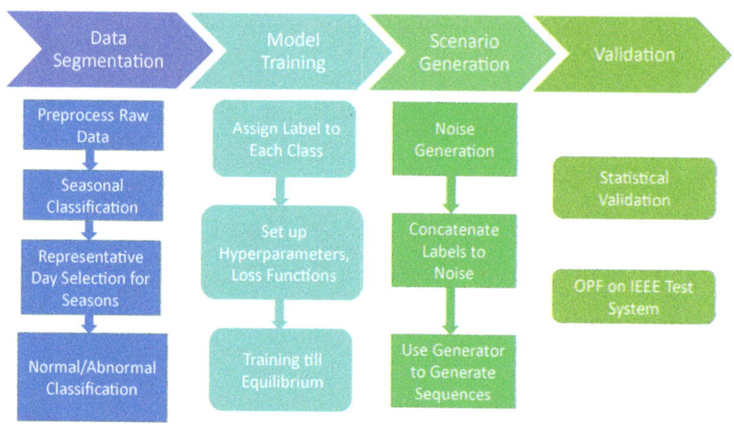

Figure 4. Flowchart of the proposed methodology. The first column captures the steps described in Section 2. The next two columns depict the training and use of the implicit generative model described in Section 3. The final column captures the thorough validation of the proposed methodology, which is described in Section 4.

4. Results and Analyses

The proposed approach was tested on a dataset provided by a power utility located in the US Southwest. The dataset comprised two years of hourly solar generation and load demand profiles at the transmission level. The nature of the data allowed for capture of temporal and cross-correlations within the variables. However, as no spatial information was provided with the dataset, spatial correlations could not be captured. After preprocessing and normalizing the dataset, it was segregated into *summer, shoulder*, and *winter* seasons, followed by classification into *normal* and *abnormal* days. The C^2RGAN was trained with these datasets. The C^2RGAN-generated scenarios were then evaluated for their similarity to the historical data in the same category. Comparison of individual generated profiles for each variable to the historical profiles showed a good match, as shown in Figure 5 for summer normal real and generated load, and Figure 6 for summer normal real and generated solar, respectively. As is evident from the figures, the seasonal segmentation results in scenarios that closely track the temporal variations of the real dataset.

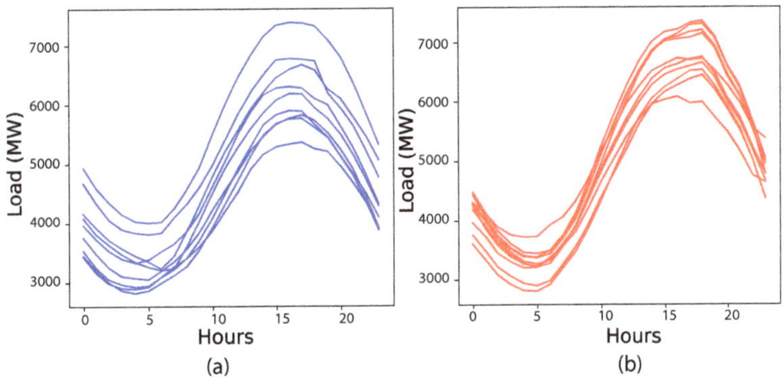

Figure 5. Selected summer season normal load profiles: (**a**) real; (**b**) generated.

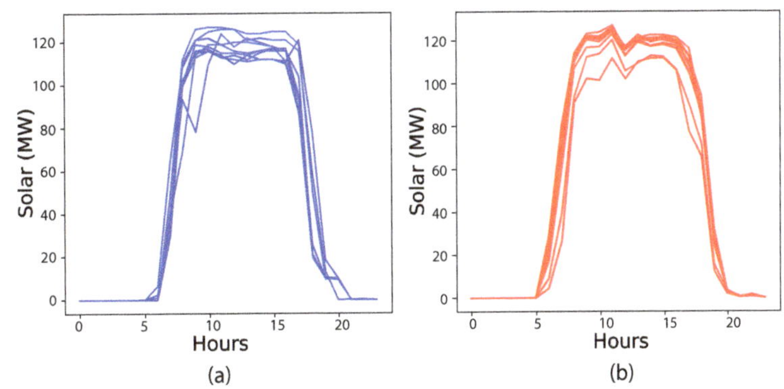

Figure 6. Selected summer season normal solar profiles: (**a**) real; (**b**) generated.

4.1. Statistical Validation of Proposed Implicit Generative Model

Going beyond visual confirmation, we performed rigorous statistical analysis to investigate the performance of the proposed scenario generation methodology. The statistical measure employed was the auto-correlation function (ACF), which defines how data points in a time series are related, on average, to the preceding data points.

Under normal conditions, the ACF shapes of the real and generated datasets for both load and solar were found to be very similar (see Figure 7a,b). The highest positive correlation at one hour for both variables confirms that the nearest temporal value has the highest correlation to any data point. However, since the normal solar peak and zero production times in summer are roughly 10 h each, the highest negative correlation occurs at a 10-h lag for the solar profile. The normal summer load pattern shown in Figure 5 is quasi-sinusoidal with peak and valley 12 h apart, which is consistent with the negative

ACF peak for the load at a 12-h lag. A similar pattern is observed for normal days in other seasons with slight variations in negative ACF peak location.

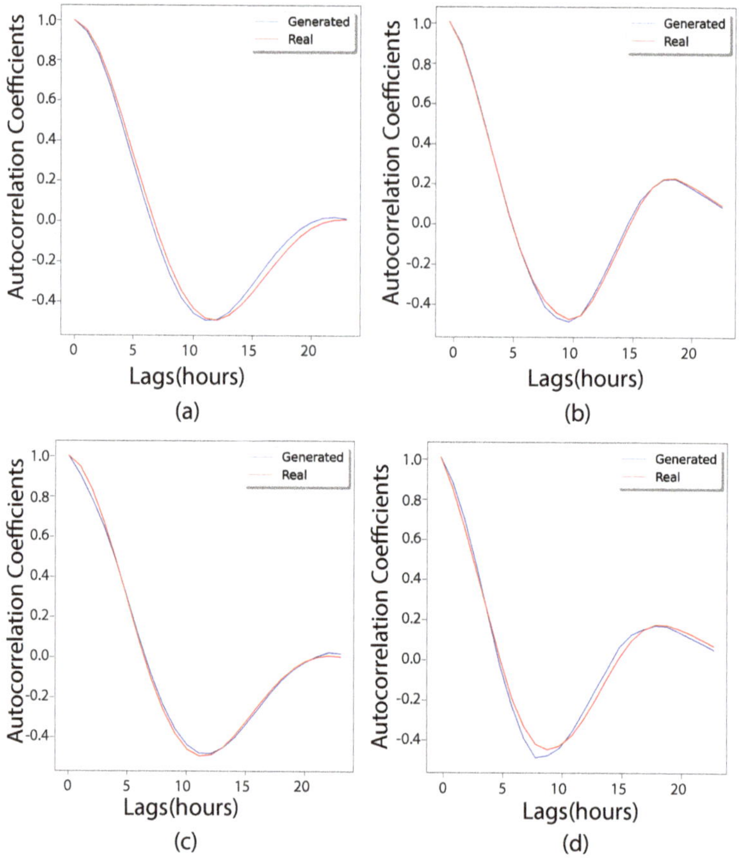

Figure 7. Summer season ACF for: (**a**) normal load; (**b**) normal solar; (**c**) abnormal load; (**d**) abnormal solar for generated and real data.

Under abnormal conditions, the load correlation shapes show a similar pattern as their counterparts under normal conditions, but a slight difference is observed between real and generated shapes for solar (see Figure 7c,d). This happens because the cross-correlated nature of the C^2RGAN can bias one or both of its outputs (solar ACF in Figure 7d), as it is trained on both the variables. Therefore, its accuracy in producing matching scenarios for any one variable might be lower. However, we demonstrate in Section 4.2.2 that for actual power system applications, creating scenarios where the cross-correlations are considered results in more realistic outcomes.

4.2. Comparison with Uncorrelated Scenario Generation for Power System Application

To highlight the value of the correlated scenario generation process, two additional GANs were trained using the same historical data—one for the independent generation of load sequences and one for the independent generation of solar sequences. These univariate uncorrelated GANs (termed load GAN and solar GAN) generate seasonal (normal/abnormal) scenarios for load and solar generation, respectively. Note that many GAN-based scenario generation techniques proposed recently are univariate and hence, uncorrelated (e.g., [22,23]). Therefore, the subsequent analysis is a comparison of the proposed methodology with the state of the art.

The selection of baseline days for uncorrelated scenarios is an important but challenging consideration. As load and solar are processed independently, there is no guaranteed or consistent overlap between the labeled training data for each set. This disjunction is more clearly pronounced for abnormal days. For example, an abnormal summer day for the load (very high load) can be vastly different from an abnormal summer day for solar generation (cloudy or rainy day). Thus, it is impossible to determine baseline days satisfying the same load and solar generation conditions.

One strategy could be to assume that the baseline days were identical for correlated and uncorrelated data. However, doing so will yield consistently favorable results for the correlated scenario generation approach since the baselines are drawn from its training dataset. Consequently, to avoid this possible (implicit) bias in favor of the proposed approach, the following strategy was devised in this paper: the baseline days for the uncorrelated scenarios were synthesized independently from the two training datasets (load and solar generation). Separate comparisons were then made between each approach's generated and baseline values.

4.2.1. Validation Using Optimal Power Flow (OPF) Analysis

To evaluate the performance of the generated scenarios for power systems applications, the generated solar and load profiles were applied to a modified IEEE 30-bus system [32]. A futuristic generation scenario was evaluated, where all the load buses also have solar generation. OPF was run under different ratios of solar generation peak to load peak, ranging from 0.3 to 1.2. Since the scenarios are derived from the historical dataset, the OPF converged for all the scenarios. To lend statistical validity to the exercise, 900 (=30 $\times$ 30) scenarios were generated for both correlated and uncorrelated methodologies for each of the 6 classes (3 seasons $\times$ normal/abnormal). This enabled application of 30 distinct and randomly assigned profiles to all the buses of the system for one OPF computation. The OPF itself was run 30 times—each time with a completely different set of profiles—to ensure consistency of the results.

The distributions for each iteration were compared with the baseline to compute the distance between the two; the Wasserstein distance was used as a measure for this comparison. Additionally, the OPF results provided costs by the hour for each iteration. Finally, the voltage-angle data based on the hour/iteration/bus were collected for further analysis. The results were evaluated from multiple perspectives. Each methodology (correlated and uncorrelated scenario generation) was compared against its baseline to identify which would generate more realistic scenarios. Furthermore, comparisons were made over iterations to evaluate the consistency and on an hourly basis to identify if the gap between generated and baseline scenarios has any time-of-day dependence. The voltage-angle distribution plots were plotted for three different hours: 07, 12, and 17.

4.2.2. Results and Discussion for Abnormal Conditions

Figure 8 shows the shoulder season hourly OPF costs, averaged over 30 OPF iterations, for the solar-to-load ratio of 0.6. The correlated generated scenarios track their baseline much more closely than the uncorrelated scenarios. In addition, the baselines for the two cases show significant differences. The abnormal conditions typically signify lower solar generation (due to cloudy or rainy conditions), which is often accompanied by a lighter load (due to lower cooling requirements). However, the baseline of the uncorrelated case shows significantly higher OPF costs that result from the unrealistic combination of independently derived abnormal conditions (high load and no to low solar generation). The generated scenarios overestimate the costs (i.e., a combination of higher load and lower solar generation), resulting in grossly unrealistic scenarios.

A similar case is presented in Figure 9 for the summer abnormal situation, where the correlated scenario's costs are tracking the baseline costs well (similar to the shoulder abnormal). The baselines for correlated and uncorrelated cases are more closely aligned compared to the shoulder abnormal (except for a few morning hours), but the uncorrelated generated scenarios are underestimating the cost by a large margin. Although not shown in the figures to ensure clarity, the behavior for the other solar-to-load ratio was consistent with these results.

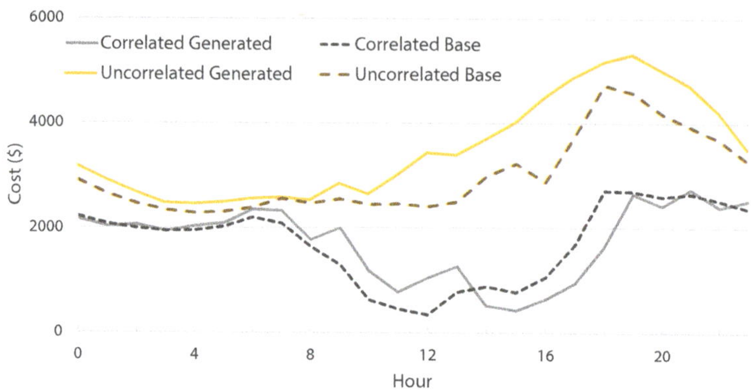

Figure 8. Shoulder abnormal hourly costs (ratio = 0.6).

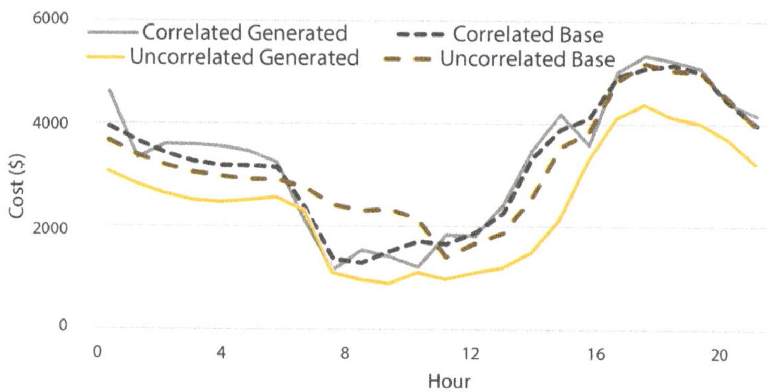

Figure 9. Summer abnormal hourly costs (ratio = 0.6).

In the case of the winter abnormal shown in Figure 10, the baseline costs between correlated and uncorrelated scenarios differ significantly—similar to the shoulder abnormal case. The correlated scenarios are much closer to their baseline than the uncorrelated ones. However, it is observed that the correlated scenarios are overestimating the costs between the hours of 8 AM and 6 PM, indicating that the generated solar scenarios are lower than the baseline. Under the winter abnormal conditions, the solar profiles are predominantly low with a few exceptions, so the GAN is getting trained to generate lower solar profiles. However, since the baseline does contain some higher solar generation profiles, there is some gap between the baseline and the correlated scenarios.

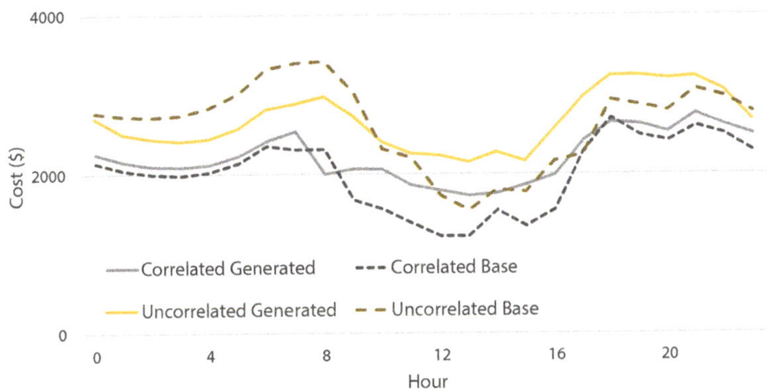

Figure 10. Winter abnormal hourly costs (ratio = 0.6).

In contrast to Figure 8, which provides averaged hourly OPF cost profiles over 30 iterations, Figure 11 depicts the average cost per hour for different OPF iterations for the solar-to-load ratio of 0.6. The correlated variations are narrower in range and closer to the baseline. This chart also underscores the baseline difference discussed above.

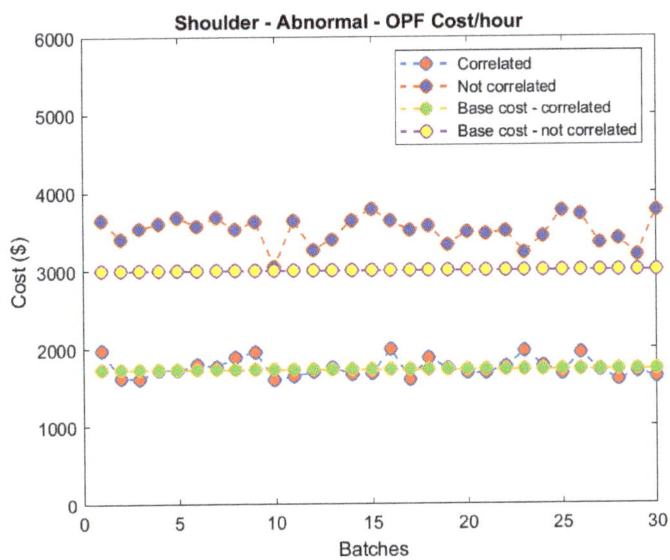

Figure 11. Shoulder abnormal OPF cost (ratio = 0.6).

Another perspective to view the differences between the correlated and uncorrelated approaches is to look at the voltage angle distributions for the 30-bus system. Figure 12 shows the probability density functions (PDFs) of the voltage angles for 5 PM for correlated and uncorrelated scenarios for all three seasons. The better overlap with the baseline distribution is clearly visible for the correlated scenarios. These plots are for one of the 30 iterations, but a similar pattern was observed for other hours and for all iterations, albeit with some variability. The larger difference between the correlated and uncorrelated scenarios in the shoulder season may be partially attributable to the data segmentation technique used in Section 2.2. However, the distinction between the two scenario generation methods is still evident in the other seasons.

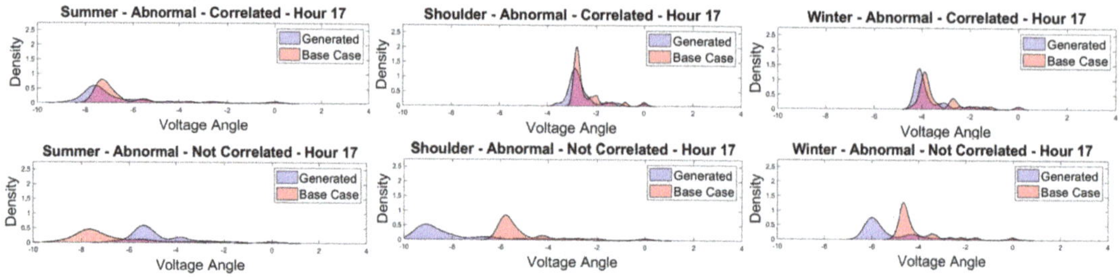

Figure 12. Voltage angle PDFs—abnormal, all seasons.

Table 3 shows the numerical results for abnormal seasonal daily OPFs. It covers solar-to-load ratios from 0.3 to 1.2 for all three seasons and correlated and uncorrelated conditions and reinforces the results and conclusions from earlier plots. The Wasserstein distances for correlated cases are lower than uncorrelated cases under most conditions, often by large margins. The Wasserstein distance should be a low number, but not 0, as we are aiming to obtain similar, but distinct scenarios. Correlated scenarios achieve this objective much better than uncorrelated scenarios, with a minor exception of high solar-to-load ratios in winter, for which the results are comparable. Moreover, the costs for the uncorrelated scenarios for the summer and shoulder seasons point to totally misleading results. For instance, even under abnormal conditions, summer costs should be highest due to high load, and shoulder costs should be lowest due to a combination of low load and good solar generation. However, the uncorrelated scenarios are showing the exact opposite behavior.

Table 3. Result summary for abnormal conditions.

		Correlated				Uncorrelated			
Solar-to-Load Ratio		**0.3**	**0.6**	**0.9**	**1.2**	**0.3**	**0.6**	**0.9**	**1.2**
Average	**Summer**	0.43	0.58	0.77	0.85	1.52	1.59	1.69	1.66
Wasserstein	**Shoulder**	0.40	0.57	0.71	0.77	1.00	1.06	1.12	1.17
Distance	**Winter**	0.31	0.47	0.67	0.84	0.58	0.59	0.62	0.65
Solar-to-Load Ratio		**0.3**	**0.6**	**0.9**	**1.2**	**0.3**	**0.6**	**0.9**	**1.2**
Average	**Summer**	3951	3308	2762	2363	2992	2452	2007	1698
Daily	**Shoulder**	2087	1749	1466	1246	3776	3515	3268	3035
Cost ($)	**Winter**	2342	2211	2084	1961	2796	2674	2556	2443

4.2.3. Results and Discussion for Normal Conditions

The difference between correlated and uncorrelated scenarios is not as significant under normal conditions. In fact, the uncorrelated scenarios showed a closer correlation to the baseline data in the summer season than the correlated scenarios, as shown in Table 4. This is understandable as the solar and load profiles for each season do not have many deviations under normal conditions, and the single variable nature of uncorrelated scenarios allows the corresponding GAN to be trained better for normal, independent signals. However, it was observed that for winter (see Figure 13), the uncorrelated scenarios are farther from their baselines (depicting lower costs) due to the overestimation of the solar generation. The Wasserstein distances for shoulder normal shown in Figure 14 indicate that the uncorrelated distances are higher than correlated ones for most hours of the day. The voltage-angle plots for three different hours for winter normal correlated scenarios (see Figure 15) demonstrate that the distributions are matching the baseline very well.

Table 4. Results summary for normal conditions.

		Correlated				Uncorrelated			
Solar-to-Load Ratio		**0.3**	**0.6**	**0.9**	**1.2**	**0.3**	**0.6**	**0.9**	**1.2**
Average Wasserstein Distance	**Summer**	0.61	0.69	0.77	0.75	0.30	0.33	0.39	0.45
	Shoulder	0.28	0.35	0.41	0.51	0.45	0.56	0.65	0.71
	Winter	0.23	0.35	0.36	0.52	0.21	0.36	0.32	0.33
Solar-to-Load Ratio		**0.3**	**0.6**	**0.9**	**1.2**	**0.3**	**0.6**	**0.9**	**1.2**
Average Daily Cost ($)	**Summer**	3499	2616	1960	1523	3417	2574	1950	1548
	Shoulder	2038	1398	993	670	1832	1208	816	498
	Winter	1978	1592	1323	1107	1940	1572	1333	1139

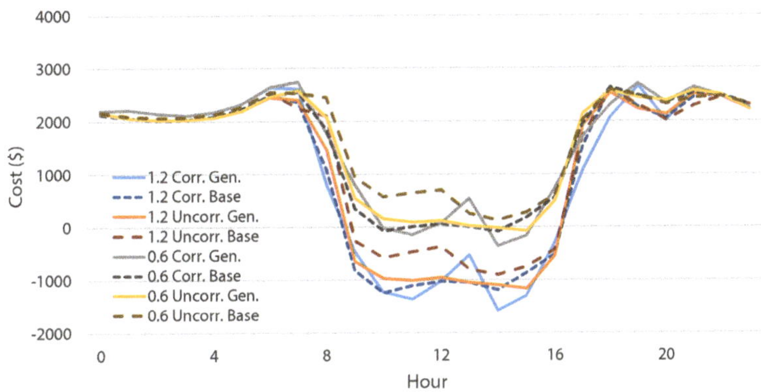

Figure 13. Winter normal hourly OPF costs (ratio = 0.6, 1.2).

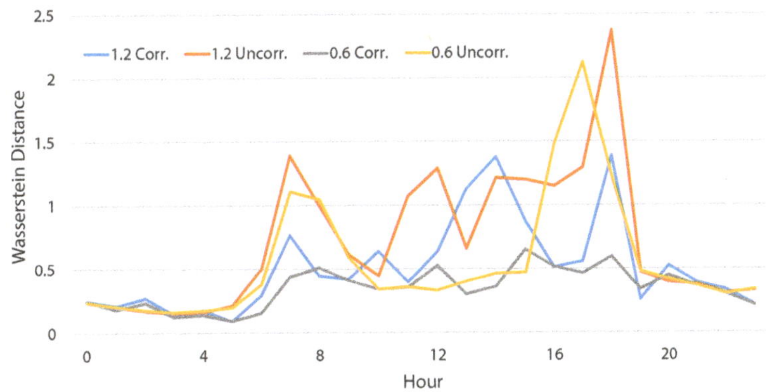

Figure 14. Shoulder normal hourly Wasserstein distances.

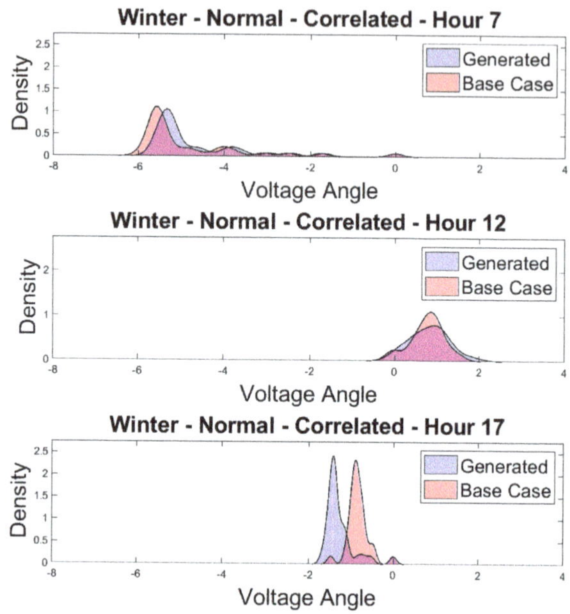

Figure 15. Voltage angle PDFs for winter normal—correlated.

Table 4 shows the normal seasonal summary results for solar-to-load ratios from 0.3 to 1.2 for all three seasons and correlated and uncorrelated conditions. It can be observed from the tables that the Wasserstein distances for normal conditions are similar (both are low) for correlated and uncorrelated scenarios. Similarly, the cost distinctions are minor under most conditions. However, the uncorrelated scenarios are consistently underestimating the costs for the shoulder season, which is in direct contrast to their behavior under abnormal conditions. As a result, the uncorrelated scenario-based OPF may demonstrate unreasonably high variations in OPF costs between normal and abnormal scenarios, leading to non-optimal outcomes from a long-term reliability planning perspective.

4.3. Practical Significance

Since many resource planning activities aim to distinguish abnormal conditions from normal conditions, it is helpful to compare how the generated abnormal scenarios differ from the generated normal scenarios. For correlated cases, the costs for the abnormal scenarios are consistently and reasonably higher than the costs of the normal scenarios due to the lower solar production on abnormal days. Winter days show the largest and most consistent gap through the day (see Figure 16), indicating the need for a longer traditional generation or battery backup requirements. For the shoulder (see Figure 17) and summer seasons, the gap between normal and abnormal is smaller and restricted to fewer hours of the day, indicating that the backup requirements may be less. For uncorrelated scenarios, consistency is absent: for the shoulder season, the abnormal scenarios grossly overestimate the net load (as shown in Figure 17); for summer, they show lower costs than the normal scenario, and no reasonable conclusions can be drawn from them. In summary, through the OPF application, we have demonstrated the ability of correlated scenario generation to create valid representative power system scenarios that are a prerequisite for long-term resource planning. In the future, we will apply the scenarios generated using the proposed approach to solve the optimal BESS sizing and siting problem [33].

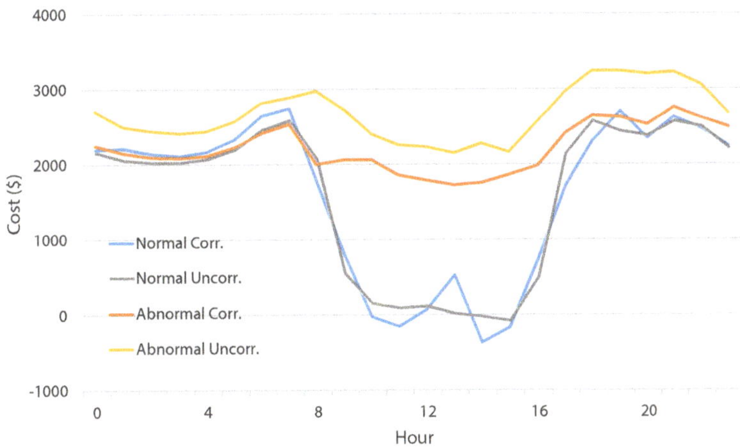

Figure 16. Winter normal–abnormal comparison (ratio = 0.6).

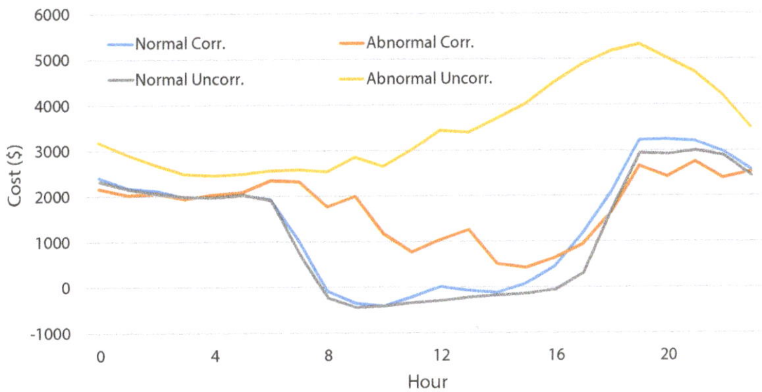

Figure 17. Shoulder normal–abnormal comparison (ratio = 0.6).

5. Conclusions

As the exploration of ways to understand and analyze the impacts of RG on grid reliability continues, synthetically generated representative scenarios will play an increasingly vital role. Due to legacy practices and/or ease of application, uncorrelated/univariate scenario generation is often used for such exploration. However, this may lead to outcomes that are not realistic. This paper demonstrates the utility of correlated multivariate scenario generation in understanding and analyzing normal and abnormal system conditions.

The proposed systematic end-to-end methodology for correlated scenario generation has the following components:

- Structured and model-free data segmentation.
- An informed selection/design of a cross-correlated conditional recurrent generative adversarial network (C^2RGAN).
- Generation of correlated representative scenarios that augment the original dataset.
- Extensive and application-oriented validation that proves the value of the proposed methodology.

Overall, correlated scenario generation was seen to create more realistic profiles due to the integration of both solar generation and load demand in the training of the C^2RGAN. From the OPF application evaluation, the following key conclusions are drawn:

- The correlated scenario generation resulted in lower and more accurate average hourly costs across the seasons (as shown in Figure 11 and Table 3).
- From the voltage angle distributions, it was observed that the correlated scenarios are more similar to the real case compared to uncorrelated scenarios (as shown in Figure 12).
- Seasonal performance analyses highlighted why inferences drawn from uncorrelated scenarios might be misleading (results from Tables 3 and 4).
- It was also shown that the results from uncorrelated scenarios are adequate for normal days, but it can lead to misplaced conviction about their applicability to abnormal scenarios (results from Tables 3 and 4).

The proposed methodology is voltage-level agnostic, scalable, and portable to different datasets, geographies, and end-application requirements. It can be used to analyze the reliability and resilience issues with various renewable energy penetration levels and come to definitive conclusions about deploying these resources. The proposed approach currently captures cross-correlations and temporal correlations that exist between RG and loads. With the right dataset and minor modifications, it can also be extended to capture spatial correlations between the different variables.

Author Contributions: D.D. and M.B. are co-first authors; conceptualization, A.P. and P.A.; methodology, A.P., D.D. and M.B.; software, M.B., D.D., H.S. and A.I.S.; validation, D.D., M.B. and A.P.; formal analysis, M.B. and D.D.; investigation, A.P., D.D., M.B. and A.I.S.; resources, A.P. and P.A.; data curation, A.I.S., H.S., D.D. and M.B.; writing—original draft preparation, D.D. and M.B.; writing—review and editing, A.P., A.I.S. and D.D.; visualization, M.B., D.D., H.S. and A.I.S.; supervision, A.P.; project administration, A.P. and P.A.; funding acquisition, A.P. and P.A. All authors have read and agreed to the published version of the manuscript.

Funding: This work was supported in part by the Salt River Project (SRP) under grant 96-180C 2021-2022 EE-04, and the National Science Foundation (NSF) under grants OAC 1934766 and ECCS 2145063.

Data Availability Statement: The data presented in this study are available on request from the corresponding author. The data are not publicly available as they were received under a non-disclosure agreement.

Conflicts of Interest: The authors declare no conflict of interest.

Nomenclature

The following abbreviations are used in this paper:

ACF	Auto-correlation Function
C^2RGAN	Cross-Correlated Conditional Recurrent Generative Adversarial Network
DTW	Dynamic Time Warping
GAN	Generative Adversarial Network
LSTM	Long Short-Term Memory
OPF	Optimal Power Flow
PDF	Probability Density Function
PV	Photovoltaic
RG	Renewable Generation
RNN	Recurrent Neural Network

References

1. Yan, R.; Wang, J.; Huo, S.; Qin, Y.; Zhang, J.; Tang, S.; Wang, Y.; Liu, Y.; Zhou, L. Flexibility improvement and stochastic multi-scenario hybrid optimization for an integrated energy system with high-proportion renewable energy. *Energy* **2023**, *263*, 125779. [CrossRef]
2. Hainsch, K.; Löffler, K.; Burandt, T.; Auer, H.; Crespo del Granado, P.; Pisciella, P.; Zwickl-Bernhard, S. Energy transition scenarios: What policies, societal attitudes, and technology developments will realize the EU Green Deal? *Energy* **2022**, *239*, 122067. [CrossRef]

3. Xie, L.; Huang, T.; Zheng, X.; Liu, Y.; Wang, M.; Vittal, V.; Kumar, P.; Shakkottai, S.; Cui, Y. Energy system digitization in the era of AI: A three-layered approach toward carbon neutrality. *Patterns* **2022**, *3*, 100640. [CrossRef]
4. Dumlao, S.M.G.; Ishihara, K.N. Weather-Driven Scenario Analysis for Decommissioning Coal Power Plants in High PV Penetration Grids. *Energies* **2021**, *14*, 2389. [CrossRef]
5. Padhee, M.; Pal, A.; Vance, K.A. Analyzing effects of seasonal variations in wind generation and load on voltage profiles. In Proceedings of the 2017 North American Power Symposium (NAPS), Morgantown, WV, USA, 17–19 September 2017; pp. 1–6.
6. Padhee, M.; Pal, A. Effect of solar PV penetration on residential energy consumption pattern. In Proceedings of the 2018 North American Power Symposium (NAPS), Fargo, ND, USA, 9–11 September 2018; pp. 1–6.
7. Mishra, C.; Pal, A.; Thorp, J.S.; Centeno, V.A. Transient stability assessment of prone-to-trip renewable generation rich power systems using Lyapunov's direct method. *IEEE Trans. Sustain. Energy* **2019**, *10*, 1523–1533. [CrossRef]
8. Buonanno, A.; Caliano, M.; Di Somma, M.; Graditi, G.; Valenti, M. A Comprehensive Tool for Scenario Generation of Solar Irradiance Profiles. *Energies* **2022**, *15*, 8830. [CrossRef]
9. Sund, L.; Talari, S.; Ketter, W. Stochastic Wind Power Generation Planning in Liberalised Electricity Markets within a Heterogeneous Landscape. *Energies* **2022**, *15*, 8109. [CrossRef]
10. Marulanda, G.; Bello, A.; Cifuentes, J.; Reneses, J. Wind Power Long-Term Scenario Generation Considering Spatial-Temporal Dependencies in Coupled Electricity Markets. *Energies* **2020**, *13*, 3427. [CrossRef]
11. Xie, Y.; Xu, Y. Transmission Expansion Planning Considering Wind Power and Load Uncertainties. *Energies* **2022**, *15*, 7140. [CrossRef]
12. Lee, D.; Baldick, R. Load and Wind Power Scenario Generation Through the Generalized Dynamic Factor Model. *IEEE Trans. Power Syst.* **2017**, *32*, 400–410. [CrossRef]
13. Hoeltgebaum, H.; Fernandes, C.; Street, A. Generating Joint Scenarios for Renewable Generation: The Case for Non-Gaussian Models With Time-Varying Parameters. *IEEE Trans. Power Syst.* **2018**, *33*, 7011–7019. [CrossRef]
14. Becker, R. Generation of time-coupled wind power infeed scenarios using pair-copula construction. *IEEE Trans. Sustain. Energy* **2017**, *9*, 1298–1306. [CrossRef]
15. Goh, H.H.; Peng, G.; Zhang, D.; Dai, W.; Kurniawan, T.A.; Goh, K.C.; Cham, C.L. A New Wind Speed Scenario Generation Method Based on Principal Component and R-Vine Copula Theories. *Energies* **2022**, *15*, 2698. [CrossRef]
16. Cui, M.; Zhang, J.; Wang, Q.; Krishnan, V.; Hodge, B.M. A data-driven methodology for probabilistic wind power ramp forecasting. *IEEE Trans. Smart Grid* **2017**, *10*, 1326–1338. [CrossRef]
17. Li, J.; Zhou, J.; Chen, B. Review of wind power scenario generation methods for optimal operation of renewable energy systems. *Applied Energy* **2020**, *280*, 115992. [CrossRef]
18. Cui, M.; Ke, D.; Sun, Y.; Gan, D.; Zhang, J.; Hodge, B.M. Wind power ramp event forecasting using a stochastic scenario generation method. *IEEE Trans. Sustain. Energy* **2015**, *6*, 422–433. [CrossRef]
19. Sideratos, G.; Hatziargyriou, N.D. Probabilistic wind power forecasting using radial basis function neural networks. *IEEE Trans. Power Syst.* **2012**, *27*, 1788–1796. [CrossRef]
20. Goodfellow, I.; Pouget-Abadie, J.; Mirza, M.; Xu, B.; Warde-Farley, D.; Ozair, S.; Courville, A.; Bengio, Y. Generative adversarial nets. In Proceedings of the 28th Annual Conference on Neural Information Processing Systems 2014 (NIPS), Montreal, QC, Canada, 8–13 December 2014.
21. Liang, J.; Tang, W. Sequence generative adversarial networks for wind power scenario generation. *IEEE J. Sel. Areas Commun.* **2019**, *38*, 110–118. [CrossRef]
22. Jiang, C.; Mao, Y.; Chai, Y.; Yu, M.; Tao, S. Scenario generation for wind power using improved generative adversarial networks. *IEEE Access* **2018**, *6*, 62193–62203. [CrossRef]
23. Chen, Y.; Wang, Y.; Kirschen, D.; Zhang, B. Model-free renewable scenario generation using generative adversarial networks. *IEEE Trans. Power Syst.* **2018**, *33*, 3265–3275. [CrossRef]
24. Zhang, Y.; Ai, Q.; Xiao, F.; Hao, R.; Lu, T. Typical wind power scenario generation for multiple wind farms using conditional improved Wasserstein generative adversarial network. *Int. J. Electr. Power Energy Syst.* **2020**, *114*, 105388. [CrossRef]
25. Yuan, R.; Wang, B.; Sun, Y.; Song, X.; Watada, J. Conditional Style-based Generative Adversarial Networks for Renewable Scenario Generation. *IEEE Trans. Power Syst.* **2022**. [CrossRef]
26. Wei, H.; Hongxuan, Z.; Yu, D.; Yiting, W.; Ling, D.; Ming, X. Short-term optimal operation of hydro-wind-solar hybrid system with improved generative adversarial networks. *Appl. Energy* **2019**, *250*, 389–403. [CrossRef]
27. Choi, J.; Lee, J.I.; Lee, I.W.; Cha, S.W. Robust PV-BESS Scheduling for a Grid With Incentive for Forecast Accuracy. *IEEE Trans. Sustain. Energy* **2021**, *13*, 567–578. [CrossRef]
28. Dalal, D.; Pal, A.; Augustin, P. Representative Scenarios to Capture Renewable Generation Stochasticity and Cross-Correlations. In Proceedings of the 2022 IEEE Power & Energy Society General Meeting (PESGM), Denver, CO, USA, 17–21 July 2022; pp. 1–5. [CrossRef]
29. Keogh, E.; Ratanamahatana, C.A. Exact indexing of dynamic time warping. *Knowl. Inf. Syst.* **2005**, *7*, 358–386. [CrossRef]
30. Fabozzi, D.; Van Cutsem, T. Assessing the proximity of time evolutions through dynamic time warping. *IET Gener. Transm. Distrib.* **2011**, *5*, 1268–1276. [CrossRef]
31. Goodfellow, I.J. NIPS 2016 Tutorial: Generative Adversarial Networks. *arXiv* **2017**, arXiv:1701.00160.

32. IEEE 30 Bus System. Available online: https://icseg.iti.illinois.edu/ieee-30-bus-system/ (accessed on 12 April 2022).
33. Padhee, M.; Pal, A.; Mishra, C.; Vance, K.A. A Fixed-Flexible BESS Allocation Scheme for Transmission Networks Considering Uncertainties. *IEEE Trans. Sustain. Energy* **2020**, *11*, 1883–1897. [CrossRef]

energies

MDPI

Article

VSG Control Applied to Seven-Level PV Inverter for Partial Shading Impact Abatement

Tiago H. de A. Mateus [1,*], José A. Pomilio [2], Ruben B. Godoy [1] and João O. P. Pinto [3]

[1] Faculty of Engineering, Architecture and Urbanism and Geography, Federal University of Mato Grosso do Sul, Cidade Universitária, Campo Grande 79070-900, MS, Brazil
[2] School of Electrical Engineering and Computer Science, UNICAMP-University of Campinas, Av. Albert Einstein, 400, Cidade Universitária, Campinas 13083-852, SP, Brazil
[3] Oak Ridge National Laboratory, 2360 Cherahala Blvd Knoxville, Oak Ridge, TN 37932, USA
* Correspondence: tiago.mateus@ufms.br

Abstract: This paper describes the use of a Cascaded H-Bridge Multilevel Inverter to deal with the problem of partial shading in a photovoltaic system connected to the grid. Combined with the topology, it proposes the use of a virtual synchronous generator for power sharing between photovoltaic arrays and to ensure the connection to the power grid even in extreme shading situations with no power buffer. The experimental results demonstrate the feasibility of the strategy adopted and the limitations to overcome.

Keywords: partial shading; cascaded multilevel inverter; droop control; active power transfer

Citation: Mateus, T.H.d.A.; Pomilio, J.A.; Godoy, R.B.; Pinto, J.O.P. VSG Control Applied to Seven-Level PV Inverter for Partial Shading Impact Abatement. *Energies* **2022**, *15*, 6409. https://doi.org/10.3390/en15176409

Academic Editor: Germano Lambert-Torres

Received: 26 July 2022
Accepted: 23 August 2022
Published: 2 September 2022

1. Introduction

Traditional MPPT techniques have a reliable performance in applications using only one PV module or in regions that operate with uniform irradiation conditions most of the time [1–5]. However, there are several causes which may generate a non-uniform irradiation condition in a PV module or PV array, such as dust, aging of the PV array, and partial shade (PS). The PS is an event that occurs when cells in a PV module or a photovoltaic arrangement suffers from obstruction of solar irradiance. Passing clouds, buildings, trees, and birds are examples of PS causes.

The PS of PV cells causes operation with reverse voltage in the shaded cells. If the reverse voltage rises too high, an avalanche breakdown may occur. With reverse polarity, these cells begin to consume energy, causing overheating and decreasing the maximum power that the group could generate. Such heating can lead to thermal rupture of the shaded cell, which is called the hot-spot effect and in some cases, this effect can lead to burning of the cell [6–9].

One solution to diminish the effects of reverse bias voltage is the use of bypass diodes in parallel with the cells and PV modules. The major drawback of this solution is that by using the bypass diode, it becomes the alternative path to the current if the respective PV panel or arrangement is shaded and, consequently, it results in multiple maximum power points in the PV curve. It is worth mentioning that recent studies indicate that the bypass diode, although attenuate, not fully prevent the damage caused by hot-spot [7–9].

2. Alternatives to Reduce PS Effects

2.1. Maximum Power Point Tracking—MPPT

With the use of bypass diodes, conventional MPPT algorithms are not able to track the global maximum point of operation [4,10,11]. In the search for a solution, or at least the reduction of the effects of PS, many alternatives have been proposed, ranging from the way in which the arrangement of PV modules is carried out, to the use of artificial intelligence techniques to find the global MPPT. However, even if the optimum point is tracked in operations under the effects of PS, or if the reconfiguration is performed in the arrangement of the PV modules, the performance achieved cannot be compared with the situation in which each PV module is capable of operating in its specific MPP [10,12].

2.2. Photovoltaic Arrangements

The best-known photovoltaic arrangements to mitigate the effects of PS are Series-parallel, Total-cross-tied, Honeycomb and Bridge-linked. Among the techniques mentioned, the interconnection through Total-cross-tied is pointed out as the best way to reduce losses by PS, however, it also does not guarantee the operation at maximum power [12,13]. Schemes for PV modules reconfiguration are extensively presented in literature; however, the main disadvantages of these techniques reside in the increased number of power switches and the need for a complex intelligent algorithm that, besides the challenge in implementation, may require high computational cost and, in some cases, an additional battery bank [11,13,14].

2.3. Arrangements for Power Transfer

The centralized converters (Figure 1a) are characterized by the interconnection of all photovoltaic modules into a single inverter. In this configuration, the photovoltaic panels are interconnected in series, resulting in a voltage high enough that no further power conversion stage is needed. To increase the electric power generated, more branches of panels can be inserted in parallel, provided that due care be taken against reverse currents that may arise if one of the strings is shaded. Common ways to protect shaded strings from reverse currents are fuses or diodes in series. Centralized converters were the first to be used, however, disadvantages such as high DC voltage, difficulty in tracking maximum power in shading situations and losses in series diodes motivated the search for alternatives to overcome such difficulties [12,15].

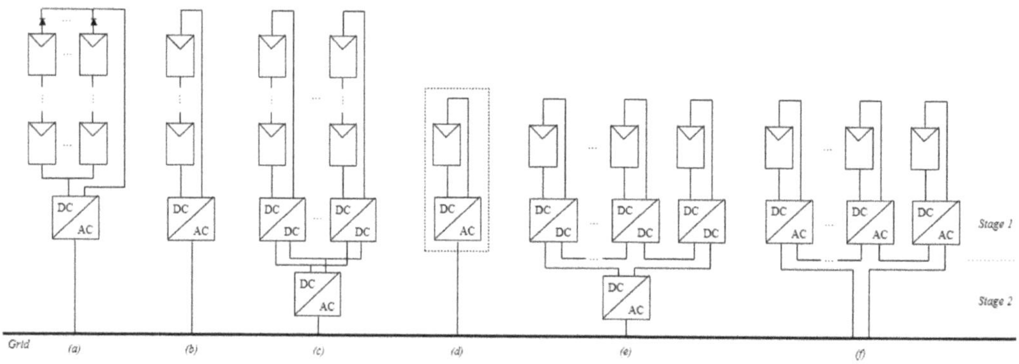

Figure 1. Power plants for grid connection: (**a**) central inverter; (**b**) string inverter; (**c**) multistring inverter; (**d**) ac-module inverter; (**e**) cascaded dc/dc converter; (**f**) cascaded dc/ac inverter.

A way to improve the efficiency in applications with centralized converters is the use of only one string of PV modules. This configuration is known as a string converter (Figure 1b) and it allows a better use of electrical energy supplied by the set of panels. In this case, there are no conduction losses from the series diodes since they become unnecessary. Thus, likewise the case with centralized inverters, power switches are needed to support the open circuit voltage of the entire string.

In order to increase the power generation without losing the string converter benefits and minimizing the PS effects, the multi-strings converter emerges as a good alternative (Figure 1c). In this topology, each branch has a DC-DC converter to carry out its respective MPPT. Such a strategy considerably improves the efficiency of tracking the global MPP.

There is a common DC bus to parallel the DC-DC converters output, followed by the inverter module that controls the current transferred to the power grid.

Figure 1d shows a representation of the AC module. This setting, commonly known as a microinverter, allows a panel (or an arrangement with few panels) to be connected to the grid. In this case, each panel will have its own MPPT algorithm, which guarantees the optimized operation in terms of power tracking. Since it is a topology that needs to be replicated to increase the injected power, it becomes more costly as compared to the centralized structures. On the other hand, it is a versatile proposal, which allows planned expansion of the system, greater reliability, maintenance without interruption of operation, and optimized adjustment of the injected power. Its cost has been reduced with the manufacture in scale [16].

Another possibility is the use of cascade DC-DC converters, as seen in Figure 1e. In this configuration, each DC-DC converter controls the MPPT of a panel (or a small PV arrangement) and associates them in series. A controlled DC bus, with voltage enough to supply the DC-AC inverter, is created. This configuration becomes less expensive when compared to the AC module (Figure 1d), since it uses a single DC-AC stage. However, it does not have the same versatility to increase or reduce the injected power.

A second possibility of using cascade converters is presented in Figure 1f. In contrast with the previous ones, the DC-AC cascade inverter uses only one stage of power conversion. Each PV module is connected to a DC-AC converter whose output is in series with the output of the other converters. Such a configuration has the good feature of being similar to the AC module converter, bringing a better use per PV module due to the individual MPPT, however, with lower cost and greater efficiency [12].

From the knowledge of the main advantages and disadvantages of each grid connection architecture, it is observed that multilevel topologies can be better explored mainly because they are a good option in PS situations. They are also interesting for operating with lower switching frequencies, for having better performance when dispensing the DC-DC boost stage, and for not having high voltage DC buses [12,15].

In view of good stability during grid-tied operation, many control techniques are presented throughout literature review. An attractive solution with relative simplicity of implementation is the virtual synchronous generator (VSG), including for a cascaded H-bridge multilevel PV converter [17]. However, facing PS situations, its poor dynamic response results in sudden voltage oscillations and imposes the use of battery banks or reserve PV modules to operate as a power buffer.

Among the topologies used for PS solution, this work chose the cascaded multilevel converter as the stage of energy conversion. It allowed for the performance of the MPPT individually or with a small PV array, making it possible to use a classic MPPT algorithm. The control philosophy for power transfer were based on the fundamentals of a VSG, however, without using any power buffer. Compared to [12], the VSG is simpler since there is no need to tune controllers. The droop dynamics is slow, which suggests the need for a power buffer as in [17]. However, this work proposes a differentiated control to regulate the output voltage of the converter. The idea is to use two control loops that update both the power reference to be transmitted and the energy distribution delivered by each PV array.

3. System Overview

3.1. Multilevel Topology

To achieve the maximum power transfer in PS situations, it is assumed that the MPPT is carried out in each PV module or in a set of a few modules, which allows the use of classical MPPT techniques. The MPPT used in this work was the perturb and observe (P&O) [2,18], however, it could be any other classic technique. Therefore, in order to deal with PS conditions and providing other advantages such as low complexity and high performance, a seven-level Cascaded H-Bridge Inverter (CHB-ML) was chosen, as shown in Figure 2.

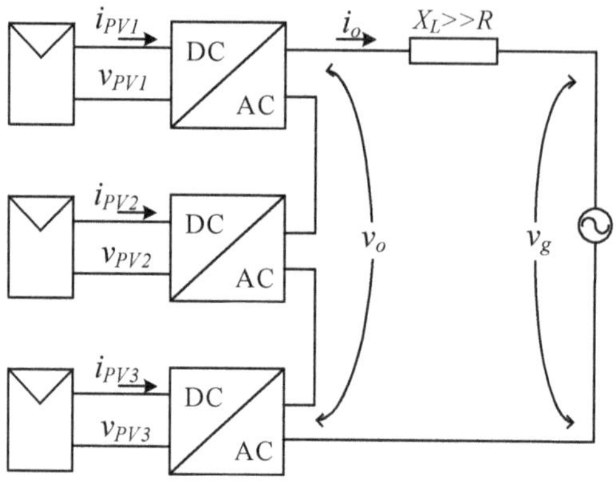

Figure 2. General schematic of a PV system grid connection using a seven levels CHB-ML inverter.

Regarding the use of multilevel inverters, there are always additional concern with the modulation techniques, as each level must respect a switching logic to obtain the desired output voltage. For this purpose, to facilitate the distribution of power to the different levels, phase shifted pulse width modulation (PS-PWM) was used. PS-PWM is the most popular carrier-based modulation technique due to its advantages in power balancing, good output harmonics and simplicity of implementation [19,20]. Much has been studied to improve its performance, especially to reduce the switching frequency, however, increasing the complexity [21,22].

With the proposed multilevel topology, the increase in the number of levels depends on the increase in cascade inverter modules (n). Consequently, aiming to guarantee an output voltage based on the modulation signal waveform, each new inverter module will have its switching logic from pulses generated by a new carrier signal. As shown in (1), for the PS-PWM the carrier signals must respect a uniform phase shift (δ) between them:

$$\delta = 180^{\circ}/n. \tag{1}$$

3.2. VSG Control

In an electric power system, the active power flow (P) can be calculated using (2), where V_o and V_g are the inverter output voltage and the mains voltage, respectively. In (2), θ corresponds to the power angle obtained from the angular difference between such voltages, and the resistance (R) and inductive reactance (X_L) make up the connection impedance.

$$P = \frac{R \times {V_o}^2 - R \times V_g \times V_o \times \cos(\theta) + X_L \times V_g \times V_o \times \sin(\theta)}{R^2 + {X_L}^2} \tag{2}$$

For situations in which the connection impedance is predominantly inductive, the variation in the transmitted power in relation to the power angle $\left(\frac{\delta P}{\delta \theta}\right)$ can be linearized to an operating point as seen in (3).

$$\Delta P = \frac{V_g \times V_o \times \cos(\theta)}{X_L} \times \Delta\theta \tag{3}$$

As noted in (3), θ was chosen to control the transmitted power. The practical and robust way to change θ is to act in the converter angular frequency (ω_r). A very useful technique to update ω_o mimics the synchronous generator and it is called droop-control. In addition to the simplicity of its implementation, this controller guarantees intrinsic stability for on-grid operation and eliminates the necessity of intercommunication loops between generators. In (4), the rule that governs the droop-control is presented. In addition to ω_o, the variables present in (4) are the grid angular frequency (ω_r), the power to be transmitted $\left(P_{ref}\right)$, the power delivered by the converter (P_{med}), and the coefficient k which defines the slope of the droop-control curve [23].

$$\omega_o = \omega_r + k\left(P_{ref} - P_{med}\right) \tag{4}$$

Assuming a generator set already connected to the power grid, from the change of P_{ref} the $P - \omega$ curve will be adjusted to change the transmitted power. In this way, a new equilibrium point will be reached since the system will balance in ω_r. According to Figure 3, dashed lines represent slope curves for situations whose reference power has changed.

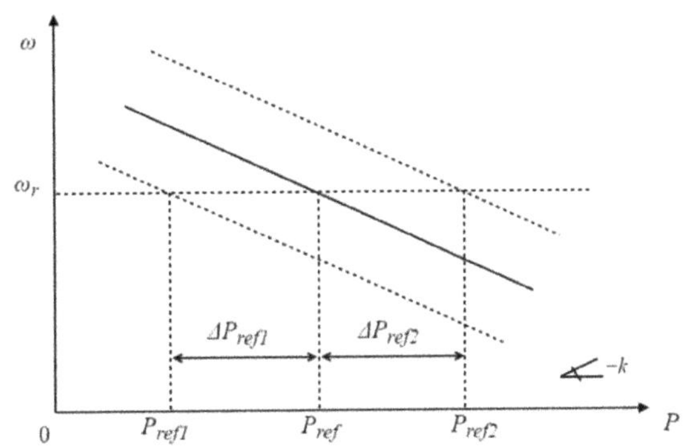

Figure 3. Droop-control curves.

3.3. Control Philosophy

The block diagram that describes the dynamic behavior of a droop-controller is represented in Figure 4. The control action occurs from perturbations in P_{med} and P_{ref}. According to the droop-control slope (k), the power error (ε) results in variations in the system frequency ($\Delta\omega$). Consequently, the power angle (θ) is perturbed and the output power is also modified. The power calculation is processed through the instantaneous product between the real output voltage (v_o) and the virtual current (i_v). The presence of the low-pass filter is necessary to obtain the average value of the instantaneous output power [24]. In Figure 5, the low-pass filter is inserted in the block called *PCalculation*.

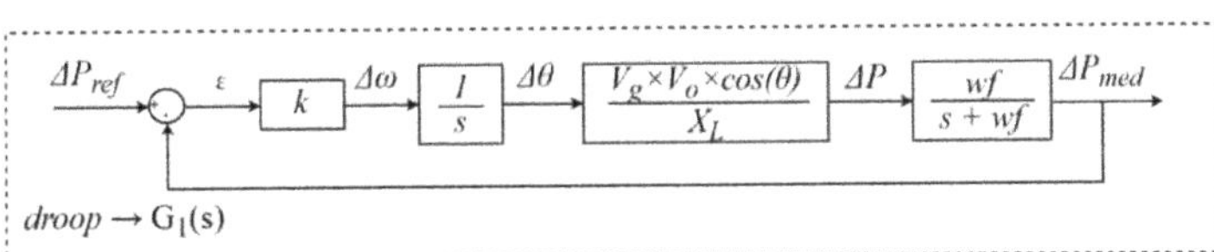

Figure 4. Block diagram of the droop controller.

As noted, the droop-control imposes a slow dynamic by acting on average power values. Furthermore, the system plant cannot collaborate for fast controller actions due to the connection reactance. Such characteristics impose an additional challenge to the transfer of power in situations of PS or sudden variations in irradiation. The controller needs to be tuned so that the power transferred to the grid occurs with a relatively quick response as the irradiation variations occur. If this dynamic response is not appropriate, the voltages in the panels can suffer sudden variations and consequently impact the connection voltage. An intermediate stage, between panels and the H-bridge inverter, could facilitate the dynamics, for it would serve as a supplier or accumulator of energy in the respective situations from a transient to a low or high irradiance. However, aiming to lower cost and better performance, this paper proposes the use of a single-stage converter.

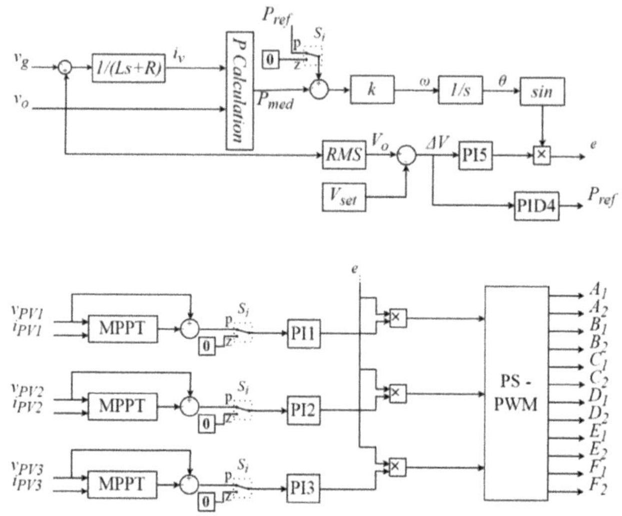

Figure 5. Proposed controller for grid-tied 7 levels CHB-ML inverter.

The solution found to minimize the effects of rapid variations in irradiance was to use the output voltage as the main power transfer control variable. To this end, it is necessary to have a controller fast enough to sustain the connection without the PV sets suffering irreversible voltage sags due to the sudden reductions in irradiation. Another challenge is to have a controller with proper tuning to ensure the connection without overloading the system when there is a rapid rise of irradiation. To solve such adversities, the controller presented in Figure 5 is proposed.

According to diagrams in Figure 5, the control system uses the reading of eight variables, being: PV arrays voltages (v_{PV1}, v_{PV2} e v_{PV3}), PV arrays currents (i_{PV1}, i_{PV2} e i_{PV3}), grid voltage (v_g), and multilevel inverter output voltage (v_o). In accordance with (5), the converter output current (i_v) is calculated through a virtual impedance, where R and L correspond to the approximated values of the connection resistance and inductance, respectively.

$$i_v(s) = \frac{v_r(s) - v_o(s)}{Ls + R} \tag{5}$$

In Figure 5 are presented the blocks that make up the droop controller. When the switch S is in the "z" position, the controller causes P_{med} to be zero, resulting in synchronism with the grid. When the switch S is in the position "p", the converter tracks the power P_{ref}. To preserve the connection stability, before the action of the power tracking algorithms (PI1, PI2 and PI3), the PID4 controller acts based on the error between the inverter RMS output voltage (V_o) and the reference voltage (V_{set}). Thus, before any tracking, the droop control acts on the power angle to reduce or increase power transfer and to preserve both PV and output inverter voltages. In parallel to PID4, the controller called PI5, also in anticipation of the tracking controllers, acts to regulate the amplitude of the signal (e) that modulates the output voltage and that ends up controlling the reactive power flow. However, mainly for low irradiation situations, some content of reactive power may be present in the system, after all these situations are more complex to regulate V_o. Although in this work the reactive power control was not addressed, an effective way to overcome this drawback depends on an additional controller and measurements or estimations of the reactive power. Thus, V_{set} and θ could be updated to minimize or cease the reactive power. Another point to be highlighted refers to the limit of voltage regulation imposed by the PV arrays. There is no link DC for voltage regulation, which means that, depending on the level of solar irradiation, the presence of reactive power becomes unavoidable, forcing the system disconnection if any amount of reactive power can be tolerated.

Now, by observing the diagram in Figure 5, from a new irradiation condition, controllers PI1, PI2 and PI3 act in the power tracking through the duty cycle of each inverter block. In this way, the indexes m_i are adjusted in proportion to the amount of power that each inverter module will process. Since the MPPT acts on the modulation indexes, this action results in the variation of the output voltage, instructing the PID4 controller to vary P_{ref} for the new power situation while PI5 adjusts the amplitude (e) to prevent reactive flow.

The controller gains are shown in Table 1. Due to the non-linear characteristic of the photovoltaic panels, there was some difficulty in modeling the plant response, leading, a priori, to select the gains based on simulations.

Table 1. Controllers' gains.

	k_p	k_i	k_d
PI 1, 2, 3, 5	0.001	0.01	0
PID 4	15	20	0.1

3.4. *Effects of Irradiation Variations*

Figure 6 presents the complete view of the system. An important point to observe in this representation refers to the PV arrays voltages. These voltages depend on the solar irradiation, and they are also affected by the delivered power variation. Consequently, PV voltage variations strongly interfere in the inverter output voltage, which makes its regulation a challenging task. It is also important to mention that the part surrounded by the dashed lines concerns the droop-control and the physical plant operation (G_1). ΔP_{medi} represent the powers delivered by each inverter bridge. Note that these powers are the result of the tracked power weighted by the modulation indexes (M_i), in this case represented by the RMS values of m_i. Looking at Figures 5 and 6, it is noteworthy that m_i are obtained from the product between two signals, which means, they depend on both the MPP tracking (p) and the modulating (e) updating.

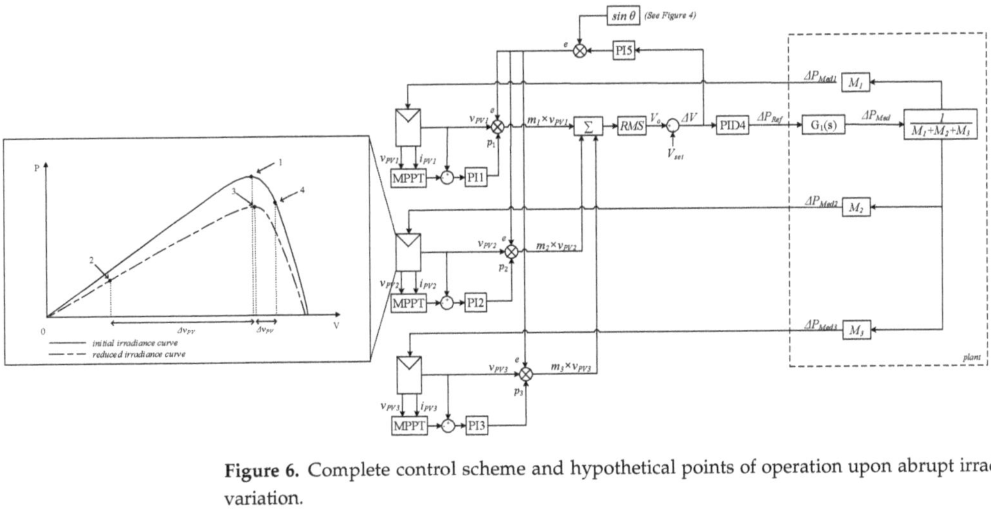

Figure 6. Complete control scheme and hypothetical points of operation upon abrupt irradiance variation.

Analyzing the curve zoomed in Figure 6, it is concluded that the main dynamic difficulty occurs with the sudden reduction of irradiation. Suppose that a given PV set is at the point of operation identified on the curve as "1". With the reduction of irradiation, the power transferred to the grid does not change instantly, requiring more power than that available on the panels at that time. As shown at point "2", this situation causes the collapse of the panel voltage and, consequently, the converter output voltage. From the PID4 controller action, the power transfer is reduced, and the voltage of the panels is restored until the plant is driven to the new MPP, represented by the point "3". If there is a return to the original irradiation, there will be an instantaneous increase in the PV voltages and available power, which will cause the system to operate momentarily at point 4. Finally, with the action of the MPPT algorithm and the other controllers, the system will be taken to the original operating point, identified in the curve by "1".

4. Experimental Evaluation

4.1. System Overview

Figure 7 contains a general schematic and Table 2 summarizes the electric parameters for the circuits used for grid connection. As noted, each inverter bridge works from a PV array formed by three photovoltaic panels in series. For the experiments, a real plant was used, composed of Sun Earth® panels, model TBP156 × 156-60P, with rated power of 235 W. In the input of each inverter bridge, an electrolytic capacitor of 4700 µF was used to minimize the inductive effect of connection cables between panels and inverters. To compose the H-bridge, since it is a prototype whose controllers would need experimental adjustments, it was decided to use power switches with over-rated characteristics, ensuring robustness even in adverse operating conditions. The switch model adopted was the IRGP50B60PD1, with blocking voltage of 600 V and continuous operating current of 33 A. For connection to the mains, a 9.74 mH connection inductor with an intrinsic resistance of about 100 mΩ was used.

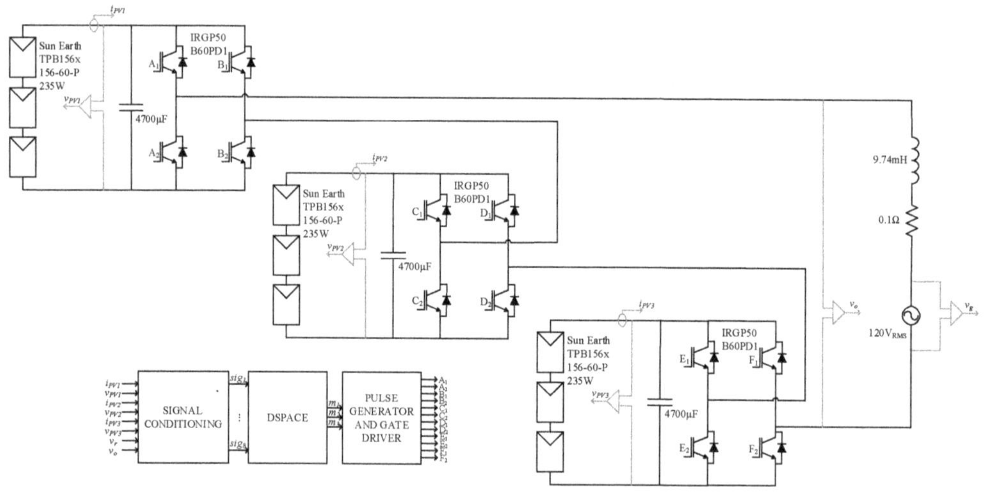

Figure 7. General schematic for the implemented prototype.

The control system consists essentially of the three blocks shown in Figure 7. At first, the sampled signals go through the conditioning hardware to adjust amplitudes and filter noise. The conditioned signals (sig_1, ..., sig_8) go to the control system embedded in the DSPACE® processor. As a result of the control actions, the modulating signals (m_1, m_2 and m_3) go to an external hardware where they are compared to the triangular carriers to generate the switching pulses. Figure 8a,b show, respectively, the implemented prototype and the set of photovoltaic panels used during the experiments.

Table 2. System and PV panel electrical parameters.

Parameters	Values
DC-link capacitor	4700 μF
Connection inductance	9.74 mH
Connection resistance	0.1 Ω
Switching frequency	2.5 kHz
Mains frequency	60 Hz
Rated grid voltage	120 V_{RMS}
Maximum power PV	235 Wp
Voltage at MPP	29.2 V
Current at MPP	7.6 A
Open circuit voltage	32.9 V
Short circuit current	7.6 A

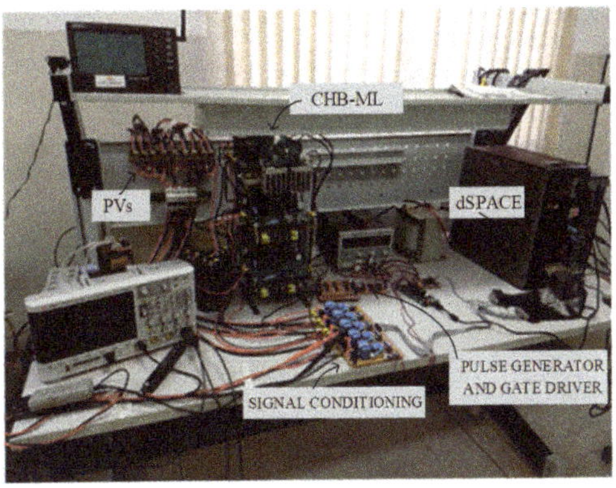

(a)

(b)

Figure 8. (**a**) Experimental setup for the seven levels cascaded H-bridge inverter; (**b**) Photovoltaic panels: Sun Earth PV-Module 235 W.

4.2. Experimental Results

Figure 9a contains the waveforms for the CHB-ML output voltage (yellow), current (red) and power (magenta) injected into the electrical grid during irradiation of 1250 W/m^2. Still for this situation, the inverter voltage and current harmonic spectra are shown in Figure 9c,e, respectively. Figure 9b contains the waveforms for the voltage (yellow), current (red) and power (magenta) during irradiation of 300 W/m^2. The harmonic spectra for both the inverter voltage and output current are shown in Figure 9d,f, respectively. For all situations, THD was calculated considering harmonics up to 3 kHz. The electrical parameters for the situations of high and low irradiance are summarized in Table 3. It is interesting to observe the equilibrium reached by the indexes m_i denoting the small differences in the power contribution of each PV array for both situations. Another point to be considered is the superiority of the indexes for high irradiation, confirming the more significant contribution of the PV arrays for this condition.

An important characteristic of PS-PWM is the capability of dealing with different indexes m_i in each inverter while maintaining an output voltage waveform with low THD. However, in situations where the contribution of one (or more) of the PV arrays is very low, the controller can lead its respective modulation index m_i to smaller values, reducing the respective inverter contribution. For these situations, if the output voltage cannot be reached or if the current THD becomes poor, the system must be disconnected.

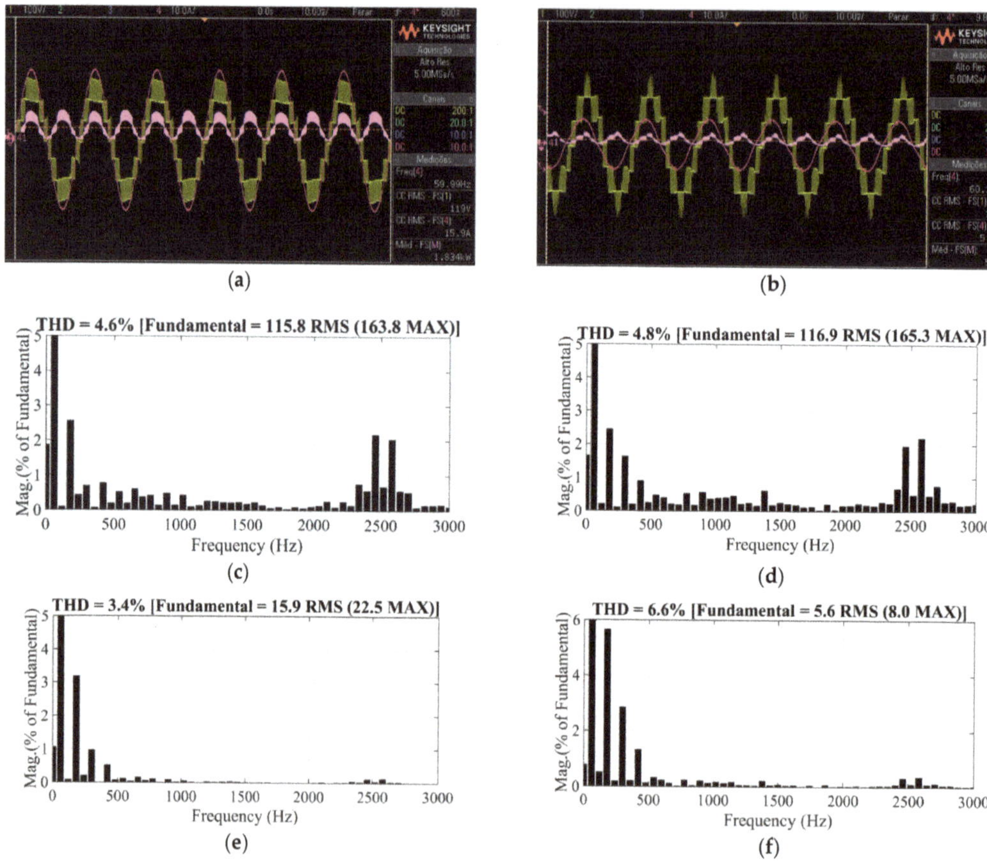

Figure 9. (**a**) Inverter output voltage and current injected to the grid for the high irradiation; (**b**) low irradiation; (**c**) Harmonic spectrum for the inverter output voltage for the high irradiation; (**d**) low irradiation; (**e**) Harmonic spectrum for the inverter output current for the high irradiation; (**f**) low irradiation.

Table 3. Comparison between steady-state parameters for high and low irradiance.

	1250 W/m^2	**300 W/m^2**
PVs Temp	~60 °C	~45 °C
V_{PV1}	71.5 V	82.5 V
V_{PV2}	71.0 V	82.8 V
V_{PV3}	70.9 V	83.3 V
I_{PV1}	9.4 A	2.6 A
I_{PV2}	9.7 A	2.6 A
I_{PV3}	9.5 A	2.4 A
V_O	119 V_{RMS}	120 V_{RMS}
I_O	15.9 A_{RMS}	5.6 A_{RMS}
P_O	1834 W	569 W
m_1	0.79	0.68
m_2	0.86	0.72
m_3	0.84	0.68

Aiming to evaluate the currents THD for the results presented in Figure 9, some considerations are noteworthy. First, this application consists of a reduced number of levels, which harden the current and voltage quality in some situations. The presence of DC level and third harmonic in the currents are slightly superior to the limits recommended by the IEEE Standard 1547-2018. Another important point refers to the absence of any high-frequency filter, resulting in the converter output voltage with significant high-order content. Based on the preliminary results, although some efforts must be concentrated to improve the quality of the voltage and current waveforms, it is reasonable to conclude that the ML-CHB and PS-PWM are a promising choice for PS conditions, especially for situations with a greater number of levels, where voltage regulation and THD will be appreciably improved.

Figure 10 shows the behavior of the system under different situations of irradiation and PS. In the scenario presented in Figure 10a, graphic 1 represents the power transferred to the electrical grid. According to the voltage curves of the photovoltaic arrays, shown in graphic 4, the increase in irradiance raises the voltage in the PV sets. As a matter of consequence, there is an increase in the output voltage of the inverter (graphic 3), which determines the essential need for increasing power transfer. Therefore, the output voltage stabilizes as the new MPPT is reached. Both in graphic 1, referring to the power transferred to the power grid, and in graphic 2, referring to the current injected into the power grid, it is evident that the system stabilizes at a new point of operation after the increase in irradiance.

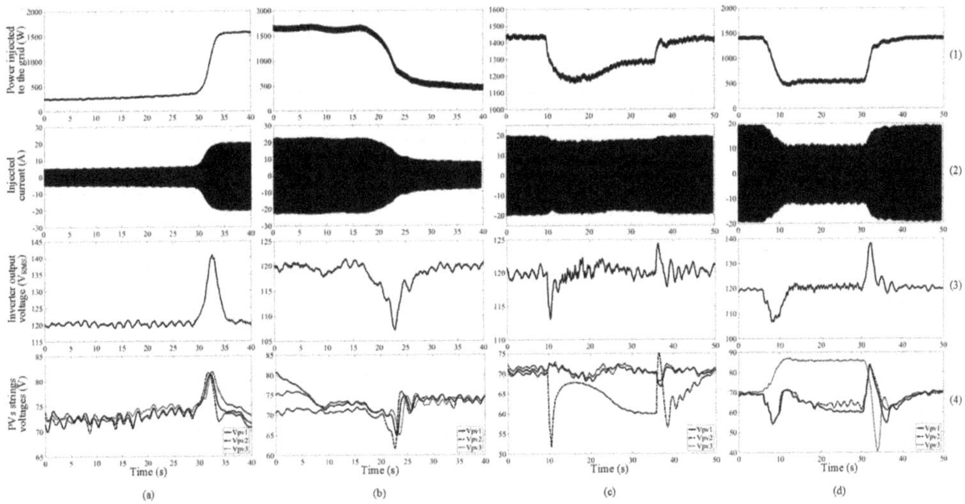

Figure 10. Evaluated scenarios: (**a**) with no PS and with transition from low irradiance (297 W/m^2) to high irradiance (1067 W/m^2); (**b**) with no PS and with transition from high irradiance (1250 W/m^2) to low irradiance (487 W/m^2); (**c**) high irradiance (1027 W/m^2) with PS of one PV set; (**d**) high irradiance (1027 W/m^2) with PS of two PV sets.

Figure 10b contains the curves corresponding to the situation of sudden irradiance reduction. Similar to the previous case, the PV sets voltage variation is reflected to the output voltage (graphic 3). Since the available power has been reduced and the transferred power is still the same, the tendency is to bring the panels to extremely low voltage levels that can be irreversible and can cause the power transfer to be definitively interrupted. Therefore, it is necessary for the control system to act from the output voltage sag, reducing the transferred power as quickly as possible until the tracking of the new MPP is achieved.

The situation presented by the graphs in Figure 10c corresponds to the shading of one PV array. As shown in graphic 1, referring to the power injected to the mains, shading occurs from the time of 10 s to the time of 36 s. According to the PV sets voltage curves (graph 4), promptly after shading, the control acts to reduce the transferred power to relieve the drained power and to recover the voltage of the shaded array. Still with a view at the curves referring to the behavior of the PV array voltages, it is noted that the tracking of the new MPP is established around the instant of 30 s. After the shading being removed, around 36 s, the control system acts to stabilize the rise in the output voltage and to track the new MPP.

Figure 10d presents a very peculiar operation scenario. In this condition, from the instant of 7 s, two PV arrangements were completely shaded, leaving only one arrangement exposed to sunlight. The shading lasted until approximately 31 s. It is worth mentioning that this condition did not allow the system to operate in its new MPP. If the new MPP were tracked, there would not be enough voltage on the panels to guarantee the regulation of the inverter output voltage at 120 V_{RMS}. Therefore, due to the reduction in the voltage of the PV arrays (graphic 4), there is an abrupt reduction in the inverter output voltage (graphic 3), which leads to the immediate need of decreasing the power transferred to the grid. As the transferred power decreases, the voltage of the arrays is restored and, consequently, the output voltage of the inverter is regulated again. However, based on the curves shown in graphic 4, once the output voltage regulation has been prioritized (graphic 3), the arrangement without shading is prevented from operating at its maximum power point. After the shading is removed, the output voltage is momentarily increased, and the maximum power point tracking is achieved. The curves corresponding to the transferred power (graphic 1) and the current injected into the grid (graphic 2) make the operating points noticeably clear before, during and after the two PV sets are shaded.

5. Conclusions

In accordance with the results achieved, the ML-CHB topology is presented as a good solution for dealing with the problem PS. Current research as well as market demands make evident the need for structures with better performance and addition of MPPT algorithms for a PV plant. In addition to ensuring a low THD output voltage, the CHB-ML is a single stage structure which allows the use of an MPPT algorithm for each power module. The number of levels can be set based on the need for the project, considering the output voltage and the number of tracking algorithms intended to be implemented. Simplification of control strategies becomes a relevant approach for the use of CHB-ML topology, and the use of the VSG control demonstrated robustness, ease of implementation and possibility of operation for islanding situations. Of course, the results show that the dynamic response of the droop control is truly slow. However, it is emphasized that in practical situations an extremely fast dynamic response is not required, since the irradiance variations during PS are not as sudden as those assessed in this paper. An important feature of the proposed control is the possibility of power tracking in two stages. Firstly, there is the prompt action of PID4 controller to adjust the power transfer based on the oscillations of the converter output voltage. Subsequently, the remaining controllers act to ensure the fine adjustment for tracking the MPP and to regulate the output voltage. It is reasonable to mention that, from plant adaptations, i.e., the reduction of the current connection impedance, the controllers can be tuned for a better dynamic response. Currently, the authors have studied the alternative of replacing the PID4 controller with a fuzzy-logic controller. This approach intends to improve the system stability and to ensure specific responses for situations of PS with sudden irradiance variations. Furthermore, it must be mentioned that the individual or global tracking factors were not evaluated, for infrastructure limitations made it difficult the achievement of V-I curves of the real PV arrays. Although the development was focused on keeping the system stability in sudden PS transitions, it was confirmed significant coherence among the delivered power for different irradiations. For future works, the authors intend not only to evaluate the PVs tracking factors but also to increase the number of levels, which will result in better output voltage quality and regulation, in addition to one MPPT per panel.

Author Contributions: Conceptualization, formal analysis, investigation, and resources, T.H.d.A.M., R.B.G., J.A.P. and J.O.P.P.; methodology, software, validation, writing—original draft preparation, T.H.d.A.M. and R.B.G.; writing—review and editing, supervision, project administration, and funding acquisition, J.A.P. and J.O.P.P. All authors have read and agreed to the published version of the manuscript.

Funding: This research was funded by São Paulo Research Foundation—FAPESP, grant number #2016/08645-9 and the National Council for Scientific and Technological Development: #303859/2020-2. This study was financed in part by the Coordenação de Aperfeiçoamento de Pessoal de Nível Superior-Brasil (CAPES)-Finance Code 001.

Institutional Review Board Statement: Not applicable.

Informed Consent Statement: Not applicable.

Data Availability Statement: Not applicable.

Conflicts of Interest: The authors declare no conflict of interest.

References

1. Zhang, P.; Sui, H. Maximum Power Point Tracking Technology of Photovoltaic Array under Partial Shading Based On Adaptive Improved Differential Evolution Algorithm. *Energies* **2020**, *13*, 1254. [CrossRef]
2. De Brito, M.A.G.; Galotto, L.; Sampaio, L.P.; Melo, G.D.A.E.; Canesin, C.A. Evaluation of the Main MPPT Techniques for Photovoltaic Applications. *IEEE Trans. Ind. Electron.* **2013**, *60*, 1156–1167. [CrossRef]
3. Villalva, M.G.; Gazoli, J.R.; Filho, E.R. Comprehensive Approach to Modeling and Simulation of Photovoltaic Arrays. *IEEE Trans. Power Electron.* **2009**, *24*, 1198–1208. [CrossRef]
4. Bollipo, R.B.; Mikkili, S.; Bonthagorla, P.K. Hybrid, optimal, intelligent and classical PV MPPT techniques: A review. *CSEE J. Power Energy Syst.* **2021**, *7*, 9–33. [CrossRef]
5. Eltamaly, A.M. An Improved Cuckoo Search Algorithm for Maximum Power Point Tracking of Photovoltaic Systems under Partial Shading Conditions. *Energies* **2021**, *14*, 953. [CrossRef]
6. Barcellona, S.; Barresi, M.; Piegari, L. MMC-Based PV Single-Phase System with Distributed MPPT. *Energies* **2020**, *13*, 3964. [CrossRef]
7. Alves, T.; Torres, J.P.N.; Marques Lameirinhas, R.A.; Fernandes, C.A.F. Different Techniques to Mitigate Partial Shading in Photovoltaic Panels. *Energies* **2021**, *14*, 3863. [CrossRef]
8. Kim, K.A.; Seo, G.-S.; Cho, B.-H.; Krein, P.T. Photovoltaic Hot-Spot Detection for Solar Panel Substrings Using AC Parameter Characterization. *IEEE Trans. Power Electron.* **2016**, *31*, 1121–1130. [CrossRef]
9. Rossi, D.; Omaña, M.; Giaffreda, D.; Metra, C. Modeling and Detection of Hotspot in Shaded Photovoltaic Cells. *IEEE Trans. Very Large Scale Integr. Syst.* **2015**, *23*, 1031–1039. [CrossRef]
10. Batzelis, E.I.; Georgilakis, P.S.; Papathanassiou, S.A. Energy models for photovoltaic systems under partial shading conditions: A comprehensive review. *IET Renew. Power Gener.* **2015**, *9*, 340–349. [CrossRef]
11. Engel, E.; Kovalev, I.; Testoyedov, N.; Engel, N.E. Intelligent Reconfigurable Photovoltaic System. *Energies* **2021**, *14*, 7969. [CrossRef]
12. Xiao, B.; Hang, L.; Mei, J.; Riley, C.; Tolbert, L.M.; Ozpineci, B. Modular Cascaded H-Bridge Multilevel PV Inverter with Distributed MPPT for Grid-Connected Applications. *IEEE Trans. Ind. Appl.* **2015**, *51*, 1722–1731. [CrossRef]
13. Chavan, V.C.; Mikkili, S.; Senjyu, T. Hardware Implementation of Novel Shade Dispersion PV Reconfiguration Technique to Enhance Maximum Power under Partial Shading Conditions. *Energies* **2022**, *15*, 3515. [CrossRef]
14. Vadivel, S.; Boopthi, C.S.; Ramasamy, S.; Ahsan, M.; Haider, J.; Rodrigues, E.M.G. Performance Enhancement of a Partially Shaded Photovoltaic Array by Optimal Reconfiguration and Current Injection Schemes. *Energies* **2021**, *14*, 6332. [CrossRef]
15. Kjaer, S.B.; Pedersen, J.K.; Blaabjerg, F. A review of single-phase grid-connected inverters for photovoltaic modules. *IEEE Trans. Ind. Appl.* **2005**, *41*, 1292–1306. [CrossRef]
16. Alluhaybi, K.; Batarseh, I.; Hu, H. Comprehensive Review and Comparison of Single-Phase Grid-Tied Photovoltaic Microinverters. *IEEE J. Emerg. Sel. Top. Power Electron.* **2020**, *8*, 1310–1329. [CrossRef]
17. Zhang, X.; Hu, Y.; Mao, W.; Zhao, T.; Wang, M.; Liu, F.; Cao, R. A Grid-Supporting Strategy for Cascaded H-Bridge PV Converter Using VSG Algorithm with Modular Active Power Reserve. *IEEE Trans. Ind. Electron.* **2021**, *68*, 186–197. [CrossRef]
18. Mohamed Hariri, M.H.; Mat Desa, M.K.; Masri, S.; Mohd Zainuri, M.A.A. Grid-Connected PV Generation System—Components and Challenges: A Review. *Energies* **2020**, *13*, 4279. [CrossRef]
19. Sochor, P.; Akagi, H. Theoretical and Experimental Comparison between PhaseShifted PWM and Level-Shifted PWM in a Modular Multilevel SDBC Inverter for Utility-Scale Photovoltaic Applications. *IEEE Trans. Ind. Appl.* **2017**, *53*, 4695–4707. [CrossRef]
20. Akhmetov, Z.; Hammami, M.; Grandi, G.; Ruderman, A. On PWM Strategies and Current THD for Single- and Three-Phase Cascade H-Bridge Inverters with Non-Equal DC Sources. *Energies* **2019**, *12*, 441. [CrossRef]

21. Kim, S.-M.; Lee, E.-J.; Lee, J.-S.; Lee, K.-B. An Improved Phase-Shifted DPWM Method for Reducing Switching Loss and Thermal Balancing in Cascaded H-Bridge Multilevel Inverter. *IEEE Access* **2020**, *8*, 187072–187083. [CrossRef]
22. Guo, X.; Wang, X.; Wang, C.; Lu, Z.; Hua, C.; Blaabjerg, F. Improved Modulation Strategy for Singe-Phase Cascaded H-Bridge Multilevel Inverter. *IEEE Trans. Power Electron.* **2022**, *37*, 2470–2474. [CrossRef]
23. Godoy, R.B.; Bizarro, D.; De Andrade, E.T.; Soares, J.D.O.; Ribeiro, P.E.M.J.; Carniato, L.A.; Kimpara, M.L.M.; Pinto, J.O.P.; Al-Haddad, K.; Canesin, C.A. Procedure to Match the Dynamic Response of MPPT and Droop-Controlled Microinverters. *IEEE Trans. Ind. Appl.* **2017**, *53*, 2358–2368. [CrossRef]
24. Godoy, R.B.; Pinto, J.O.P.; Canesin, C.A.; Coelho, E.A.A.; Pinto, A.M.A.C. Differential-Evolution-Based Optimization of the Dynamic Response for Parallel Operation of Inverters with No Controller Interconnection. *IEEE Trans. Ind. Electron.* **2012**, *59*, 2859–2866. [CrossRef]

energies

Article

Active Distribution Networks with Microgrid and Distributed Energy Resources Optimization Using Hierarchical Model

Thaís M. Blasi [1,*], Cyntia C. C. B. de Aquino [1], Rafael S. Pinto [1], Mauro O. de Lara Filho [1], Thelma S. P. Fernandes [1], Clodomiro U. Vila [1], Alexandre R. Aoki [1], Rodrigo B. dos Santos [2] and Fabricio H. Tabarro [2]

[1] Department of Electrical Engineering, Federal University of Paraná, Curitiba 81531-990, Brazil; cyntiacristinne@ufpr.br (C.C.C.B.d.A.); rafael.pinto@ufpr.br (R.S.P.); mauroobladen@ufpr.br (M.O.d.L.F.); thelma.fernandes@ufpr.br (T.S.P.F.); clodomiro.vila@ufpr.br (C.U.V.); aoki@ufpr.br (A.R.A.)

[2] Copel Distribution, Curitiba 81200-240, Brazil; rodrigo.braun@ufpr.br (R.B.d.S.); fabricio.tabarro@copel.com (F.H.T.)

* Correspondence: thais.blasi@ufpr.br

Abstract: Distribution networks have undergone a series of changes, with the insertion of distributed energy resources, such as distributed generation, energy storage systems, and demand response, allowing the consumers to produce energy and have an active role in distribution systems. Thus, it is possible to form microgrids. From the active grid's point of view, it is necessary to plan the operation considering the distributed resources and the microgrids connected to it, aiming to ensure the maintenance of grid economy and operational safety. So, this paper presents the proposition of a hierarchical model for planning the daily operation of active distribution grids with microgrids. In this case, the entire grid operation is optimized considering the results from the microgrid optimization itself. If none of the technical constraints, for example voltage levels, are reached, the grid is optimized, however, if there are some violations in the constraints feedback is sent to the internal microgrid optimization to be run again. Several scenarios are evaluated to verify the iteration among the controls in a coordinated way allowing the optimization of the operation of microgrids, as well as of the distribution network. A coordinated and hierarchical operation of active distribution networks with microgrids, specifically when they have distributed energy resources allocated and operated in an optimized way, results in a reduction in operating costs, losses, and greater flexibility and security of the whole system.

Keywords: active distribution networks; distributed energy resources; microgrids; optimization

Citation: Blasi, T.M.; de Aquino, C.C.C.B.; Pinto, R.S.; de Lara Filho, M.O.; Fernandes, T.S.P.; Vila, C.U.; Aoki, A.R.; dos Santos, R.B.; Tabarro, F.H. Active Distribution Networks with Microgrid and Distributed Energy Resources Optimization Using Hierarchical Model. *Energies* **2022**, *15*, 3992. https://doi.org/10.3390/en15113992

Academic Editor: Alberto Geri

Received: 13 April 2022
Accepted: 9 May 2022
Published: 28 May 2022

1. Introduction

An active distribution network is a new concept associated with distribution networks that present distributed energy resources (DERs) as distributed generation, controllable loads, and storage systems, as well as new monitoring, communication, and controls, which allow the supervision and management of the resources placed [1].

These new concepts no longer correspond to future scenarios, but rather to what has been seen in different utilities in Brazil and around the world. In the Brazilian scenario, the growth in the connection of distributed generation (DG) systems has occurred exponentially. Even with the COVID-19 pandemic, the search for distributed generation systems intensified, so that by the end of 2020, the number of systems was more than double the previous year, totaling 393,573 consumer units with distributed generation, corresponding to an installed power of 4786 MW. In 2021 this growth continues to happen, presenting at the end of September an installed capacity of more than 7164.4 MW, with more than 630,663 systems installed in all country regions [2].

It is possible to see this growth as a trend and, consequently, the inclusion of new equipment in the network that allows greater use of this generation, as well as maximizing its benefits and reducing its impacts. One of these types of equipment is the storage systems, mostly completed through batteries. BESS is being used at distribution grids on the power-utility scale or even on small scales connected to rooftop PV systems.

With PVs and BESS connected at the consumer units or close to them, there is the possibility of forming and operating microgrids (MG), which can operate connected or disconnected from the grid. In this case, when this part of the system is disconnected from the main grid they are supplied by distributed generation and storage systems belonging to the microgrid.

When connected to the main grid, microgrids can operate as a load or even as a generation providing active and reactive power. Due to the possibility of providing power in different moments, they can realize some ancillary services, such as voltage and frequency control, congestion management, reduction in power losses, and power quality improvements [3,4].

Nowadays, in Brazil, there are some MGs under development, and simultaneously, the development of regulatory aspects for them. One of the initiatives is the implementation of a real microgrid in Curitiba City (Brazil) at the Barigui Park. This system was built inside of the research and development project PD-02866-0511/2019, being composed of 36 kW of solar generation and 7.2 kW/28.8 kWh of an energy storage system with lithium-ion batteries.

It is important that the control system of the microgrids optimize and manage the operation of the resources available in the MG, as well as should operate coordinate, and communicate with the distribution system operator. In this context, this paper presents a hierarchical method for the operation planning of active distribution grids with microgrids integration.

The hierarchical structure proposed here aims to act at the tertiary control level to provide set points for a day ahead based on the status of the microgrid and the primary network to which it is connected [5,6]. Using a centralized structure, the utility has access to control actions that interfere in the operation of microgrids in order to ensure quality indices of the primary network.

According to [5], the implementation of this type of control can be performed via decomposition techniques to reduce computational effort. Thus, the optimization problem involving the primary network and the microgrids is decomposed into several sub-problems that are solved iteratively until convergence.

Therefore, based on this idea of decomposition, the approach of this work is to decompose the problem into two levels of optimization of the daily operation. Level 1: optimizing the operation of the active distribution network and Level 2: optimizing the operation of each microgrid. These two levels are optimized separately, in a hierarchical, iterative, and coordinated way in order to exchange information until a convergence point is reached that reconciles the objectives of each one of them.

In this work, the layout considered for connecting the microgrids to the distribution grid is the multiple microgrids connected to the active distribution grid as a single equivalent bus [5,7]. As for the design of the hierarchical model, which is based on the decomposition of the problem, one can cite the work of [8], which first performs a Power Flow (PF) of the entire active distribution network or primary distribution network to indicate an electrically feasible operating point. The results of this PF are forwarded to each microgrid that performs its optimization process to establish power transactions between the microgrids. This process is repeated until a convergence point is established between the grid and the microgrids, as in a master–slave scheme.

In this paper, the goal is to establish a point or state of operation where the operational and cost constraints of the grid and the microgrid are met or find a compromise solution where these operational constraints are met within an acceptable tolerance criterion. For this purpose, the active distribution grid is analyzed using the non-linear Multiperiod Optimal Power Flow (MOPF), responsible for the optimization of the primary grid (designated as the master model) and for the optimization of the microgrid (designated as the slave model) which is also performed through a Multiperiod Optimal Power Flow but using the mixed-integer linear model.

The main points that will be delineated in this hierarchical model are:

- Maintain primary grid voltage profile within safety range;
- Minimize the operating cost of the primary grid and the microgrids;
- Control the active and reactive power flow between the microgrid and the primary grid, to maintain the load-generation balance;
- Manage the power injections of the microgrids to ensure operational and economic quality;
- Perform tertiary control so that the controller (distribution utility), from hourly forecasts of power injection from the microgrid and distributed generation, suggests actions to the microgrid connected to the grid to maintain the power balance and voltage control of the active distribution network of the primary branch;
- Decomposition of the problem into master–slave optimization subproblems, where the master part optimizes the operation of the primary active grid (which involves voltage adjustment, adjustment of flexible loads, operation of storage systems, and closing the active and reactive power balance) and feeds back the optimization process of the microgrid operation as a slave problem. This slave part separately optimizes the operation of the microgrid (which involves setting flexible loads, operating storage systems, closing the active power balance, and dispatching controllable loads);
- Feedback the results of the slave problem to the master until a convergence point is reached that reconciles the objectives of each problem, which are coordinated through a centralizing unit;
- In the master model, the microgrid is represented as a single bus equivalent that injects or absorbs power, while in the slave problem the modeling of the microgrid is performed in detail, considering the internal network of each microgrid;
- Use a planning horizon of one day ahead (24 h), and each period contemplates an interval of Δt equal to 1 h.

Paper [9] also proposes a multi-period optimization for a distribution grid, considering the integration of multiple microgrids simultaneously. The approach of the authors is similar to the one proposed in this present paper. However, in [9] the microgrids have their operation optimized by the energy trading with other MGs, as well as the electric vehicles charging behavior, by the way, the current paper proposes the optimization of the microgrids and the distribution grid simultaneously to bring the best operational scenario.

Considering also a multi-microgrids scenario [10] proposes the transformation of the distribution network into a network of microgrids, with multiple agents. In this case, different objectives are considered to be optimized, according to the control level that is being considered. For the primary control, which happens inside of the MG, the objective is to perform a droop control, for secondary control the goal is the frequency/voltage restoration and arbitrary power-sharing and for tertiary control, the objective is to reduce the power losses.

Different hierarchical control approaches can be founded for microgrid integration into the distribution grid. Ref. [11] proposed an enhanced microgrid power flow considering a Newton approach to power-sharing and voltage regulation and a modified Jacobian matrix for secondary control, implementing a droop control. Ref. [12] proposes a multiple ac and dc community microgrid integration considering an economic dispatch that allows the energy flow between microgrids and from microgrids to the main grid, considering the voltage and frequency constraints. This approach is similar to the one proposed in the present paper, which considers the internal microgrid optimization, as well as the integration with the main grid, considering technical and economic aspects.

Ref. [13] considers the real-time pricing strategy for PV system integration into the distribution grid considering the active and reactive power dispatch. In this case, feedback is considered from the distribution system operator to the power conditioning systems from PVs to optimize the entire system operational cost, being this a similar approach that was also considered in the present paper, since the main goal is to reduce the operational cost of grid operation considering a multi-objective approach.

So, the main contribution of this paper is the development of a hierarchical model that plans the operation one day ahead of an active distribution network with a microgrid connected directly to the active primary distribution, using the master-slave methodology. The evaluation of the proposed methodology is made using a real distribution grid in the city of Curitiba/Brazil.

This paper is organized as follows: the second section presents the Material and Methods, comprehending the design of a hierarchical model for joint operation of microgrids and active distribution grid, the multiperiod optimum power flow methodology for grid optimization, the methodology for day ahead optimum planning of microgrids, and the details of the hierarchical model implemented; in the sequence, the scenarios definition and the results of the simulations are evaluated; in the end, the main conclusions are presented.

2. Materials and Methods

2.1. Design of a Hierarchical Model for Combined Operation of Microgrid and Active Distribution Network

This section aims to describe the basic concepts of the methodology that should perform the daily scheduling of microgrids to optimize and coordinate them with the operation of the active distribution grid to which the microgrids are connected. The basic conception of the model is based on the development of a coordinated and hierarchical optimization methodology:

- Of the radial active distribution network in medium voltage (from buses of the 13.8 kV substations) that contemplate: distributed generation, demand-side management (DSM), and coordination of microgrids and batteries connected in medium voltage;
- Of microgrids that contemplate DSM of flexible loads, distributed microgeneration, and optimal dispatch of batteries. For this, it considers the flexibility of the loads in the microgrid seeking to reduce the operational costs of the microgrid, avoiding the connection of loads at peak times of the feeder that correspond to more expensive energy prices and increasing the energy efficiency of the system as a whole.

Figure 1 illustrates the structure of an active distribution network, with multiple distributed energy resources connected, as well as a microgrid, able to operate connected or disconnected from the main grid.

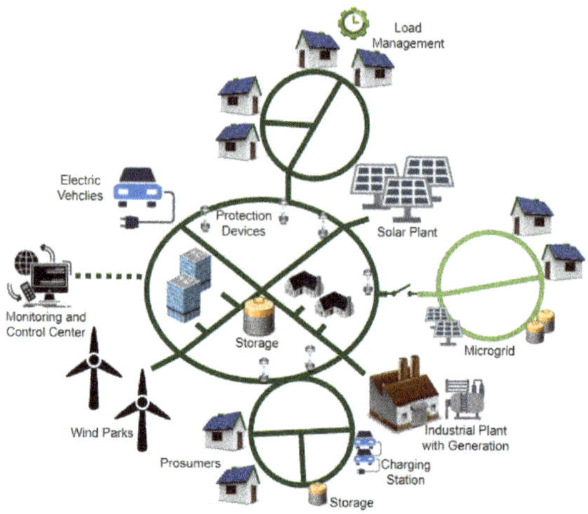

Figure 1. Illustration of an active distribution network.

The proposed architecture considers centralized/distributed control according to levels of action (master-slave) in order to optimize the overall performance of the system.

At the master level (active distribution grid optimization model using a Multiperiod Optimal Power Flow Model—MOPF), there is a detailed operation of the active distribution grid in medium voltage, and at the slave level (Microgrids Optimization Model—MGOM), there is a detailed operation of the microgrid.

2.2. Multiperiod Optimal Power Flow to Active Distribution Networks Methodology

To perform the daily operation at the master level, the reduced distribution radial feeder is modeled in medium voltage, in which are connected DG generation (fixed according to the solar incidence throughout the day), batteries, DSM, and the management of the power injections from the microgrids.

To operate these elements aiming to maintain this system in a permanent regime, part of the MOPF formulation proposed in [14] is used, which is a non-linear optimization problem that minimizes costs and maximizes load supply of microgrids. The operational constraints are defined by equalities and inequalities expressions representing the operative limits of the grid and the considered equipment, as briefly described below. The MOPF considers np periods, simultaneously resolved.

For the proposed problem, the non-linear equations that correspond to the active and reactive power balances of the grid seek to balance between the power injections from the loads, batteries, microgrids, and DGs in order to satisfy the operational characteristics of the grid, such as limits of voltage magnitudes, power flows through the lines, power, and energy of the batteries; for each of the 24 h of the study horizon.

The objective function (OF) corresponds to a multi-criteria function, composed of three parcels to be minimized: operational cost (1), battery degradation cost (2), and cost of flexible loads (or microgrids) deviations from the predicted power load (3):

$$foper = wc.c(\sum_{t=1}^{np}\sum_{i=1}^{nb} Pg_i^t) \tag{1}$$

$$fbat = wbat.c_{bat}(\sum_{t=1}^{np}\sum_{i=1}^{nb} Pbat_i^t) \tag{2}$$

$$fmicro = wmicro.c_{micro}(\sum_{t=1}^{np}\sum_{i=1}^{nb}(gamma_i^t - 1)^2) \tag{3}$$

where np is the number of periods of the horizon time of planning, nb is the number of buses, $c(\,)$ is the cost function of the system operation; $c_{bat}()$ is the cost function of the BESS, $c_{micro}()$ is the cost function of the deviation from the injections power of microgrids calculated by MGOM; Pg_i^t is the total generation power at each bus i and period t (solar + substation); $Pbat_i^t$ is the power injection of the BESS connected at bus i and period t, $gamma_i^t$ is the percentage of predicted power injection of the microgrids (connected at bus i and period t) to be cut and wc, wbat, and wmicro are the weights of the OF objectives.

The values of $gamma_i^t$,that multiply the injection power of microgrids (FPd_i^t, calculated from MGOM formulation) (5), decrease their predicted power injection and must deviate minimally from the unit value (which corresponds to the ideal values).

The operation cost of the power utility is based on the Brazilian white tariff [15]. The operation cost of BESS is associated with its degradation cost due to its usage in each cycle [14]. The cost of power injections deviations from the predicted load of microgrids obtained from MGOM is defined as fines for not meeting the loads.

The complete problem formulation is given by:

$$min(fmicro + foper + fbat) \tag{4}$$

s.t.

$$Psub_i^t + P_{PV}{}_i^t - Pbat_i^t - Pd_i^t - gamma_i^t.FPd_i^t = P_i^t(\dot{V}, areg) \quad (5)$$

$$\begin{aligned} Qsub_i^t + P_{PV}{}_i^t.tan(acos(pf)) - Pbat_i^t.tan(acos(pf)) - Pd_i^t - \\ gamma_i^t.FPd_i^t.tan(acos(pf)) + |\dot{V}_i^t|^2.B_{sh}{}_i^t = Q_i^t(\dot{V}, areg) \end{aligned} \quad (6)$$

$$V_{min}{}_i^t \leq |\dot{V}_i^t|^2 \leq V_{max}{}_i^t \quad (7)$$

$$-fl_{max}{}_j^t \leq fl_j^t \leq fl_{max}{}_j^t \quad (8)$$

$$a_{min}{}_j^t \leq a_{reg}{}_j^t \leq a_{max}{}_j^t \quad (9)$$

$$gamma_{min}{}_i^t \leq gamma_i^t \leq gamma_{max}{}_j^t \quad (10)$$

$$Pbat_i^{min} \leq Pbat_i^t \leq Pbat_i^{max} \quad (11)$$

$$Ebat_{acum}{}^{min} \leq Ebat_i^t \leq Ebat_{acum}{}^{max} \quad (12)$$

$$i = 1, ..., nb, j = 1, ..., nl \text{ and } t = 1, ..., np$$

Equations (5) and (6) are the equality constraints that represent the active and reactive power balance, respectively. In these equations: $Psub_i^t$ is the active power provided by the substation bus, $P_{PV}{}_i^t$ is the active power of the PV systems at bus i and period t, $Pbat_i^t$is active power of BESS injected at bus i and period t, Pd_i^t is the power demand at bus i and period t; FPd_i^t is the predicted injection of active power of the microgrids at bus i and period t (obtained from MGOM), and $P_i^t(\dot{V}, areg)$ is the total active power injection at bus i and period t, this power injection depends on the voltage phasor (($\dot{V}$) in module and angle) and if there is any voltage regulator or transformer connected to the bar, it will depend on the tap (areg) of this equipment. The injections of reactive power are calculated by the power factor (pf) of PVs and BESS, the reactive power from substation ($Psub_i^t$), the reactive power contribution from capacitor banks at bus i and period t ($|\dot{V}_i^t|^2.B_{sh}{}_i^t$), making the total reactive power injection at bus i and period t ($Q_i^t(\dot{V}, areg)$) ; $gamma_i^t$ represents a percentage of the predicted power injection of the microgrids (FPd_i^t); and $|\dot{V}_i^t|$ is the voltage magnitude of bus i and period t.

The inequality constraints represent the operational limits of the grid or the equipment. Equation (7) is related to maximum and minim voltage limits ($V_{min}{}_i^t, V_{max}{}_i^t$) where $|\dot{V}_i^t|$ is the voltage magnitude of bus i and period t. Equation (8) is related to the limits of the power flow ($fl_{max}{}_j^t$) through the nl lines of the network; fl_j^t is the power flow through the line j and period t. Equation (9) represents the maximum and minimum limits ($a_{min}{}_j^t, a_{max}{}_j^t$) of the taps positions of voltage regulators and a_j^t is the tap position of a voltage regulator installed at line j and period t. Equation (9) is related to the maximum and minimum limits ($gamma_{min}{}_i^t, gamma_{max}{}_i^t$) of $gamma_i^t$. Additionally, Equations (11) and (12) are related to the limits of power and energy of the BESS ($Pbat_i^{min}, Pbat_i^{max}, Ebat_{acum}{}^{min}, Ebat_{acum}{}^{max}$). The sub-indices i represent the buses, j the lines, and t each time step.

The values $Pbat_i^{min}, Pbat_i^{max}$ is limited according to the sizing of the BESS and how much power it can absorb (charge, $Pbat_i^t$ assumes a positive value) or provide (discharge, $Pbat_i^t$ assumes a negative value) along 24 h.

All the parameters and variables of the MOPF are represented by the system per unit (pu).

The optimization problem (4)–(12) is solved by the Interior-Points Method [16] completely implemented in Python language, without use any solver. More derails about grid and its equipments modeling in MFOP approach can be founded at [14].

The input data of this problem are the power demand, solar generation, voltage limits, taps positions limits of voltage regulators, power lines flow limits, limits of $gamma_i^t$, values of susceptance capacitive of capacitor banks, system parameters, reference bus, and location of the equipment. For the BESS model, the input parameters are power and energy limits, energy at the begging and final of the evaluation period, system efficiency, and the location of the BESS. For the microgrids, the values of the power injections of the microgrids, FPd_i^t,are calculated by the MGOM problem.

The control variables are the power provided by the substation bus at each time step, the voltage magnitudes, taps positions of each voltage regulator also at each time step and the power and energy of the BESS at each time step. Mainly, the values of $gamma_i^t$ indicate the deviations that must be made in the previous calculation obtained by MGOM.

2.3. Methodology for Day-Ahead Optimal Operation Planning for Microgrids

The microgrids are modeled in the slave stage, and they are connected with the active distribution network at one or more nodes called points of common coupling (PCC). The microgrids can be modeled similarly to active distribution networks since they are small portions of networks that can be operated by the distribution system operator or by private owners. Furthermore, microgrids can operate isolated from the main grid and they have specific characteristics of operation, such as greater penetration of intermittent renewable distributed energy resources in microgrids compared to distribution networks; more frequent installation of storage systems to control the intermittence and randomness of renewable sources; typically, low voltage operation, some larger microgrids at medium voltage; capability of operation isolated in relation to the main network; demand response; and, uncertainties and unavailability of renewable generating sources.

The internal networks of microgrids, in low or medium voltage, are mathematically represented as a system of non-linear equations. However, computationally, nonlinear problems are very challenging and often even intractable. Thus, most of the works in the literature on microgrid optimization use linearized models, which simplify the network in exchange for an acceptable loss of precision or single-bus models. Single-bus models completely neglect the existence of an internal network inside the microgrids, this approach can be applied to small microgrids at low voltage.

In this context and considering that a utility distribution feeder can contain several microgrids with different sizes and configurations, the MGOM model proposed in this work considers the single-bus model for smaller microgrids at low voltage.

In summary, the computational methodology of the daily operation of microgrids aims to find the lowest operating cost for the microgrid through the management of its distributed energy resources while satisfying operational constraints.

The proposed model includes energy resources in microgrids, composed of DG, energy storage systems, and demand response through flexible or controllable load management. Island operation of the microgrid is also considered, along with network maintenance windows and other scheduled events. Additionally, the optimization problem is defined for a 24-h window, with hourly steps.

After the optimization of the lowest operating cost, the master stage of the algorithm receives the information about the daily operation schedule for each microgrid at the common connection point between the microgrid and the active distribution network. Thus, the master stage performs its optimization using the MOPF model. If the operation of all microgrids is feasible and does not set operational problems for the main grid, the operational planning for the 24-h window is validated. Otherwise, the MOPF model makes suggestions or includes operational constraints on the microgrids, which must perform a new optimization process within the imposed limits.

As formulated by the MOPF model, the objective of the microgrid is also to minimize operating costs during the considered time horizon. Thus, the problem is defined as:

$$min_x \sum_{t=1}^{N}(Cb_t) + LS_t.Tls_t + \eta \tag{13}$$

$$\eta = \sum_{t=1}^{N} Grid_t.T_t + (|Pd_{ideal,t} - |Pd_{realized,t}|).T_{pen} \tag{14}$$

where N is the number of periods, Cb_t is the cost related to battery degradation in a period t (BRL); LS_t is the load shedding in a period t (kWh); Tls_t represents the load shedding cost (BRL/kWh); η is the auxiliary variable that represents the energy costs; x is the set of variables to be optimized; $Grid_t$ is the net energy to be purchased (positive values) or injected (negative values) into the active distribution network in a period t (kWh); T_t is the price for energy purchase/sale in period t (BRL/kWh); $Pd_{ideal,t}$ is the ideal active power transaction for MFOP in a period t and $Pd_{realized,t}$ is the active power transaction realized from MGOM optimization in a period t (kWh); $Tpen$ is a penalty cost for not meeting the value defined by the MOPF (BRL/kWh). The ideal power is calculated with an auxiliary variable $gamma$, which is shared from the hierarquical process to make the adjustment in the energy transaction from the microgrid to the network:

$$Pd^{i}_{ideal,t} = Pd^{i-1}_{realized,t}.gamma^{i}_{t} \tag{15}$$

where i is the iteration from the hierarchical process. The formulation for demand response and other microgrid operations is adapted from [17].

The load shedding is calculated through the energy balance that depends on critical, non-controllable, and non-controllable loads, as will be shown below. Due to operational constraints for the case of the practical case of the microgrid in the IMAP–Barigui complex, the battery cannot directly inject energy into the main grid, so the energy balance is modeled in two parts, defined in Equations (16) to (19). These equations show that the battery can supply controllable and critical loads. In the case of an island operation, there is no energy injected by the grid into the system.

$$Grid_t = Er1_t + Er2_t \tag{16}$$

$$Er1_t = L^{NP}_t + L^{CP}_t - Gpv_t + A^{c}_t - LS1_t \tag{17}$$

$$Er2_t = L^{P}_t + L^{CONT}_t - A^{d}_t - LS2_t \tag{18}$$

$$Er2_t.X^{d}_t = 0 \tag{19}$$

$$t = 1, \ldots, N$$

where $Er1_t$ is the energy balance that cannot be supplied by the battery in a period t (kWh); $Er2_t$ is the energy balance that can be supplied by the battery in a period t (kWh); L^{NP}_t, L^{P}_t are, respectively, the non-controllable critical load and the non-controllable priority load in a period t (kWh); L^{CP}_t is the load of the carport in a period t (kWh); L^{CONT}_t is the controllable load at period t (kWh); Gpv_t is the photovoltaic generation in a period t (kWh); A^{c}_t, A^{d}_t are respectively the amount of energy stored and discharged from the battery in a period t (kWh); $LS1_t$, $LS2_t$ are, respectively, the load shedding without and with batteries in a period t (kWh).

The load shedding $LS1_t$ and $LS2_t$ are formulated as shown in Equations (20) to (22):

$$LS_t = LS1_t + LS2_t \tag{20}$$

$$0 \leq LS1_t \leq L^{NP}_t \tag{21}$$

$$0 \leq LS2_t \leq L^{P}_t \tag{22}$$

The amount of controllable load in a period t is defined from Equation (23):

$$L^{CONT}_t = \sum_{cl=1}^{Ncc} Xcc_{t,cl}.Pcl_{cl}.\frac{W_{t,cl}}{60} \tag{23}$$

where Ncc is the number of controllable loads present; $Xcc_{t,cl}$ is a binary variable that indicates whether a controllable load cl is operating in period t; Pcl_{cl} is the power of a controllable load cl (kW); $W_{t,cl}$ is the operating time in minutes of the controllable load cl in a period t.

In order to operate the controllable loads in a way that is useful to the consumer, but also economically efficient, the following data must be included as parameters of the problem:

- $Ti(cl)$ and $Tf(cl)$ represent the time interval in which the load can be operated (e.g., during business hours);
- $Ncycles_cl$ is the number of charge cycles that must be operated within the proposed time interval;
- MTO_{cl}^{real} indicates the cycle time of a controllable load (for example, a washing machine may have an operating cycle longer than one hour, thus incorporating multiple periods).

With these parameters, it is possible to implement the constraints (24) to (28):

$$Xcc_{t,cl} - Xcc_{t-1,cl} = Yoncc_{t,cl} - Zoffcc_{t,cl} \quad (24)$$

$$Yoncc_{t,cl} + \sum_{l=1}^{MTO_{cl}^{aux}} Zoffcc_{t+l,cl} \leq 1 \text{ when } Ti(cl) \leq t \leq Tf(cl) - 1 \quad (25)$$

$$W_{t,cl} \leq 60.Xcc_{t,cl} \quad (26)$$

$$\sum_{t=Ti(cl)}^{Tf(cl)} W_{t,cl} = 60.Ncycles_{cl}.MTO_{cl}^{real} \quad (27)$$

$$Yoncc_{t+MTO_{cl}^{aux},cl} - Xcc_{t+MTO_{cl}^{aux},cl} \geq 2.Yoncc_{t,cl} - Xcc_{t,cl} - 1 \quad (28)$$

where $Yoncc_{t,cl}$ and $Zoffcc_{t,cl}$ are binary variables that indicate the beginning and the end of a cycle for a controllable load cl in a period t, respectively; MTO_{cl}^{aux} is an auxiliary variable that rounds up the cycle time of controllable loads for use in specific constraints.

Constraint (24) uses the time variation of Xcc to detect the beginning and end of an operating cycle of a controllable load. Constraint (25) ensures that if a controllable load is turned on, it cannot be turned off until the end of its operating cycle. Constraint (26) defines that the operating time of a load in a period t of one hour is a maximum of 60 min. Constraint (27) represents the total number of operating minutes required for a controllable load to be met. Additionally, Constraint (28) ensures the inclusion of consecutive operating cycles of a controllable load, avoiding the consideration of just one cycle. Equations (24) to (28) also apply for the controllable load of the carport L_t^{CP}.

2.4. Hierarchical Model

This section aims to present the formulation of the hierarchical model for daily microgrid scheduling coordination that contemplates: an active medium voltage distribution network composed of distributed generation (DG), coordinating microgrids and batteries connected at medium voltage; and microgrids that contemplate the demand response through the dispatch of flexible loads, distributed microgeneration, and the optimal dispatch of batteries.

The architecture proposed in this paper considers centralized control according to actuation levels (master-slave) to optimize the global performance of the system. The master level uses a Multiperiod Optimal Power Flow (MOPF) that operates the active distribution network in medium voltage, and the slave level uses the Microgrid Optimization Model (MGOM) that operates the microgrid.

The proposed chaining to interconnect the information between the MFOP Model and the MGOM Model is described in the following Algorithm 1:

Algorithm 1: Hierarchical Model

Step 1: Declare input parameters.
Step 2: Make $kiter_{masterslave} = 0$.
Step 3: Run first simulation of MGOM, which provides optimal power injection value by the microgrid view, namely as FPd_{MGOM}.
Step 4: Initialize the variables, load the ones already calculated in pre-simulation and run MOPF to obtain FPd_{MOPF}, which is the optimal power injection value to be performed by the microgrid, but prioritizing the feeder. If there is a convergence of the MOPF, go to **Step 5**, otherwise go to **Step 9**.
Step 5: If $kiter_{masterslave}$ is greater than the maximum number of resupply (kmax) go to **Step 8**. If there is dispatch deviation of FPd_{MOPF} from FPD_{MGOM}, go to **Step 6**. Otherwise, make $kiter_{masterslave} = 100$ and go to **Step 9**.
Step 6: Run MGOM and recalculate new FPd_{MGOM}. If the new value of FPd_{MGOM} satisfies the value of FPd_{MOPF} (calculated in **Step 5**), go to **Step 9**. Otherwise, make $kiter_{masterslave} = kiter_{masterslave} + 1$.
Step 7: Perform MFOP optimization process and obtain new of FPd_{MOPF} and go to **Step 5**.
Step 8: If microgrid injection cannot be adjusted after kmax iterations, simulate MOPF with minimum voltage reduced to 0.9 pu and using last load value set by MGOM (FPd_{MGOM}).
Step 9: END

If after kmax iterations between Step 2 to 5, no load shedding of the microgrid is achieved, there will be two options: the mandatory easing of power injection proposed by MOPF is recommended, as it trumps feeder operational safety, or the MOPF is executed with relaxation of the minimum voltage levels (Step 9) to meet the power injection proposed by MGOM.

Figure 2 presents the flowchart of the proposed algorithm in the master–slave scheme.

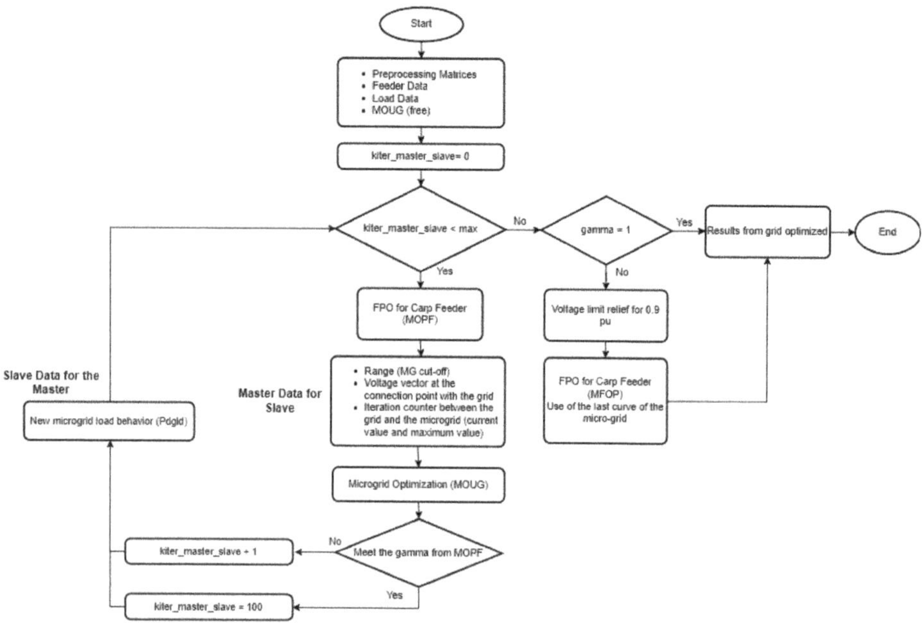

Figure 2. Flowchart of the proposed algorithm in the master–slave scheme.

2.5. Simulation Scenarios

To evaluate the grid behavior with the integration of the microgrid using the developed hierarchical model, sets of scenarios were defined to allow a comprehensive evaluation of the grid, considering different load configurations, microgrids, as well as the allocation of this system in the distribution grid.

For the simulations, a real distribution feeder, from Curitiba, Brazil, was considered. This feeder presents 2119.08 kW of installed power, of which 38% corresponds to low voltage consumers and 62% to medium voltage consumers. It has a total number of 359 buses and 358 lines, which were modeled for the power flow simulations.

Figure 3 shows the schematic diagram of the active network feeder, with the substation represented by the pink rectangle at the base of the diagram. In the figure legend, ssdbt means low voltage lines; ssdmt are medium voltage lines, sub corresponds to the substation; ucbt is the low voltage consumers; ucmt are medium voltage consumers; unsemt are medium voltage switches, and untrd are distribution transformers that connect LV to MV (there are a total number of 64 in the circuit).

Figure 3. Real Feeder schematic diagram.

The IMAP Barigui complex is a microgrid being implemented in the city of Curitiba, Brazil, deployed under the ANEEL R&D project PD-02866-0511/2019. The complex constitutes on premises of the Curitiba City Hall and features batteries, photovoltaic generation, controllable loads, and an electric vehicle charger (carport). Figure 4 shows a single-line diagram of the IMAP–Barigui microgrid. The complex has a total of 36.64 kWp in roof-mounted photovoltaic panels. More data and other specifications of this microgrid can be found in [18].

Thus, the scenarios evaluated comprised:

- Scenario 1: original load for the feeder and the Barigui microgrid;
- Scenario 2: condition of programmed microgrid isolation (schedule islanding);
- Scenario 3: feeder and microgrid with increased load;
- Scenario 4: consideration of the most complex grid scenario, with the presence of distributed generation and batteries, as well as the microgrid, with load increased.

It should be noted that the Barigui microgrid is allocated at the bus of the circuit referring to the IMAP complex (bus 350 in the active network modeling), the month used as a base was December, with the highest solar generation.

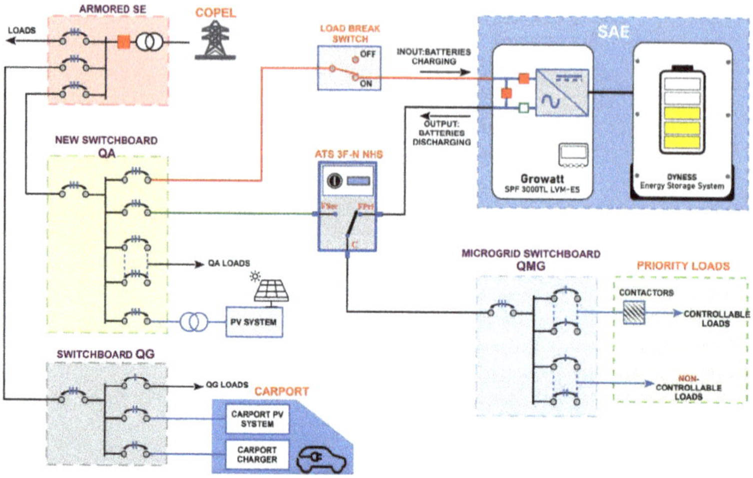

Figure 4. Schematic diagram of the IMAP–Barigui microgrid.

2.6. Materials

To validate the proposed formulation, all experiments and simulations were carried out in Python 3.8 language, with Gurobi Optimizer on a laptop with an Intel Core™ i7-5500U CPU @ 2.40 GHz with 8 GB RAM.The feeder data were obtained from the geographical database of the power utility, as well as the data for characterization of the loads, which was obtained from the treatment and analysis of historical measurement data.

3. Results

3.1. Scenario 1: With Original Loads

In this scenario, the simulation uses the original loads for the microgrid and active network. The behavior of the load in the feeder is shown in Figure 5.

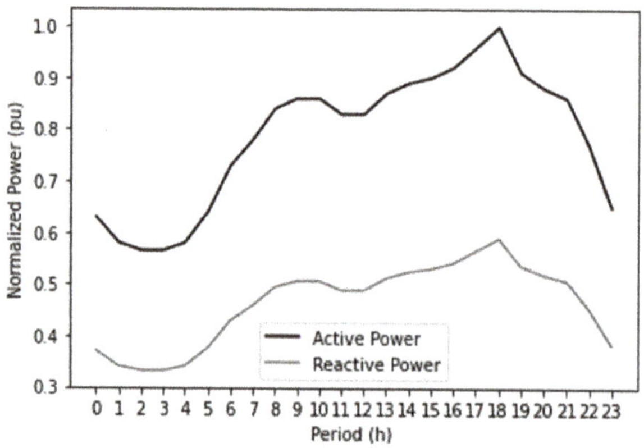

Figure 5. Behavior of the load curve in the feeder.

The initial dispatch proposed by the MGOM model is presented in Figure 6a. For this dispatch, it is possible to observe that the charging (in pink) of electric vehicles is performed during the period from 11:00 a.m. to 4:00 p.m., minimizing the operating cost of the microgrid. The controllable loads are allocated from 9:00 a.m. to 12:00 p.m. and 3:00 p.m., following the operating constraints presented in Table 1. The discharges of batteries (in red) are in the periods of 6:00 p.m. and 7:00 p.m. since the photovoltaic generation is ending and the price of energy is at peak demand, showing that DERs are used to equalize the reduction in generation and reduce the microgrid total cost. In addition, the battery charges during hour 23 (green color), when the off-peak power price is cheaper. No load shedding (gray color) is observed in this scenario.

It is important to highlight that in this step of the hierarchical process there has been no feedback from the main grid yet. This first run ($kiter_{masterslave} = 0$) provides the optimal power injection value from the microgrid point of view. In this case, the total cost of dispatching the microgrid is −26.27 BRL, which means that at the end of the day, there is more power injection into the network than consumption. There is no need for the feedback process ($kiter_{masterslave} = 100$), because no restrictions were violated, and the main grid operation took place within the operational limits. Thus, the microgrid behavior does not need to be changed, and the operation proposed by the MGOM is accepted by the MOPF calculation. Figure 6b shows the dispatch of the microgrid from the point of view of the network, which shows that the load/injection performed, and the load/injection desired by the microgrid is the same, without the need for changes.

Table 1. Controllable loads of the microgrid.

Circuit	Equipment	Power (kW)	Number of Cycles per Day	Cycle Time (minutes)	Start Time	Ending Time
A22	Electric Stove	5	2	30	11	13
A23	Microwave	0.75	5	12	11	13
A12	Common Outlets	1	4	60	8	18
A14	Common Outlets	1	4	60	8	18

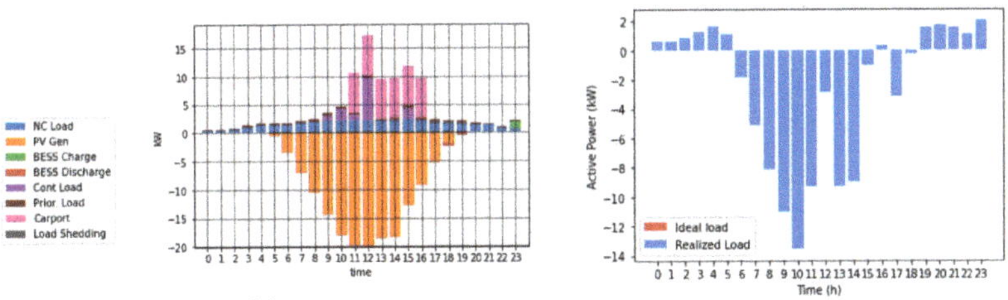

(a) (b)

Figure 6. Proposed Dispatch for the Pilot microgrid. (**a**) Microgrid dispatch. (**b**) Dispatch from the point of view of active network.

For the voltage profile, it can be observed in Figure 7 that the voltage levels follow what is specified according to the load behavior (light, medium, and heavy), showing a small variation between feeder buses in relation to the specified voltage at the substation. In this case, it is also verified that the voltage magnitudes of all the buses of the feeder are far from the voltage limits of 0.93 and 1.05 pu. The voltage behavior for all the feeder buses is shown in Figure 7a, with the emphasis on the bus where the microgrid is connected Figure 7b.

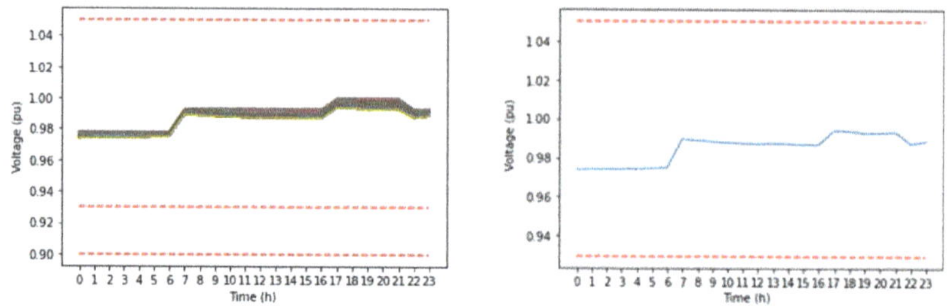

(**a**) (**b**)

Figure 7. Voltage behavior and highlight for voltage on the Bus in which the microgrid is connected. (**a**) All buses from the feeder. (**b**) Bus with microgrid connection.

3.2. Scenario 2: Scheduled Islanding

In this case, it is considered a planned island operation of the microgrid during the periods from 1:00 p.m. to 3:00 p.m. Note that, in the case of islanding, the battery can only supply critical loads on the microgrid. The microgrid dispatch is shown in Figure 8. It is possible to observe that during the islanding period, there is no charging of electric vehicles, nor allocation of controllable loads. The battery discharges to supply critical loads, while non-critical uncontrollable loads are cut off. Once the islanding operation ends, the battery is charged again in the next period. By not using photovoltaic generation, as it is not connected in the same frame as the priority loads, the costs of the microgrid increase to 32.59 BRL.

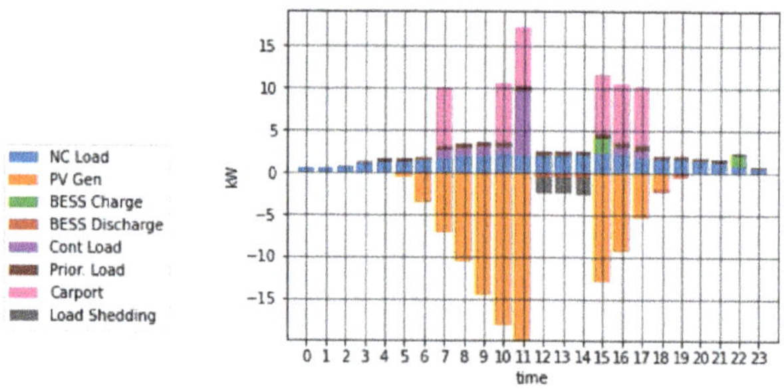

Figure 8. Dispatch proposed by the microgrids for scheduled islanding.

The behavior of the power injection by t load observed by the main network is shown in Figure 9, in which during the programmed islanding period, there is no energy transition with the main grid. Even so, in other planning periods, it was not necessary to define load shedding on the main grid.

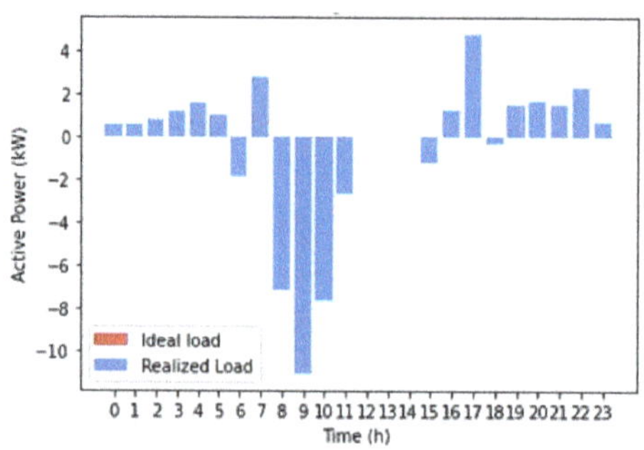

Figure 9. Behavior of the injection/load power of the microgrid observed by the network in the scenario with programmed islanding.

3.3. Scenario 3: Feeder with Load Increase

Since the feeder load and the Barigui microgrid load are composed of low values in the practical case, there are no problems with the voltage profile, and no operational and violated constraints. So, no iterative process occurs within the hierarchical model.

To force feedback and, therefore, verify the hierarchical model, the load values of the active network feeder were increased 9 times and the load values of the microgrid were increased 20 times. These values were defined after exhaustive tests with several other lower values, which did not cause problems for the operation of the network. Here is the sequence of steps performed for this case:

- Initialization of the counter for the iteration between the methodologies, $kiter_{masterslave} = 0$;
- Run the first simulation of the MGOM model, which provides optimal power injection value through the microgrid point of view ($PdGL_{MGOM}$). In this case, the dispatch corresponds to the behavior of load and generation in the summer, with the power increased by 20 times;
- Run the MOPF model. In this step, all the constraints are checked and a new microgrid operation is proposed by the MOPF ($PD_{GLDMOPF}$), as shown in Figure 10a. The proposed load shedding takes into account the voltage violation that occurs, especially between 7 p.m. and 8 p.m., as shown in Figure 10b. There was a reduction in the microgrid's load, which tends to improve the voltage magnitude profile, thus avoiding undervoltage.

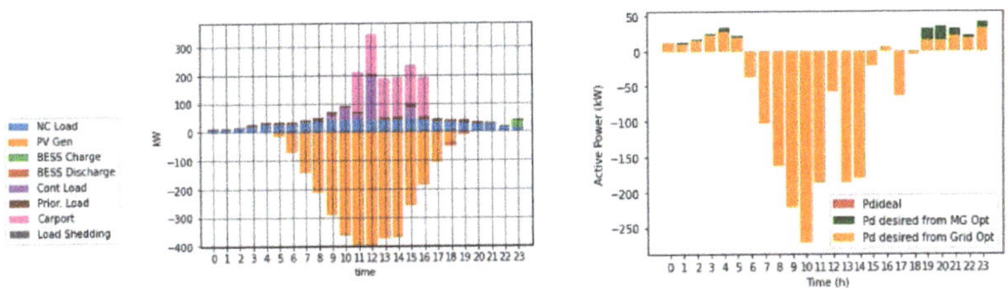

(a) (b)

Figure 10. Proposed dispatch for the microgrid before the feedback process. (**a**) Dispatch of the microgrid (MGOM). (**b**) Dispatch from the point of view of the active network (MOPF).

A new microgrid dispatch (MGOM) is run using the proposed dispatch of MOPF (Figure 10b). There is a load reduction at periods suggested by the main grid, but not enough as suggested by the main grid. Figure 11a shows the load shedding that occurs from 7 p.m. to 8 p.m. by the proposed new dispatch. Furthermore, to better adapt to the suggestion of the MOPF, the MGOM also reduces part of the battery charge and discharge. From the point of view of the main grid, it would be necessary to further reduce the load during the 7 p.m., 8 p.m., and 11 p.m. periods. Therefore, there is a new feedback process, thus sending a new range so that the microgrid can adjust its energy resources to the requests made by the main grid.

After the increase in load shedding and reduction in the use of batteries, a new dispatch is proposed. However, even with the adjustments, the load shedding is very costly for the microgrid during peak hours, which is exactly when the MOPF requests the reduction in load. Therefore, even after three feedbacks from the hierarchical process, the microgrid cannot adapt to the power requested by the MFOP, as shown in Figure 12.

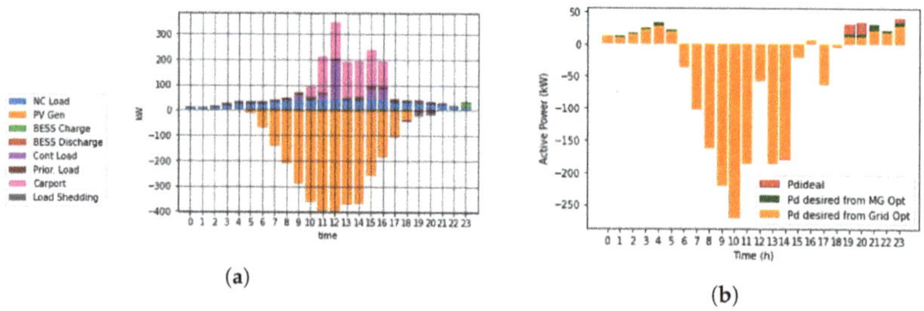

Figure 11. Proposed dispatch for the microgrid in the first iteration of the hierarchical model. (**a**) Dispatch of the microgrid. (**b**) Dispatch from the point of view of the active network.

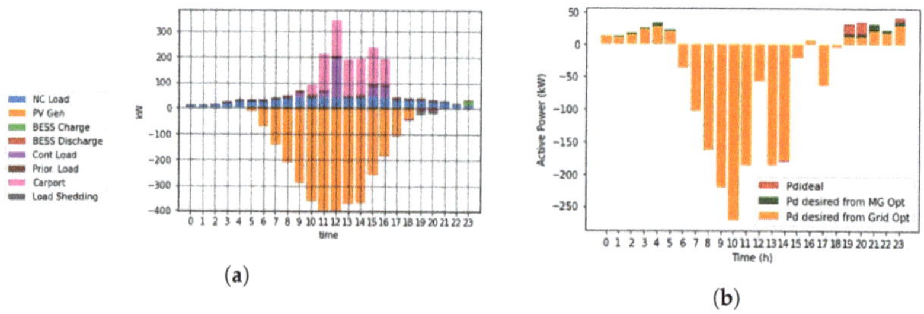

Figure 12. Proposed dispatch for the microgrid in the third iteration of the hierarchical model. (**a**) Dispatch of the microgrid. (**b**) Dispatch from the point of view of the active network.

Therefore, the main grid reduces the minimum voltage to 0.9, and the dispatch from the microgrid is accepted by the main grid under these conditions. The behavior of the load at the end of the hierarchical process can be seen in Figure 13, in which the difference between the dispatch initially proposed by the microgrid (ideal load) and the dispatch carried out (realized load) is shown, both from the point of view of the main grid.

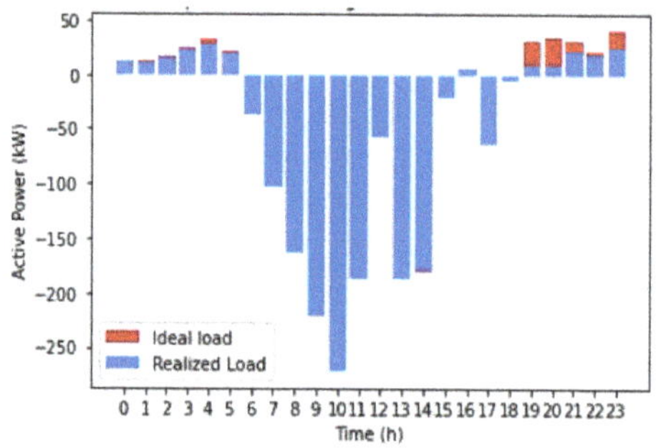

Figure 13. Proposed dispatch for the microgrid at the end of the hierarchical process.

Regarding the voltage behavior of all buses at the end of the hierarchical model iterations, it shows that with the new microgrid dispatch it was yet possible to guarantee the voltage of all buses within the operational limits, as shown in Figure 14, where the voltage for all buses of the feeder is within the specified voltage limits.

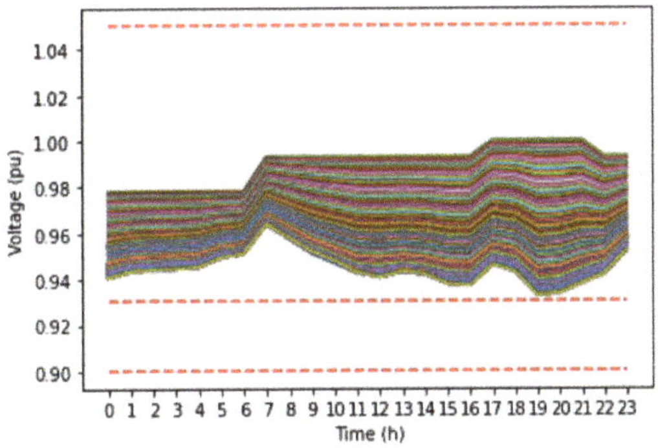

Figure 14. Voltage profile of all buses of the feeder after the hierarchical control for the mg dispatch.

In the first iteration, the total costs of the microgrid were −525.40 BRL, in which the controllable loads are allocated within the commercial period and the battery is discharged at 6 p.m. and 7 p.m., and it charges at 23 h. In the second iteration, when there was the feedback of the MFOP and the definition of a new range for readjustment in the microgrid dispatch, the total cost was −380.41 BRL, since there was a need to reduce the load between the periods 6 p.m. to 8 p.m., decreasing the financial benefit. For this scenario, after the iteration of the hierarchical process, the microgrid carried out a 6.93% reduction in its load compared to the initial proposition.

The complete behavior of the network presents losses of 3.2% (8.088 MWh) and operating costs of 141,319.57 BRL. It should be noted that in this scenario the costs are higher since the demand supplied by the feeder is 9 times higher than the load in the original scenario, which represents a total of 252 MWh throughout the day.

3.4. Scenario 4: Active Distribution Network with Microgrid + Photovoltaic + Battery, with Load Increase

In this scenario, the operation of the network was considered with the allocation of distributed solar generation systems and batteries operating simultaneously along the feeder, as an active distribution network. In other words, this scenario looks at future cases in which the feeder also has distributed energy resources. The analysis was carried out considering the penetration of 30% of distributed generation in this feeder. The allocation of the distributed energy resources was defined proportionally to the load buses, considering that part of the loads allocated may have solar photovoltaic generation systems (PV).

Regarding the allocation of distributed energy resources, a battery storage system (BESS) with 1 MW/2 MWh was allocated at bus 183, since 80% of the total feeder load is located downstream from this point. For this scenario, the loads of the main grid and of the microgrid were increased in order to evaluate the behavior of the hierarchical model under conditions of stress for the system. The pre-processing follows the behavior of scenarios 2 and 4. However, for this case, there is a slack due to the insertion of BESS and GD; thus, the MOPF did not require readjustment, providing a cost in the microgrid of −525.37 BRL. Figure 15 shows the microgrid dispatches, which is similar and proportional to the microgrid dispatch without increased load. Similarly, Figure 15b also shows the load behavior from the point of view of the network and it is also proportional to the behavior of scenarios 1, 3, and 5.

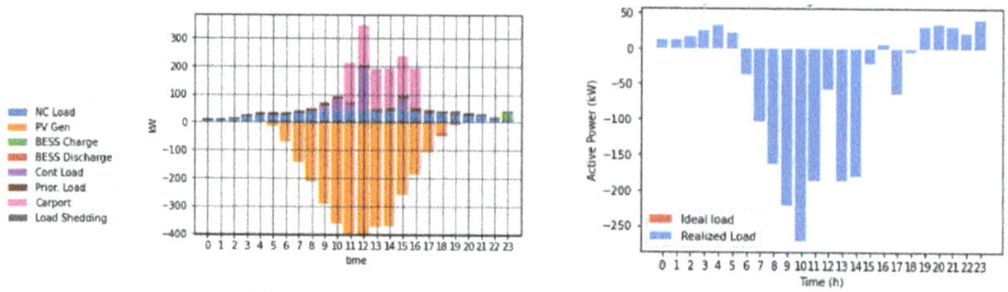

(**a**) (**b**)

Figure 15. Microgrid's dispatch to active network with increase load. (**a**) Dispatch of the microgrid. (**b**) Dispatch from the point of view of the active network.

Regarding other characteristics of the network operation for this scenario with a larger number of elements, it appears that the losses were 3.31% (8.35 MWh), remaining the levels obtained in the previous simulations. Regarding the costs of operation of the network, these are 134,761.88 BRL including the cost of battery depreciation of 6089.52 BRL.

Figure 16 shows the active power balance from the point of view of the substation for the scenario with increased load. In this case, it appears that most of the load is served by the substation since the solar generation power installed in the feeder remained unchanged in relation to scenario 5 (no-load increase).

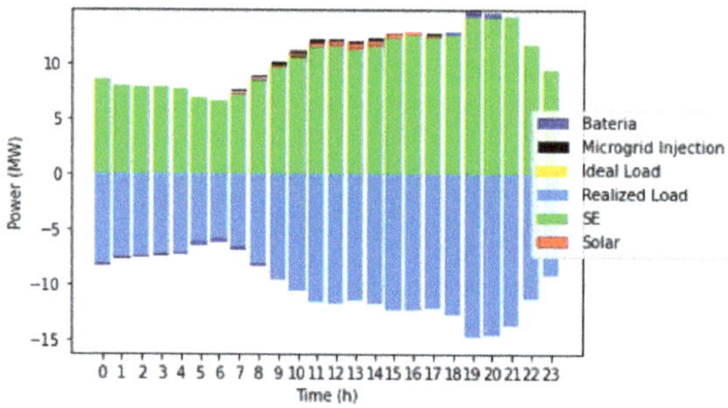

Figure 16. Active power balance of the point of view of the substation for the case with active distribution network and with increased load.

The battery operation stores 1.537 MWh throughout the day and provide 1.309 MWh of energy for the same period, note that the chosen capacity for the system is 2 MWh. In Figure 17, the behavior of power (upper part of the figure) and energy (lower part) is presented. It can be observed that the charge occurs in the first hours of the day, which has little impact on the feeder operation since the load is at low levels. The discharging, on the other hand, occurs at the peak load time of the feeder, contributing to the fulfilment of the load.

Finally, regarding the voltage behavior, since this scenario has a heavier load, there is a tendency for voltage violation. However, with the battery dispatch, it is possible to perform the network operation in a satisfactory manner, without the occurrence of under or over voltage in any of the feeder buses (Figure 18). Note that for this case, the voltages of all the buses do not reach the minimum limit, perhaps because of the slack that GD and BESS provide to the system.

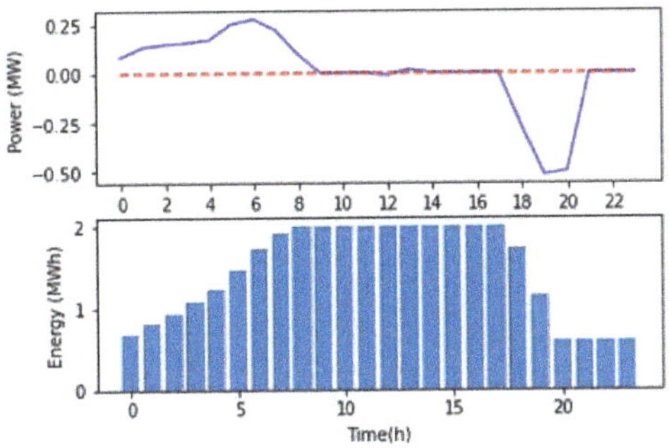

Figure 17. Behavior of the battery for the active distribution network scenario with increase charge.

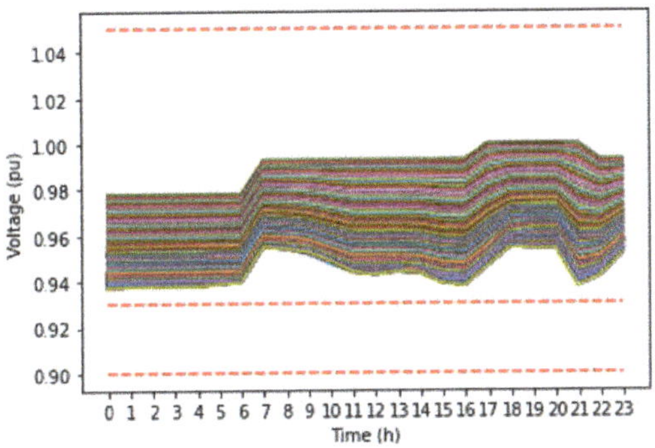

Figure 18. Behavior of voltage at all buses for the scenario of active distribution and load increase network.

3.5. Scenarios Comparison

Table 2 shows a comparison of the main characteristics evaluated in all simulation scenarios. Scenario 0 (zero) was added to compare the numeric results and corresponds to the scenario of the feeder with loads only, i.e., without microgrids, distributed generation, and battery. This table presents microgrid costs, total operational costs, total energy consumption, computational time, number of iterations, total power losses, and variation of the original MG and the optimized.

From the results presented in Table 2, it can be seen that adding microgrids to active distribution grids, in general, is beneficial in terms of economic and technical aspects because the costs are not greatly increased (scenarios 1 and 2). The microgrids have enough energy and try to contribute to the grid by injecting energy. In scenarios where there is an increased load (3 and 4), there is a clear reduction in the total costs from the microgrid due to the injection of energy from the microgrid and the DERs to the grid. This reduction in costs and losses for the microgrid is more significant in scenario 4 due to the incorporation of more distributed energy resources in the active distribution grid. The computational costs were effective and appropriate for the size of the problem. The variation of energy injection/consumption by the microgrid for the different scenarios was very little due to the great flexibility and self-sufficiency of the microgrid.

Table 2. Comparison of the results of the scenarios analyzed.

Scenarios	0	1	2	3	4
Microgrid Costs (BRL)	-	−26.27	32.59	380.41	−525.37
Total Operational Costs (BRL)	14,662.91	14,836.34	14,860.52	137,326.58	137,461.88 BESS: 6089.52
Total Energy Consumption	27 MWh	27.37768 MWh	27.4294 MWh	245.6516 MWh	245.73525 MWh Charge BESS: 1.537 MWh Discharge BESS: 1.309 MWh
Computational Time	2.52 min	8.613 min	6.592 min	27.981 min	14.784 min
Number of iterations	5	12	9	33, 4, 4, 1	21
Power Losses	0.094 MWh (0.34 %)	0.092 MWh (0.33 %)	0.093 MWh (0.34 %)	8.088 MWh (3.29 %)	8.350 MWh (3.40 %)
Variation of the original MG and the optimized	-	0.02%	0.50%	6.93%	0.10%

4. Conclusions

This paper presents the integration of two distinct methodologies: (1) an internal optimization of the dispatch of microgrids, and (2) responsible for the optimization of the operation of the distribution network. The integration of these methodologies was accomplished through a hierarchical problem, that joint them as an integrated optimization of the methodologies. From this model, it was possible to test different scenarios considering the application of a real distribution grid and a pilot microgrid implemented in the city of Curitiba/Brazil. Different feeder load levels, and microgrid operations, were evaluated. In addition, only one microgrid input was evaluated, as well as a future scenario of the feeder behaving as an active distribution grid, with distributed generation and energy storage systems.

For the scenarios with the entry of the pilot microgrid in the feeder, the operating costs decrease in relation to the scenario without the microgrid. This happens is since the microgrid contributes to the injection of power into the grid, as well as reducing the load since part of it is supplied by the microgrid's own resources. When considering the operation of an active grid, the injection of distributed generation, as well as the battery operation contribute to meeting the feeder demand and consequently reduce the operation costs. However, in scenarios of increased load, it is necessary to perform the joint operation of the grid with the microgrid to ensure operation within operational limits, especially in relation to voltage. Even so, thanks to the operation of the hierarchical model it is possible to communicate the optimal dispatch from the grid's point of view to the microgrid, allowing a redispatch and, therefore, a new grid optimization.

Additionally, for the scenarios with load elevation on the feeder, more expressive losses occur, but they still do not correspond to operational problems, since they do not exceed 5% of the feeder's daily demand. Furthermore, in relation to the voltage profile behavior, for the light load scenarios, it can be seen that there is little voltage variation on the feeder's buses in relation to the magnitude of the substation's output voltage.

However, for the higher load scenarios, the impact on voltage behavior is verified in such a way that the voltage decreases with greater intensity along with the buses farther from the substation or that present a higher load. So, in general, both microgrids and other distributed resources that can be incorporated into the active grid, if their operation and the DERs were appropriately optimized/allocated, tend to decrease power losses and operation costs of active grids with microgrids and other DERs.

Furthermore, with more complex compositions of grids, with greater penetration of distributed energy resources and loads growth, it can be observed greater computational requirement to the model achieves convergence, demanding more computational time, and a greater number of iterations.

Author Contributions: Investigation, C.C.C.B.d.A. and R.B.d.S.; Methodology, T.M.B., R.S.P., M.O.d.L.F., T.S.P.F. and C.U.V.; Project administration, A.R.A. and F.H.T.; Resources, F.H.T.; Supervision, T.S.P.F., C.U.V. and A.R.A.; Writing—original draft, T.M.B., C.C.C.B.d.A. and R.S.P.; Writing—review and editing, T.S.P.F., C.U.V. and A.R.A. All authors have read and agreed to the published version of the manuscript.

Funding: This research was founded by Companhia Paranaense de Energia—COPEL research and technological development (RTD) program, through the PD-02866-0511/2019 project, regulated by ANEEL.

Institutional Review Board Statement: Not applicable.

Informed Consent Statement: Not applicable.

Acknowledgments: Authors thanks the grant for Technology Development of CNPQ—National Council for Scientific and Technological Development within the Ministry of Science, Technology, Innovations, and Communications. Authors thank the support of CAPES—Brazilian Federal agency for Support and Evaluation of Graduate Education with the Ministry of Education of Brazil. The authors also thank Companhia Paranaense de Energia—COPEL for the funding from the research and technological development (RTD) program, through the PD-02866-0511/2019 project.

Conflicts of Interest: The authors declare no conflict of interest.

References

1. Cigre. *C6.19 Planning and Optimization Methods for Active Distribution Network*; Working Group Technical Report; Cigre: Paris, France, 2014. Available online: https://e-cigre.org/publication/ELT_276_7-planning-and-optimization-methods-for-active-distribution-systems (accessed on 12 April 2022).
2. ANEEL. Geração Distribuída Aneel; 2021. Available online: https://app.powerbi.com/view?r=eyJrIjoiY2VmMmUwN2QtYWFiOS00ZDE3LWI3NDMtZDk0NGI4MGU2NTkxIiwidCI6IjQwZDZmOWI4LWVjYTctNDZhMi05MmQ0LWVhNGU5YzAxNzBlMSIsImMiOjR9 (accessed on 12 April 2022).
3. Kurundkar, K.M.; Vaidya, G.A. Ancillary Services through microgrid for Grid Security and Reliability. *Electr. India* **2017**. Available online: https://www.electricalindia.in/ancillary-services-through-microgrid-for-grid-security-reliability/ (accessed on 12 April 2022).
4. Cigre. *Technical Brochure 435—Ancillary Services: An overview of International Practices*; Technical Brochure; Cigre: Paris, France, 2010. Available online: https://e-cigre.org/publication/435-ancillary-services-an-overview-of-international-practicess (accessed on 12 April 2022).
5. Yazdanian, M.; Mehrizi-SANI, A. Distributed control techniques in microgrids. *IEEE Trans. Smart Grid* **2014**, *5*, 2901–2909. [CrossRef]
6. Guerrero, J.M.; Chandorkar, M.; Lee, T.L.; Loh, P.C Advanced Control Architectures for Intelligent Microgrids—Part I: Decentralized and Hierarchical Control. *IEEE Trans. Ind. Electron.* **2013**, *60*, 1254–1262. [CrossRef]
7. Bullich-Massague, E.; Diaz-Gonzalez, F.; Aragues-Penalba, M.; Girbau-Llistuella, F.; Olivella-Rosell, P.; Sumper, A. Microgrid clustering architectures. *Appl. Energy* **2018**. [CrossRef]
8. Zhao, B.; Wang, X.; Lin, D.; Calvin, M.; Morgan, J.; Qin, R.; Wang, C. Energy management of multiple microgrids based on a system of systems architecture. *IEEE Trans. Power Syst.* **2018**, *33*, 6410–6421. [CrossRef]
9. Liu, Y.; Wang, Y.; Li, Y.; Gooi, H.; Xin, H. Multi-Agent Based Optimal Scheduling and Trading for Multi-Microgrids Integrated With Urban Transportation Networks. *IEEE Trans. Power Syst.* **2021**, *36*, 2197–2210. [CrossRef]
10. Wang, Y.; Nguyen, T.; Xu, Y.; Tran, Q.; Caire, R. Peer-to-Peer Control for Networked Microgrids: Multi-Layer and Multi-Agent Architecture Design. *IEEE Trans. Smart Grid* **2020**, *11*, 4688–4699. [CrossRef]

11. Feng, F.; Zhang, P. Enhanced Microgrid Power Flow Incorporating Hierarchical Control. *IEEE Trans. Power Syst.* **2020**, *35*, 2463–2466. [CrossRef]
12. Che, L.; Shahidehpour, M.; Alabdulwahab, A.; Al-Turki, Y. Hierarchical Coordination of a Community Microgrid With AC and DC Microgrids. *IEEE Trans. Smart Grid* **2015**, *6*, 3042–3051. [CrossRef]
13. Hirata, K.; Akutsu, H.; Ohori, A.; Hattori, N.; Ohta, Y. Decentralized Voltage Regulation for PV Generation Plants Using Real-Time Pricing Strategy. *IEEE Trans. Ind. Electron.* **2017**, *64*, 5222–5232. [CrossRef]
14. Blasi, T.M.; Fernandes, T.; Aoki, A.; Tabarro, F. Multiperiod Optimum Power Flow for Active Distribution Networks With Provisioning of Ancillary Services. *IEEE Access* **2021**, *9*, 110371–110395. [CrossRef]
15. COPEL. White Tariff; 2021. Available online: https://www.copel.com/hpcopel/root/nivel2.jsp (accessed on 12 April 2022).
16. Granville, S.; Mello, F.C.; Mello, A.C.G. Application of Interior Point Methods to Power Flow Unsolvability. *IEEE Trans. Power Syst.* **1996**, *11*, 1096–1103. [CrossRef]
17. Lara Filho, M.; Pinto, R.; Campos, A.; Vila, C.; Tabarro, F. Day-Ahead Robust Operation Planning of Microgrids Under Uncertainties Considering DERs and Demand Response. In Proceedings of the 2021 IEEE PES Innovative Smart Grid Technologies Conference— Latin America (ISGT Latin America), Lima, Peru, 15–17 September 2021; pp. 1–5.
18. Lara Filho, M.O. Modelo Robusto Orientado a Dados para Programação diáRia de Operação de Microrredes Considerando Recursos Energéticos Distribuídos Sob Incertezas. Master's Thesis, Federal University of Paraná, Curitiba, Brazil, 2021.

energies

MDPI

Article

Integrating Smart Grid Devices into the Traditional Protection of Distribution Networks

Bruno Silva Torres [1,2,*], Luiz Eduardo Borges da Silva [1,2], Camila Paes Salomon [1] and Carlos Henrique Valério de Moraes [1]

1 Electrical Engineering Graduate Program, Itajuba Federal University, Itajuba 37500-903, Brazil; leborgess@gmail.com (L.E.B.d.S.); camilapsalomon@unifei.edu.br (C.P.S.); carloshvmoraes@gmail.com (C.H.V.d.M.)
2 R&D Department, Gnarus Institute University, Itajuba 37500-052, Brazil
* Correspondence: bs_torres@hotmail.com

Abstract: Smart grids are a reality in distribution systems. They have assisted in the operation, control, and most of all, the protection of urban networks, significantly solving the contingencies of these networks. This paper treats the initial stage of implementing smart grid switching devices in distribution networks. In this stage, smart grid technologies need to operate with the traditional protection elements (such as fuses, reclosers, and sectionalizers). This fact can create trouble in the protection schemes because there are two distinctive philosophies. In some companies, especially those without substantial capital, these two protection philosophies can run together for many years. The most popular intelligent electronic devices (IEDs) available in the market are studied to verify their features and the possibility to incorporate techniques to allow the two philosophies to work together. After that, the proposed approach shows how the existing IEDs can interact with the traditional devices. Special functions can also be incorporated to inform the control center of an operational problem, increasing the observability of the network. With the proposed approach, the IEDs are transformed into intelligent agents. Practical examples using real distribution systems are presented and discussed, proving the efficacy of the proposed methodology.

Keywords: smart grids; power distribution; traditional urban protection; contingencies; IED; intelligent agents; rule-based system; multiagent systems

Citation: Torres, B.S.; Borges da Silva, L.E.; Salomon, C.P.; de Moraes, C.H.V. Integrating Smart Grid Devices into the Traditional Protection of Distribution Networks. *Energies* **2022**, *15*, 2518. https://doi.org/10.3390/en15072518

Academic Editor: Ahmed F. Zobaa

Received: 9 March 2022
Accepted: 25 March 2022
Published: 29 March 2022

1. Introduction

Electrical distribution networks have been structured and operated since their inception to meet the needs of their loads, which are constantly evolving and presenting new demands and operational problems [1]. Distribution systems have become increasingly more important due to the increasing number of branches and growing complexity. New challenges stand in the way of delivering quality energy, which more and more consumers demand.

In traditional distribution networks, the protection system has played a significant role in preventing problems at one point in the network from spreading and causing disturbances in other parts of the circuit [2]. In addition, they create a necessary operational guarantee because they act autonomously, aiming at the security of the network and the people who are close to it.

With the advent of the new devices and philosophies of smart grids, there has been a gradual exchange of traditional devices for smart devices in all distribution systems. Two types of devices and philosophies exist at the moment. Depending on the power distribution company's investment capacity and operational needs, deployment of new technology can take a few months to several years. Moreover, the growing presence of distributed energy resources (DERs) is increasing the challenges of the protection philosophies and the integration of different devices. Thus, in this context, one should notice that several recent works in the literature have presented innovative protection philosophies for distribution systems. The traditional protection schemes and operations are not dealt with in most of these works.

1.1. Overview of the Smart Grid Strategies

The main idea of this section is not to present a complete and systematic overview of all smart grid strategies, but only to show that the previous works do not deal with traditional protection. They contribute with new control and protection strategies using smart grid devices only. They do not deal with the integration of the traditional protection elements (such as fuses, reclosers, and sectionalizers). This paper uses the terms "smart grid device" and "smart device" interchangeably. They refer to switches existing in the network that have some degree of automatism or intelligence.

In [3], a new strategy for adaptive overcurrent protection was prepared for short-circuit currents when distribution generation exists. This proposed strategy is based on intelligent electronic devices (IEDs) only, establishing procedures for islanded circuits. In [4], active network management is presented, defining operational and network structure changes according to the load and areas without energy. IEDs work together, without other information, to determine the best performance according to data from the overcurrent relays. In [5], a new strategy for optimal protection coordination using dual-setting, directional over-current relays was installed in switches connected in IEDs. There was communication among the IEDs. The reference [6] proposes a meta-heuristic approach for feeder reconfiguration after a short-circuit. The process is based on a discrete particle swarm optimization technique. It considers the load of each branch, the branch capacities, and the IED switch positions to define the best new reconfiguration of the distribution network. In reference [7], fast islanding detection after a protection scheme operation is studied in many different situations with several points of view. Microgrid protection, operating in a grid-connected or an island situation, is also studied. Any single line is devoted to traditional protection elements in all these references. The proposed strategies are established for IEDs only. Equipment such as reclosers and sectionalizers without IEDs features and fuses are not dealt with in those papers.

Some papers related to distribution protection in smart grids follow. In [8], many analyses about protection devices, including distribution generation, are studied with control methodologies to change power flow in protection operation and reconfiguration schemes. In [9], some relay protection strategies are also examined, and the traditional relay protection tuning technology is incorporated in the proposed approach, aiming for wide-area protection. Additionally, in [10], an adaptive instantaneous overcurrent protection scheme is proposed, which sets parameters of IEDs using a real-time Thevenin equivalent circuit. Non-linear equations and Gauss solutions are involved in the IED settings. Only IEDs are implicated in the problem's resolution, and no traditional protection is mentioned.

Recently, more papers have been published on smart grid devices' implementations and their acting in the contingency problems in distribution networks. In [11], an interesting approach is introduced using two layers, one for power and another for communication. Several catastrophic contingencies are studied, and the smart grid device features and acting are studied. Experiments were carried out. In [12], several operational conditions and contingencies are studied to locate distributed generation units and shunt capacitors in the distribution network. Those studies took into consideration the actuation of smart devices only. In [13], the system reconfiguration strategy is based on smart devices and external energy sources to supply the critical load of the network. Mixed-integer linear programming runs a stochastic model to create reconfiguration solutions for the network. In [14], a new strategy for forming microgrids in the reconfiguration problem is described. The issue of reverse power flow, the frequency/voltage regulation provided by inverters, and grid support capabilities are studied. The contingency is isolated by the smart grid devices only. Finally, reference [15] presents an interesting complete overview of the smart grid computational methods. In those methods, only smart devices are used to solve protection and contingency problems. Traditional protection is not mentioned.

Therefore, there is a research gap in the integration of smart grid devices into the traditional protection devices presented in distribution networks, which neglects the reality of these systems in many places around the world. Thus, there is a need to study the implementations of the functions in IEDs and in the traditional protection schemes. It is crucial to notice that the term "intelligent" in "intelligent electronic devices" is not appropriate. The reason for that is presented in a section of this paper.

This paper deals with the integration of smart grid devices into the traditional protection devices of distribution networks more practically. Our main contribution is to present a proposal for precisely integrating a rule-based system into IEDs, transforming them into "real" intelligent devices (named intelligent agents) interacting with the existing devices in the protection systems of current distribution networks.

1.2. Operation in an Incomplete Smart Grid

This consideration is crucial, because these two operational philosophies can remain simultaneously active for months or years. It depends on the power distribution company's financial capacity and the network's necessities. In fact, in most developing countries, this incomplete integration will remain for years without a deadline. For example, in Brazil [16], as in many developing countries, such as India, Russia, China, Thailand, Mexico, and South Africa, the most critical aspect to accelerating the substitution of one structure for another is the consumers' power supply continuity indexes.

The integration analysis for the two philosophies presented in this paper is fundamental mainly for crowded cities with millions of inhabitants. These cities in developing countries have distribution systems in several stages of evolution. For instance, a city with two million inhabitants has around twenty substations spread throughout. Each substation has 5 to 10 circuits, and each circuit is 50 to 70 km and has 500 transformers for low voltage supply. Usually, the distribution protection system is using reclosers in some critical points of the circuit (around 10 per circuit), sectionalizers (approximately 2 to 4 per reclosers), and fuses at each branch of the circuit and in all transformers.

Table 1 shows an aerial photo of a region supplied by two substations (SS1 and SS2), and only three 13.8 kV circuits are represented. SS1 and SS2 have 8 and 10 circuits, respectively. In these three circuits, there are 283 transformers, more than 5000 consumers, 79 branches, and 18 km of extensions, supplying approximately 15 MVA [17–19]. The red squares represent the locations of nine existing reclosers (three per circuit with three sectionalizers each), and the green squares are the three tie switches (Figure 1).

Table 1. Typical load values for branches before 6 pm on a Friday.

S8–S9	S9–S10	S10–S11	S11–X
15A	15A	15A	5A

Regarding these circuits in a smart grid implementation process, the first switches that would be transformed in IEDs are some of the nine reclosers and twenty-seven sectionalizers, and primarily the three tie switches [17]. A significant number of the circuits would continue with the traditional protection (the reclosers and sectionalizers not transformed in IEDs, and the fuses, would be around 350). Thus, it is clear that the implemented IEDs need to interact with the fuses and the other traditional devices (not transformed into IEDs) to seek collaboration in the network trying to integrate the two protection philosophies.

Figure 1. Example of a real distribution network [17].

This paper offers a step in this direction, presenting how simple rules can be incorporated into the IEDs to integrate operations of the traditional devices. It starts with an overview of the main concepts of the traditional protection elements for urban and rural networks and the main elements of the smart grid and intelligent systems. After that, the general structure of an IED is shown, and the most popular devices used in the automation distribution systems are presented along with an analysis of their features and a comparison among them. Then, possible integration of IEDs with fuses is proposed, after introducing the concept of load partitioning each branch. Then, the integration of intelligent agents with reclosers and sectionalizers is presented. Finally, case studies using an existing distribution network are presented, and the computational implementation results prove the efficacy of the proposed methodology.

2. Presentation of Important Concepts

2.1. Main Concepts of the Traditional Protection for Distribution Networks

When protection structures in urban and rural distribution networks are observed, it is possible to verify that they have some switches to open short-circuits along with their circuits, such as fuses, reclosers, and sectionalizers.

A fuse is an element that interrupts the short-circuit current by burning its filament, which then has to be physically replaced by a utility maintenance team. This repair can sometimes take hours (between fault detection and travelling to the location), although the exchange can be done in just a few minutes.

Reclosers are devices that can eliminate short-circuits, including the entire reconnection procedure, by the verification of non-permanent short-circuits.

The sectionalizers act in conjunction with the reclosers, but cannot eliminate short-circuits. Suppose a sectionalizer observes a short-circuit current at one of the reconnection openings. In that case, it opens and remains open by isolating the power from the short-circuit and allowing the rest of the circuit to continue to operate.

Other topics in the protection of traditional urban networks are their coordination and selectivity. These two concepts are fundamental [20]. Coordination is the act of disposing of two or more pieces of protective equipment in series, according to a specific order. Selectivity is the ability of protective devices to act before their backup devices. For example, only one protection device should work as close as possible to the short-circuit when a short-circuit occurs. When this happens, it is said that there was coordination between the protection devices and that there was selectivity between them. Another example of this situation is when some (or all) protective devices that have seen the short-circuit current act. In this case, it is said that there was no selectivity, but that coordination occurred. Finally, neither selectivity nor coordination happened when both the nearest protective device (or devices) did not act, and Figure 2 shows these situations.

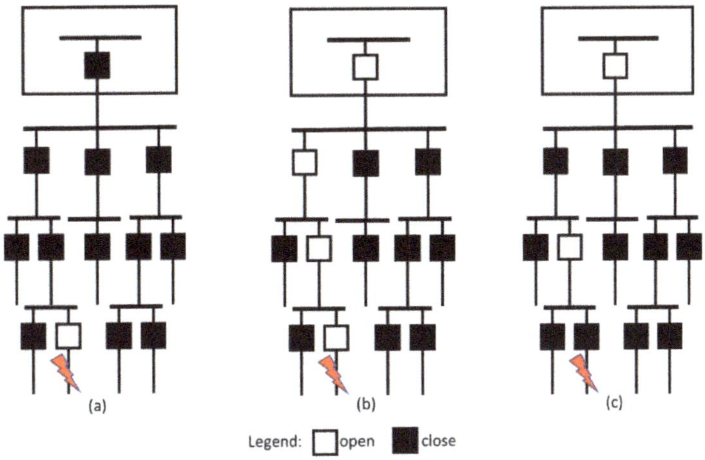

Figure 2. Situations of protective action: (**a**) with coordination and selectivity, (**b**) with coordination but without selectivity, and (**c**) without coordination and selectivity.

2.2. Differences between Smart Grids and Intelligent Systems Concepts

The aim of writing this section was not to present a complete overview of the smart grid concepts and their structures, but only to show some points related to the proposed developments. Firstly, it is important to distinguish between two concepts: smart grids and intelligent systems. Some smart grids can be intelligent systems, but not every smart grid is intelligent; they only are if intelligent techniques are applied to solve a problem. Thus, it is fundamental not to confuse systems with automatisms with intelligent systems. Many manufacturers often falsely sell their products as intelligent products, but they do not include intelligent techniques. Example of these types of devices which have in their names the word intelligent are the IEDs (intelligent electronic devices). They each have a set of structures (analyzed in the next section) to produce an answer automatically, but they are not intelligent devices.

Figure 3 presents a Venn diagram, showing that smart grids do not always include intelligent systems and that intelligent systems are not always applied in smart grids. The important point for this work is the elements that can be present in both: the intelligent elements of the smart grids.

In the region of interest, there are devices with intelligent techniques to solve certain problems. These devices have communication systems to exchange information with other devices or a central system for the most part. If they do, they have all the features necessary to make them "intelligent agents" [21].

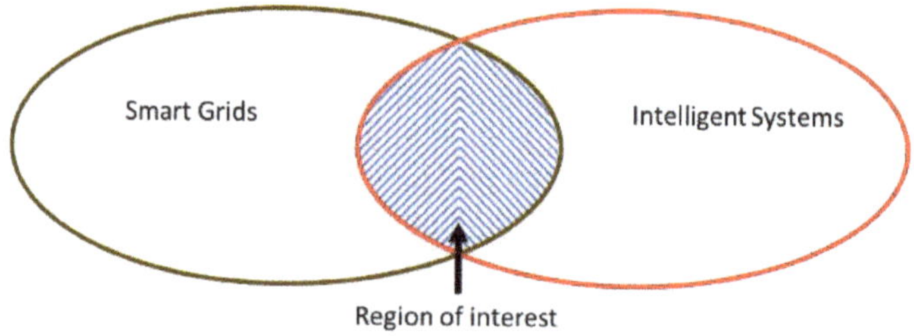

Figure 3. Venn diagram showing smart grids and intelligent systems worlds.

Other important differences introduced by smart grid technology compared with traditional protection are two-way communication, self-monitoring, self-healing, and pervasive control. In traditional protection systems, there is no communication, manual monitoring, manual restoration, and limited control [22]. Additionally, a smart grid has sensors throughout the network, increasing the observability, and the communication allows interactive activity among devices and with multiple stakeholders [23].

3. Overview of IEDs

The term intelligent electronic device refers to an extensive range of electronic devices that automate a system, be it a power plant, a substation, or a distribution system. Several functions can be incorporated into IEDs, such as protection, monitoring, and local control. The main aims of this section are to address whether there is any mention in the literature about the interactions of these IEDs and traditional protection elements and to verify if the IEDs have features to support the proposed approach. This section initially discusses the structure of an IED, showing its features and how an external program can be incorporated into it. Then, the most popular IEDs used currently in distribution system automation are shown. Finally, we show how a set of IEDs can be transformed into a multi-agent system (MAS), producing an intelligent multi-agent system.

3.1. General Structure of an IED

A typical IED architecture consists of microprocessors, local memory, digital and analog inputs, outputs, a communication system, and a power supply. The functions of the microprocessors include the execution of the protection algorithms. Most software used in IEDs works through a set of pre-programmed tasks, which only require adjustments to their parameters. These functions include displaying voltage and current values and limits, data recording, local database control, switch status, interlocking, self-reconnection, and adjusting groups of protections. Within this set of tasks, an essential component is the real-time operating system (RTOS), whose role is to ensure that all other tasks are performed appropriately and in the established priority order [24].

Concerning the software's logic, it is usually made by a human–machine interface or a dedicated communication port. These logics can be written in simple languages, such as Ladder, or directly in more elaborate programming languages, such as C.

Nowadays, for communication, IEDs use IEC 61850 [25], which is a standard for communication networks and systems in substations. Additionally, the IEC TC57 [26] standard has been developed for distribution network automation, and it is based on IEC 61850 guidelines. These standards allow all IEDs from different manufacturers to communicate, exchanging data and messages. The GOOSE protocol [27] provides an efficient way for IEDs to send and receive messages.

3.2. Overview of the Most Popular IEDs Systems Used in Distribution Systems

This section includes the analysis of some of the technological solutions of IEDs systems used in distribution systems available in the market, which are: Schneider Electric's Intelligent Loop Automation, S&C Electric Company's IntelliTeam SG, Eaton's Yukon

Feeder Automation, NovaTech's Distribution Automation, SEL's DAC, Hitachi Energy's Distribution Feeder Automation, and the Siemens Self-Optimizing Grid. The intention is to show the possibility of incorporating functions in all these systems, especially those proposed later in this article, and to map these devices' functional characteristics, applications, and strengths.

3.2.1. Schneider Electric's Intelligent Loop Automation (ILA)

ILA [28] uses a distributed architecture based on logical schemes embedded in the ADVC reclosers' controllers and the disconnector switches' controllers. Depending on the mode of operation, each IED has one of three possible pre-programmed functions: feeder, tie, or mid-point. A feeder IED usually is closed and opens when it detects an overcurrent in its terminals. The tie IED is the point of the normally open scheme that closes when it catches a loss of voltage in the source terminal or the load terminal. The mid-point IED is located along the feeder, anywhere between a feeder IED and a tie IED. The mid-point IED changes protection group and enters before the tie device closes to maintain the radiality of the circuit.

The loop automation schema can contain two feeder IEDs, multiple mid-point IEDs, and one tie IED. It is necessary to integrate the various LA schemes configured within a distribution system to use a remote terminal unit (RTU) to control the hybrid system. The most current versions of firmware support overload control, which allows the system to avoid transferring load to a feeder that cannot support it. ILA still has an editor to create and supervise the network architecture, in which some additional rules can be incorporated.

3.2.2. S&C Electric Company's IntelliTeam SG

IntelliTeam SG [29] uses a decentralized architecture with an RTU (called IntelliNode) installed at the top of each participating IED (relays, reclosers, circuit breakers, and disconnector switches) or directly on S&C's reclosers. IntelliTeam SG monitors voltages and currents in real-time using sensors built into fault-sectioning devices. Each IED is configured to know its normal function in the system and other essential parameters, such as the amount of load the device is authorized to receive, the role of the device in the set, and the priority of the source.

The system divides the network topology into groups called teams, hence the system's name. The manufacturer states that there is no limit to the size of the system. S&C offers a topology configurator that also acts as a SCADA to edit and supervise what IntelliTEAM does. It can edit the topology almost freely by inserting the IEDs and connecting them within the schema. This tool also has a replay option, allowing one to visualize in a graphic diagram what happened before, during, and after self-recovery procedures. The S&C system can prioritize sources, handle multiple short-circuits, and avoid feeder overloading. Additionally, it is possible to restore the system to its normal configuration state or decide on a new normal configuration state after the self-healing process has been completed.

3.2.3. NovaTech's Distribution Automation

The Orion DA-Master is a NovaTech Orion system [30] with specialized software that functions as an independent master controller in a single or multi-feeder, single or multi-substation distribution automation system. The DA-Master scans measurements, analyzes the data, and initiates the opening and closing commands in the various IEDs installed in the distribution system to perform user-defined system reconfiguration in response to abnormal system conditions. The DA-Master supports several communication protocol options that are available in the various IEDs currently found on the market.

If remote monitoring or control is required, the DA-Master can also communicate with an existing SCADA master system in the distribution system. NovaTech Communications Director (NCD) is the logical configurator and has a vast library and protocol profiles of the various modern IEDs. A computational package called DA-Simulator is also available that allows the user to model the distribution system, simulate different short-circuit conditions in each participating IED, and predict the actions of the self-healing algorithm before the equipment is installed in the field.

3.2.4. Eaton's Yukon Feeder Automation (YFA)

YFA [31] uses a centralized architecture with a self-healing algorithm configured in an RTU, usually installed in the substation. A dynamic system integrates real-time data from the distribution system to detect disturbances and automatically reconfigure the system to isolate the short-circuit, minimizing the total number of affected customers. Several IEDs can be integrated, such as relays, disconnector switches, capacitor bank controllers, voltage regulators, and short-circuit indicators.

YFA scans the participating IEDs of the self-healing system with various communication protocols, managing all automation functionalities of the distribution system, and can also be integrated into existing SCADA systems. The YFA editing program is a dynamic and configurable platform that uses object-oriented programming. It has a topology editor and a simulator, offering the user the possibility to test the configured programming and its response under normal and abnormal conditions in the distribution system.

3.2.5. SEL's Distribution Automation Controller (DAC)

The DAC [32] uses a hybrid architecture with a self-healing algorithm configured in an RTU, usually installed in a substation. The RTU responsible for data concentration, execution of logical schematics of self-healing, and data communication processing is a real-time automation controller. The system has a primary real-time automation controller that communicates with other substations and the IEDs participating in the distribution network. In this controller, the computer program responsible for the automatism of the network exists. The controller performs algorithms automating isolation and power restoration procedures during short-circuit situations, while additional algorithms optimize network's performance through voltage control and reactive power control. Overload reduction techniques are also used in cases of abnormal electrical system operation.

The system is independent of the IED manufacturer. It can also be integrated into the existing SCADA system, and the configuration is based on libraries and protocol profiles of the various IEDs available on the market. The various IEDs, such as relays, disconnector switches, capacitor bank controllers, voltage regulators, and short-circuit indicators, can be integrated.

3.2.6. Hitachi's Distribution Feeder Automation

Hitachi's Distribution Feeder Automation [33] uses a decentralized solution installed in the field. The RTU500 series are controllers with an integrated predefined algorithm that allows integrations in medium and low-voltage grids. Sensors and functions work using motorized switches to quickly identify a faulty cable, isolate the fault, and restore the supply. This application uses an Automatic Transfer System (ATS) for simple reconfiguration functionality and includes a high-level solution with peer-to-peer communication. The trigger happens by a failure detection in the primary active source.

This system contains all necessary applications connected to a distribution system to small-scale or mobile power stations. This RTU-based application includes a communication link, automatic speed, power factor, and voltage control. The system registers long-term or periodic overload stresses, and the asset's maintenance plan algorithm can support them.

3.2.7. Siemens Self-Optimizing Grid

The Siemens Self-Optimizing Grid system [34] is based on a regional controller on the substation level, ensuring automatic fault localization, isolation, and restoration. Special controllers make the interface with the control center. They measure data through the distribution network and host the grid's regional and centralized applications. Self-Optimizing

Grid solutions integrate the different grid levels, and depending on integration level, can be: centralized, hybrid, or decentralized.

The Self-Optimized Grid solution comprises the most relevant applications for hybrid solutions. These applications accurately monitor the grid and remotely control stations to ensure high supply reliability and improve system performance. The solution integrates these intelligent automation functions: self-healing, load management, automatic source transfer, overload reduction, and area voltage control.

3.2.8. Comparative Analysis of the Described Technologies

It is known that the structure of a smart grid must be elaborated so that its equipment allows interoperability with different systems. Thus, different algorithms can be deployed for system optimization. The seven systems offered have applications in distribution networks with different evolutionary levels of automation. Altogether, they cover each level of automation and every available network. It is important to note that the evolutionary level of each system is not linked to the technical capacity of the company that owns it, nor the ability of its technical development team. This evolutionary level is related to a company's market strategy, aiming to target one type of system more than another.

The usefulness of the described systems is easily verified by their automation of the identification, isolation, and restoration operations in response to the short-circuits. However, none of them confirmed the existence of functions that deal with traditional protection systems. They would need functions to interact with reclosers, sectionalizers, and fuses. Nevertheless, all shown IEDs have a way to write and manage these functions and analyze the data coming from the network. It is essential to note that traditional devices do not communicate with IEDs. Still, the current that passes through an IED makes it possible to make inferences, as shown later in this article.

3.3. *Transforming a Set of IEDs into a Multi-Agent System*

The shown IEDs can also be called agents [35] because they have all the main features of them: autonomy, adaptivity, contextuality, and sociability. They have a degree of autonomy which can be chosen based on the system they are to be used in. They also have a certain level of adaptivity to choose the best actions to take. They can contextualize their activities to define if an action must be taken. Finally, they have the ability to interact with others IEDs (sociability). The main elements of the agent are represented in Figure 4. In this figure, the intelligent controller is in a recloser; however, it can be put in another distribution network switch. The intelligent controller receives the data (such as current and voltage measurements and the status of the switch, among others) though its sensors. These data are given to the processing unit of the control, which contains algorithmic, numerical, and intelligent parts. Each part has a specific mission. The algorithm controls the entire process of the IED. The intelligent part provides inferences for each operational state of the network and the controlled switch. It uses the data from the sensors and the data saved in the local memory to choose an action for the controlled switch and to communicate information with other IEDs. The intelligent part usually triggers the numerical part, but the algorithmic part also uses the numerical part a few times. The local memory contains data of the observed part of the network observed by the agent. Finally, the communication system is the way to send and receive messages from and to the other agents.

An agent can be of two types: a common agent or an intelligent agent. The common agent is the one that reacts (or acts) according to a certain logic but without the basis of techniques of an intelligent system. In the world of artificial intelligence (AI), an intelligent agent is a computational system that is able to perform independent actions for the benefit of itself or its controller. Agents are expected to act rationally, operate autonomously, perceive their environment, adapt to changes, and achieve their goals. When uncertain, a rational agent obtains the best or expected outcome.

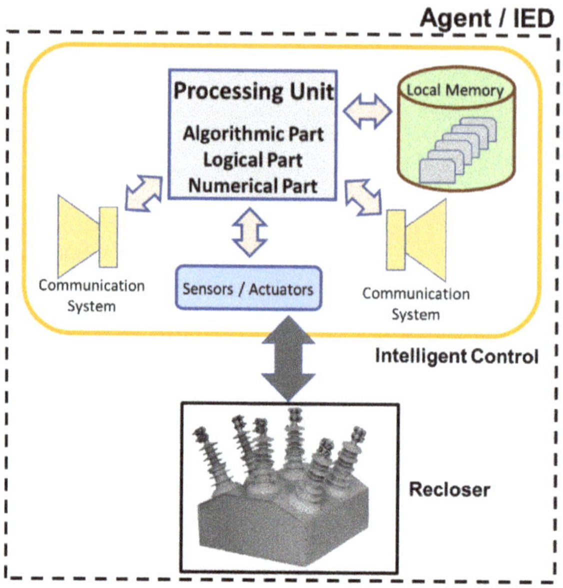

Figure 4. Intelligent agent structure.

Agents can interact among them to find a better solution to a problem, creating a multiagent system (MAS) [36]. The MAS may have one more qualifier, the word "intelligent." There are two situations in which an intelligent multiagent system can be found. The first is when the system has intelligent agents that interact and perform functions capable of accomplishing the objective proposed by the system in which they are inserted. The second type is one in which there are only "common" agents (which do not use AI techniques), but that "intelligence" emerges from their interaction. In this work, the intelligent multiagent system is composed of intelligent agents; however, the acronym MAS is maintained.

In a MAS, all intelligent agents should be able to communicate. Each must have the knowledge and skills to perform a particular task and may or may not cooperate to achieve a global goal. It is noteworthy that each agent in a MAS has incomplete information, and agents form a system with sufficient knowledge and ability to solve the problem. Nowadays, MAS is the most important technique to operate distribution networks [37], helping especially during the contingencies, such as short-circuits. In a complete MAS, all actions to eliminate a short-circuit, isolate the branch in which it occurred, and restore the branches without energy, are done by the intelligent agents distributed along with the electrical distribution network [38].

The MAS can be classified into three main types: centralized, decentralized, and hybrid. In the centralized MAS, there is a central process control that maintains communications with all agents, performs the main operation strategies of the network, and sends which actions each agent must produce. Only a few inferences are made locally, and the most part is done in the central unit. Examples of this type of MAS are presented in [39–41].

There is no central unit in the decentralized MAS, which means the agents must make all performances and inferences locally. The agents communicate with other agents to provide a solution for a problem. There is no global system control, decentralized data, and asynchronous computation. Examples of the decentralized MAS are presented in [42–44]. In this work, the MAS considered is of this type, and all the processing is done locally in the agents.

There is a balance between the central and local inference in the hybrid MAS. Central control performs the computation linking all data from the network, such as numerical programs (like load flows) or main strategies (like load shedding). In contrast, the local control performs by the agent taking care of the switching strategies and local verification of the overload, over (or under) voltage, or short-circuit detections. Examples of the hybrid MAS are presented in [45–47].

4. Proposed Methodology

This section presents the proposed methodology for integrating smart grid devices into the traditional protection of distribution networks. The section presents functions to work with fuses, reclosers, and sectionalizers, and then a summary of the intelligent agent rules.

4.1. Functions to Work with Fuses

In a decentralized MAS, all functions must be in the agents to be performed through inference processes. This section presents a function that incorporates the intelligent agent that allows interactions with fuses. The inclusion of this function allows a more gradual introduction of smart grid elements next to traditional urban networks. Over time and with financial availability, one could replace fuses with intelligent agents.

Implementing this function in the smart agent is not enough to detect the time to break the fuse links before it acts. The intelligent agent should monitor this loss of load and have a supervisory function; i.e., it should inform the operating plant about the fuse link disruption and the consequent loss of load. This allows the utility to send the maintenance team to change the fuse before receiving calls from consumers without power, reducing the durations of power outages and improving the reliability of the System Average Interruption Duration Index (SAIDI). This is one of the advantages of a smart grid.

Agents always supervise fuses downstream and monitor a load loss resulting from opening one of these fuses. The strategy is easy to understand, but its implementation needs a set of calculations that to be performed, depending on the current reduction view by the agent, involving daily load behavior and load forecasting. The load reduction perception function detects "abnormal" variations in the current and places the agent at the attention level.

The system shown in Figure 5 can be helpful to illustrate how this function performs. This figure contains a hybrid protection system merging intelligent agents, fuses, reclosers, and sectionalizers. Each switch S is one protection device. For instance, the elements S7 and S8 are intelligent agents; and S9, S10, S11, S12, and S13 are fuses. Agent S7 supervises fuses S12 and S13, while agent S8 supervises fuses S9, S10, and S11.

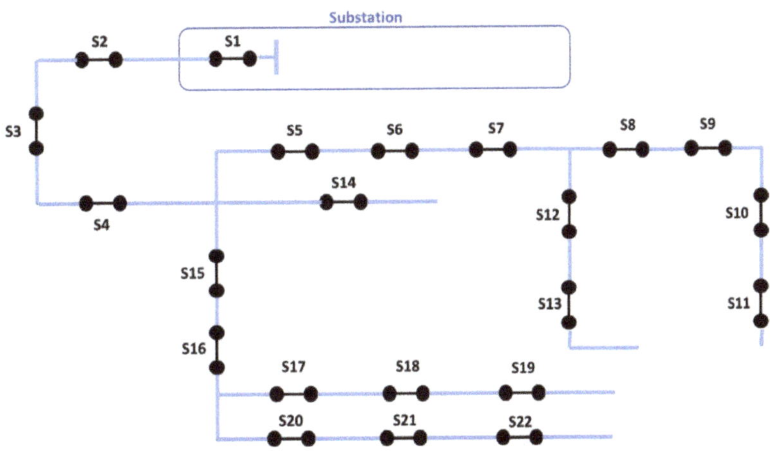

Figure 5. Typical distribution network with hybrid protection devices.

Thus, when the S11 fuse opens, S7 and S8 check the load reduction. They must communicate to determine if this occurred in the fuses supervised by S7 or S8. If both noticed, the loss is on agent S8's line, but if only S7 realized, it was one of the fuses it supervises. Again, if both detected the load reduction, then it is one of the fuses supervised by S8, and it should start an analysis process to know if it was the fuse of S9, S10, or S11 that broke.

4.1.1. Verifying Which Fuse Has a Problem

Once an agent has established that it is likely responsible for the fuse that acted, this verification procedure is initiated. The intelligent agent infers whether the abrupt load reduction occurred due to the opening of a fuse or the simple removal of a large load block (for example, a factory). This check is done through a procedure that investigates whether a short-circuit current occurred immediately before the reduction of a large load block. If this happened, there is a strong chance that the fuse link would have broken; otherwise, there must have been a load reduction. However, the question remains: are these statements entirely factual? The answer to this question is no, because the short-circuit may have occurred within a factory, and in this case, its main circuit breaker has removed the factory without burning the fuse link. In the second hypothesis, the fuse link has been burned due to an overload in the circuit.

It should be remembered that the fuse link has a "memory" of the events that passed through it. This memory exists because it does not necessarily merge at once. It is degraded whenever an overload or a short-circuit current passes through (which is not primarily responsible for opening). Remember that this degradation process is thermal and irreversible.

Although this procedure is not foolproof, it increases the observability of the electrical network, because an agent closer to the event will have greater sensitivity to direct network maintenance efforts than the total lack of observability that occurs in networks with traditional protections.

After checking for a short-circuit current, the agent calculates the load drop, in percentage, by the difference between the pre-fault current and the current divided by the pre-fault current value. The result is used in the next steps of this procedure.

The agent verifies whether the value of the percentage load drop was higher than a pre-established Δ value: if yes, the continuous inference process moves to the next step; if not, the process is completed, and the agent returns to normal. This Δ value allows minor typical load variations to not be confused with load dropouts.

With the process continuing, the next step is to check the memory of the intelligent agent for the load partitioning of each branch. It is placed there when the agent is the initial setup.

This load partitioning is done through the expected workload in the branches, and it represents the typical workload of the branch for one week. Usually, purely residential branches have similar load profiles, only vertically moving the data curve. In contrast, other load profiles (industrial or commercial, or composed of various load types) have different profiles. Power utilities have this data because they carry out measurement campaigns for other purposes.

Thus, when the inference is being made (or during an event, which is the same), it searches for the nearest whole time, and calculates the load of each switch for this time and the partitioning values in percentages. Hence, these partitioning values are compared with the percentage value of load drop, checking the nearest. The nearest value defines the fuse that acted.

Then, the agent must send information to the substation informing it of the broken fuse link to expedite the maintenance process and return energy to the branch (or branches).

4.1.2. Example of Integration of Fuses with Smart Grid Devices

Consider Figure 5, where, as reported above, agent S7 supervises fuses S12 and S13; and agent S8 supervises fuses S9, S10, and S11. Suppose a short-circuit occurs in branch S10–S11. By the coordination criteria, the proper response is the fuse of S10 interrupting the power supply of branch S10–S11 and the branch after S11.

As the short-circuit occurred on the S10–S11 branch, the intelligent agents S7 and S8 observed the short-circuit current. Other agents may also have seen the short-circuit current. Still, due to the simplicity of the explanation, the inferences are reported only by agents S7 and S8, since the other agents will verify that they should not take any action, as would occur in the case of agent S7.

Agent S7 notes that at least one of the agents of each team also observed the short-circuit current, and for this reason infers that it should not take any action because the short-circuit did not occur in its observation area.

Agent S8, on the contrary, alone in its team received this information, and hence infers that the short occurred in its observation area. Hence, this agent differentiates between the pre-fault current and the actual current. Assuming that these values are 45 and 32A, respectively, the percentage value of load drop is calculated: 29% (=(45 − 32)/45 × 100%).

Assuming that the Δ value is 4%, the agent infers that one of the fuse links it oversees has been opened. Thus, its rules observe the current hours and take the nearest whole hour. Adopting in this example 17:42 h on a Friday, the typical load data of 18:00 h of each branch are accessed. For example, consider the data shown in Table 1. Table 2 shows the percentage value of the load drop of 29%, showing that the S10 fuse link broke.

Table 2. Typical loads on agent S8 and the fuses S9, S10, and S11 at 6 pm on a Friday.

S8	S9	S10	S11
50A	35A	20A	5A
-	70%	40%	5%

4.2. Functions to Work with Reclosers and Sectionalizers

Other devices that have been used in the traditional protection of urban networks are reclosers and sectionalizers. The function of the reclosers, as stated above, is similar to that of circuit breakers in substations. They can also eliminate a short-circuit and are provided with automatism that allows the timed interruption of the short-circuit current with two main objectives: to verify if the short-circuit is transient and enable the operation of the other protection devices closer to the fault. The sectionalizers also observe the short-circuit current but cannot interrupt it. This opening only occurs in one of the interruptions caused by the recloser. There can be multiple sectionalizers together with a recloser. For example, consider the network of Figure 5, where the S15 protection device is a reconnector, and the S17 and S20 devices are sectionalizers. The main idea behind this configuration is that faults in the branches after S17 (or S20) are interrupted by the sectionalizer responsible for the branch (not by S15). It causes the faults to be restricted to a minor portion of the network.

It is essential to mention that these two devices may not be simultaneous. Still, the number of times the reconnector takes action can be counted by the sectionalizer agent.

4.2.1. Integration Function Procedure with Reclosers and Sectionalizers

To exemplify the integration function with reclosers and sectionalizers, consider the network of Figure 5, where the protection device S15 is a recloser; devices S17 and S20 are sectionalizers; S4, S5, S14, S16, and S18 are switches for intelligent agents; and S19, S21, and S22 are fuse switches. It is important to note that S15, S17, S19, S20, S21, and S22 are traditional devices without intelligent agents. Thus, the communication happens in the following teams (formed in the area of interest of the example): S4–S5–S14–S16, S16–S18.

The procedure of this function is similar to that presented in the previous section. It uses the concept of load block loss. This loss occurs with the opening of the fuse (breaking of its link); this loss is caused by the opening of the recloser or a sectionalizer. Thus, it is essential to know which agent monitors which device.

In this example, agent S4 supervises S15; agent S16 supervises S17, S20, S21, and S22; and agent S18 supervises S19. There is no difference between monitoring a fuse (S21 or S22) and a device (S17 or S20—sectionalizers) to the agent. What matters is simply knowing the load partition on each branch so that the calculation described above can be performed by the agent that monitors these protection devices (which is S16).

Three illustrative cases are studied in this section, with short branches between S17 and S18, between S21 and S22, and between S16, S17, and S20.

In the first case, the short-circuit occurs on branch S17–S18. Hence, the protection acts as follows: S15 verifies whether the short-circuit is temporary or permanent, and if it is permanent, in one of its openings, the S17 sectionalizer opens. Then, after eliminating the short, the agents begin their verification process and possible restoration of the branches. Agents S4 and S16 observed the short-circuit current, and S18 continued.

Thus, agent S4 concluded that nothing should be done because the short-circuit is not in its team's area (S16 also observed the short-circuit current). Agent S18 did not notice anything, then continued right away. Agent S16, which also noticed the short-circuit current, contacts agent S18 and verifies that the short ran through it and is not beyond agent S18.

As agent S16 monitors S17, S20, S21, and S22 protection devices, it should verify which device acted. This is done by the same load partitioning process shown earlier. Thus, it is determined that S17 was the device that acted.

One could also have an additional rule that would indicate whether there are protective devices between S16 and S18 (which ran out of power), and if this second agent ran out of energy, it is because a protective device activated between them. This rule verifies the veracity of completion performed by agent S16. Moreover, when the agent infers that a device has acted, it sends a signal towards the substation informing it of the fact (in this case, device S17).

In the second case, the short-circuit occurs on branch S21–S22. Hence, the protection acts analogously: S15 verifies the type of short-circuit, and being permanent, in one of its openings, the fuse link S21 opens. Then, after eliminating the short, the agents begin their verification process and possible restoration of the branches. Agents S4 and S16 observed the short-circuit current, and S18 did not observe it. Thus, agent S16 infers that the short ran through it and that it happened no further than agent S18.

Agent S16 contacts agent S18, informing it of its actual current. In this way, the agent, as it has the protection devices S17, S20, S21, and S22 under it, performs the calculation of the load loss with the load partition process and checks which device is closer to the load loss, which in this case should be S21. It will send this information to the substation.

In the middle of this process, the verification of the veracity of conclusion is performed, using the rule, which confirms that if S18 also observes the current, it is because S17 did not act.

These two cases involve similar actions and are based only on the partitioning of loads between the branches. The following case is a little different.

In the third case, the short-circuit occurs on branch S16–S17–S20. Hence, the protection acts similarly to the past: S15 checks whether the short-circuit is permanent; it opens, as no other device acted. Then, after eliminating the short-circuit, the agents begin their inference process. Agents S4 and S16 observed the short-circuit current, and S18 did not maintain it. Thus, agent S16 infers that the short ran from after it and was from no further than agent S18. It also verifies that it is out of charge current and that the short-circuit insulation process should begin.

Then, agent S16 starts this process by opening, locking, and sending a message so that S18 also opens and locks. If there were relief circuits, the restoration process would continue in the same way as this document presented.

When opening, S16 sends a signal, in this case, to S4, stating that the circuit breaker can close. As there is a load current in S4 and it monitors a traditional protection device, it turns this message into an acting message for the S15 protection device. There is a problem with the branch after S16 and before S18 (branches S16–S17–S20, S17–S18, S20–S21, S21–S22, and S22–X). When the S15 recloser reaches its standby time, it automatically closes, and there is no longer a short-circuited power supply, and the branch S15–S16 powers back up.

4.2.2. Incorporation of an Intelligent Agent in Reclosers and Sectionalizers

The proposed agent is structured to act on any type of switch, with or without the ability to eliminate the short-circuit current. Until this point in this work, the agent has been used in two ways: in the switches of the substations (circuit breakers—which open short-circuit current) and in the branch switches (disconnectors—which do not open short-circuit current). However, it important to remember that the agent's structure is the same for each of these switches, regardless of their type.

Thus, the inclusion of the proposed agent, reclosers, and sectionalizers does not bring about any problem, because reclosers are similar (in their performance) to circuit breakers. At the same time, sectionalizers are identical (in their performance) to disconnectors. The philosophies of coordination and selectivity should be kept from the traditional protection structure.

With the agents placed over the protection devices S15, S17, and S20 of Figure 5, the new teams formed in the area of interest are: S4–S5–S14–S15, S15–S16, S16–S17–S20, and S17–S18.

4.3. Intelligent Agent Rules for Integration with Traditional Protection Devices

This section summarizes the key actions (rules) of the intelligent agent (IEDs plus knowledge-based system—KBS) for integration with traditional protection devices. All inference processes start when the IED notes a short-circuit current. After the short-circuit area's elimination, the KBS starts to be executed. It can be divided into two main parts: verifying the performance of a traditional protection device and validation of the integrity of the conclusion, and the KBS is shown here through production rules.

4.3.1. Verification of the Integrity of the Supervised Switches

This rule is used to confirm (or change) the checklist of devices that have provided the loss of a load block. This occurs when these devices are between two intelligent agents. There are two possible situations for using these rules, which are always activated after setting the short location by the agent on one of its sides.

The first case is when the agent has load current and the adjacent agent does not. The second case is when both have load current after the short's elimination. With this, in the first case, it turns out that the device that operated is between the two agents and that all the devices between the two are the ones that make up the list of devices to be checked. In the second case, it was concluded that neither device between the two agents put up with it, so neither device should be on the list to be checked.

Notwithstanding these facts, all devices must be present in the composition of the load partition calculation. The rules for this check of the completion veracity are as follows:

Check the load currents of the agents if both are nonzero. Action: Delete devices between these agents from the solution list.

If I of the agent $\neq 0$ and I of the adjacent agent $\neq 0$, then delete the devices between these agents from the solution list.

Check the load currents of the agents, and one of them is zero. Action: The solution list consists of the devices between these agents.

If I of the agent $\neq 0$ and I of the adjacent agent = 0, then the solution list comprises the devices between these agents.

4.3.2. Verification of the Performance of a Traditional Protection Device

The first part of the ruleset is to verify that the intelligent agent is responsible for searching for the possible action of a traditional protection device (fuse, recloser, or sectionalizer). The IED must communicate with the adjacent agents (adjacent IEDs) and verify that they have also recognized the short-circuit current (I_{SC}). If at least one of the agents on each side recognized this short-circuit current, it is not the agent responsible for doing this check. However, suppose neither of the agents adjacent to it observed the short-circuit current. In that case, it is responsible for searching for a possible action by a traditional device supervised by it. The rules for this are:

Check the short-circuit current, verify that at least one adjacent agent observed the short current, and verify that no agent on the other side noted the short-circuit current. Action: Search for the device that acted.

If I_{SC} was recorded, if one of the teams did not observe $I_{SC,}$ then seek action for the device.

If I_{SC} was recorded by any of the agents of the two teams who observed I_{SC}, then return to the normal state of operation.

The second part of the rule, which is applied only to the agent who went through the first part, is to verify if the traditional protection device has been active, with a consequent loss of a load block. Thus, when the current returns to "normal," it should be verified whether the difference between the pre-fault current (I_{PF}) and the actual current (I) is greater than the Δ value (pre-fixed, which is smaller than the load partition of the branches). If yes, it should be assumed that one of the devices supervised by the agent operated; if not, it is taken that there was no operation of supervised protection, and the agent returns to a normal state. The rules for this are:

Check if the difference between the pre-fault current and the actual current was greater than Δ. Action: continue the search for the device that acted.

If $I_{PF} - I > \Delta$, then continue searching for the device that acted.

The third part of the rule is the search for the device that operated, and this is related to the loss of a load block (difference between the pre-fault current and the actual current in the agent) and the time of day. There is a pre-registered form in the agent's memory that supervises the load partitioning among the traditional devices. This load partitioning is registered on an hourly basis. Thus, the agent should approximate the current time to the nearest hour and compare the percentage of load block with the partitions at this time, and infer that the traditional protection that acted was the one with the most immediate partition. The rules for this third part are:

Transform the difference between the pre-fault current and the actual current to a percentage value and check the current time (assuming the nearest whole time). Action: Search for the closest percentage value.

If $d = I_{PF} - I$, assuming H:00 time, then check the partition closest to d to H.

4.3.3. Application of the Proposed Approach

This section shows the entire proposed approach where the intelligent agents handle the traditional protection operations. Figure 6 presents all necessary steps of this approach. At the start, the agent assumes all the required data from the observed (traditional) switches: the switch loads for each hour during the week. Additionally, the agent measures its actual current. The algorithm starts with validating the traditional switches observed by the agent. The process continues when a short-circuit current (I_{SC}) is detected. Initially, the agent needs to verify if it should take part in the process or not. If any adjacent agents did not observe the short-circuit current, the agent takes part in the process. Otherwise, if at least one adjacent agent on either side observed the short-circuit, it ends the process. This is performed using the first part of the rule presented above.

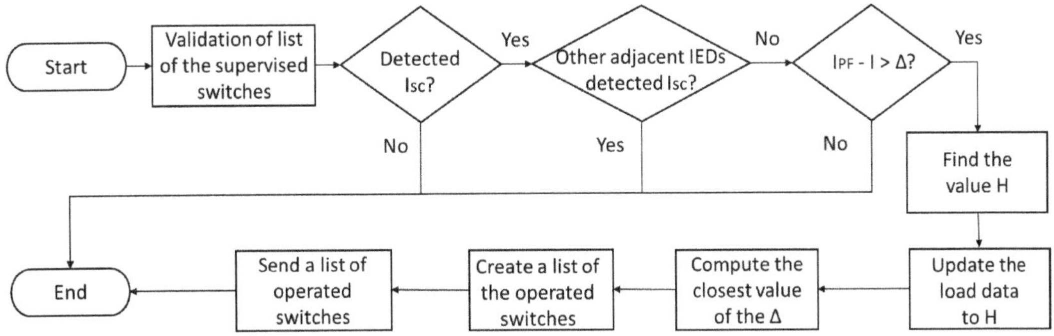

Figure 6. Flowchart of the proposed approach.

The next step is to verify the value of Δ, a pre-fixed value, and compute the difference between the pre-fault current (I_{PF}) and the actual current (I). This value must be smaller than the small load of the observed switch. If this difference value is bigger than Δ, the process continues; otherwise, the process ends. This step starts only after the system operation returns because the agent needs to know the (post-fault) current (I, named here actual current). The focus of this step is performance and is addressed by using the second part of the ruleset presented above.

After these previous steps are performed, the agent is ready to determine a list of possible operated switches. The first step of this part is to determine the value of H—the integer value of the time. For the time between T hours and zero minutes and T hours and twenty-nine minutes, the assumed value of H is T; and between T hours and thirty minutes until T hours and fifty-nine minutes, the implied value of H is T + 1. This part of the approach uses the third part of the ruleset.

In the next step, the updating of the load data occurs. The pre-saved hourly load of each switch for the H value must be updated using a proportionality between the pre-saved value of agent load (I_0) for H and pre-fault current (I_{PF}). Then, for each switch, the pre-saved load value (I_{SWx}) must be updated to a new load value ($I_{SWx\text{-}new}$) for the following relation: $I_{SWx\text{-}new} = I_{SWx} \times I_{PF}/I_0$. This step's main idea is to promote an update of the values as if it were a system of forecasting load in the very short term.

The next step is the computation to determine a list of the closest values to the difference between the pre-fault current (I_{PF}) and the actual current (I). This list can contain a pre-defined number of switches or selected switches by a given criterium, like a pre-defined interval or percentage of the difference value. It is essential to notice the values used for this computation, the $I_{SWx\text{-}new}$ values.

Finally, the message with a list of possible operated switches is sent to the distribution control center, ending the process.

4.3.4. Final Remarks

It is worth mentioning that other characteristics than coordination and selectivity, such as characteristic curves, coordination issues, and other concepts of the traditional protection schemes, are not necessary to understand the proposed approach and are out of the scope of this paper. Moreover, in the proposed approach, the protection studies and the protection element settings continue unchanged. They are not modified by the introduction of smart devices to the network.

5. Application of the Proposed Strategy for Intelligent Agents

This section presents the proposed approach applied in case studies to illustrate the use of the proposed methodology in a didactic way. For that, the network shown in Figure 7 was used. It shows a single-line diagram of a real network of a Brazilian distribution company, which is described in [1,48]. Each bus has a protection device: fuse, recloser, sectionalizer, or intelligent agent. This network's data contains one-hour load curves divided into working days, Saturdays, and Sundays/holidays. This network is a smart metering pilot project with more than 7000 residential/commercial low-voltage consumers.

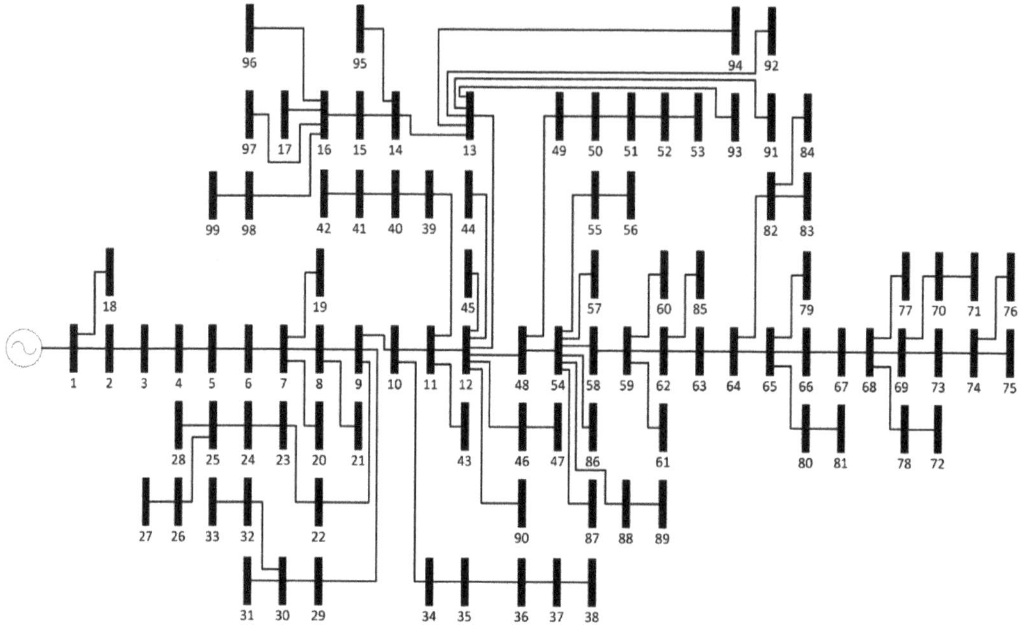

Figure 7. A diagram of a real network of a Brazilian distribution company [1,48].

In this case study, the part of the interest of this network is located from switch 64 onward, i.e., switches 64–84. These switches can be classified as intelligent agents (64, 68), a recloser (64), and sectionalizers (69, 70, 78). For switch 64, the agent was built in the recloser. All other devices are fuses (65, 66, 67, 69, 71, 72, 73, 74, 75, 76, 77, 79, 80, 81, 82, 83, 84). According to the configuration of Figure 7, agent 64 supervises the devices 65, 66, 67, 79, 80, 81, 82, and 83; and agent 68 supervises the following devices: 69, 70, 71, 72, 73, 74, 75, 76, 77, and 78. Notice that the terminal branches are not represented in this figure; however, they exist. For instance, after switch 76, there is a branch with consumers and a load.

Three case studies were created by causing short-circuits in branches 82–83, 70–71, and 74–76 to evaluate the performance of the proposed approach. A fourth case study is presented wherein the protection did not occur correctly.

5.1. Case Study 1: Short-Circuit in Branch 82–83

In this case study, the short-circuit happened in branch 82–83. First, the traditional protection eliminated the short-circuit, and the fuse 82 broke. As agent 64 found the short-circuit current, it established communication with agent 68 and verified this agent did not see this current. Agent 64 inferred the short-circuit happened in its supervised area. Then, it started the computation of the load partitioning of each branch under its supervision. These branches are 64–65, 65–66, 66–67, 67–68, 64–82, 82–83, 82–84, 65–79, 65–80, 80–81, 79–X, 81–X, 83–X, and 84–X.

The agent measured the actual current, computed the difference between the pre-fault current and the actual current, and computed the percentage value of load drop. In this data-driven solution, the current time was verified, and the data were extracted from the agent memory (computed by a previous computation). The entire database used in this example can be found in [19,49].

The load in each bus was transformed into percentual values using the data extracted from the memory. The agent, comparing the percentage value of load drop and the percentual values of each bus, detected the operation of switch 82. Immediately, it sent a message to the substation informing it that switch 82 was broken.

5.2. Case Study 2: Short-Circuit in Branch 70–71

In this second case study, the short-circuit happened in branch 70–71. The traditional protection eliminated the short-circuit using recloser 65 and sectionalizer 70. Then, after three attempts to close the circuit by recloser 64, sectionalizer 70 opened, interrupting the energy in branches 70–71 and 71–X. If the sectionalizer 70 had a problem opening, in the fourth attempt to close the circuit, sectionalizer 69 (the backup protection of 70) could open, but this was not the case.

Then, agents 64 and 68 noticed the short-circuit current. They established communication among them. Agent 64, as agent 68 noticed the short-circuit current, inferred that the short-circuit was not located in its supervised area. Agent 68 inferred the short-circuit happened in its supervised area. Thus, it was one of the branches 69–70, 70–71, 69–73, 73–74, 74–75, 74–76, 71–X, 75–X, and 76–X.

Agent 68 followed the same strategy presented in the previous case study, measuring the actual current, computing the difference between pre-fault and actual current, and computing the percentage value of load drop. After that, data were searched for in the memory (according to the current time), and the load in each bus was composed and transformed into a percentual value. By comparing the percentage value of load drop with the percentual values of each bus, agent 68 inferred the operation was of switch 70, so it sent a message with this information to the substation.

5.3. Case Study 3: Short-Circuit in Branch 74–76

This case study presents a short-circuit in branch 74–76, and via traditional protection, the fuse 74 acted, interrupting the short-circuit. The inference processes in agents 64 and 68 were similar to those in case study 2. Agent 64 inferred the short-circuit happened out of its supervised area. Then, agent 68 started the load partitioning of each branch, concluding, in the end, that fuse 74 acted, so it sent this information to the substation.

5.4. Case Study 4: Wrong Actuation of the Traditional Protection

Suppose that fuse 74 did not act in the previous case study and its backup protection (fuse 73) broke. This fact is not a problem for the inference process of intelligent agents, and the agents would promote the inference process in the same way. Agent 64 would do nothing because the short-circuit is not in its supervised area. In contrast, agent 68 would compute the load partitioning of each branch, conclude that fuse 73 acted, and send this information to the substation.

The answer is correct because fuse 73 acted, but only the maintenance team will verify the malfunction of the protection system in this part of the network.

6. Computational Validation of the Proposed Approach

Computational programs were prepared to validate the proposed approach. This section presents the results of the developed computational programs after many tests were performed.

The first sub-section below presents the structure of the computational program developed to test the validity of the proposed approach. In the second sub-section, we describe how the developed program with the proposed algorithm was executed once to determine a list of operated switches, creating a separate single short-circuit for each branch. The third sub-section shows validation of the proposed approach; thousands of tests were executed, and the quality of the answer is analyzed therein.

6.1. Presentation of the Validating Computer Program

The developed computational program was written to be run thousand times to validate the proposed approach. This section describes what happens when the program is run once. Figure 8 shows the flowchart of the computer program. At the start, the agent is provided with the structure of the monitored network, the positions of the switches, and the positions of the adjacent agents. It starts with a random set of loads (expressed in current) for each branch. The next step is to compute the initial current (I_0) through the agent using the random load of each branch. The agent will use these loads in the following steps. This part of the algorithm is equivalent to the proposed approach when the agent knows each branch's hourly load registered in the local memory.

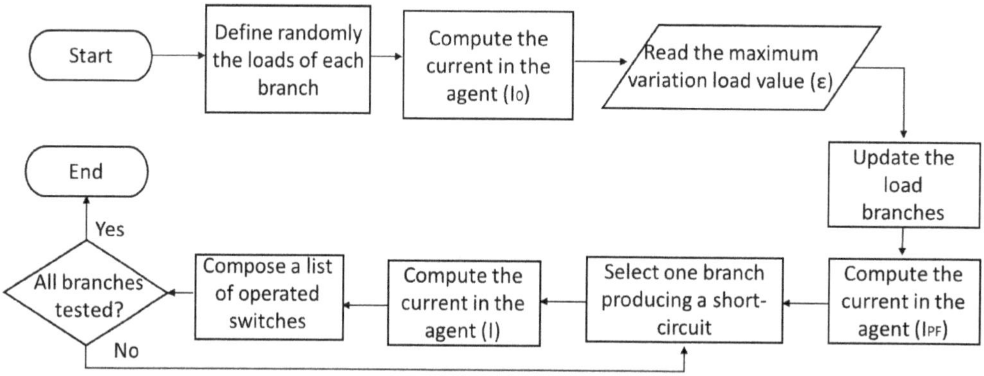

Figure 8. Flowchart of the computer program developed for validating the proposed approach.

After that, the program reads a value of ε (a percentage), the maximum variation, to modify the load branches. In the next step, each load branch is modified using random values between $-\varepsilon$ and $+\varepsilon$ drawn individually for each branch. It is equivalent to the load variation during regular operation. Then, the current in the agent is computed using these new loads. This current is the pre-fault current (I_{PF}) at the proposed approach. Notice that these new loads are used to determine the operated switch, but only to compute the pre-fault current, and in some steps ahead, the actual current (I).

The algorithm starts the test in each branch, simulating a short-circuit and the consequent operation of the switch responsible for eliminating it. This process is performed individually for all observed branches of the agent. After selecting one branch, the actual current (I) is computed without a switching operation, eliminating the load branches.

The list of operated switches is then composed following these actions: (1) update values of the initial values of branch load with the relation I_{PF}/I_0; (2) using the initial configuration of the system, determine the current in each switch; (3) compute the difference between $I_{PF} - I$, determining of the loss of a load block; and (4) compose a list with the possible switches, observing the closest values of the load switch. All these actions are represented in the flowchart in a single box.

The next step verifies if all branches were tested or not. If no, the process continues in the next branch; if yes, the process is ended.

This flowchart presents a single execution of the validation algorithm. However, in practice, the algorithm is run a thousand times, and for each execution, a new set of branch loads is chosen, but the value of ε is not modified. Maintaining a ε constant is done to know the algorithm's performance for a given load variation. Notice that changing the load individually for each branch continues to occur.

6.2. Executing the Computational Program Once

In this example, part of the circuit presented in Figure 7 is used, as shown in Figure 9. In this section, two examples are shown. In the first one, the agent is located in switch 64; in the second example, two agents are situated in switches 64 and 68.

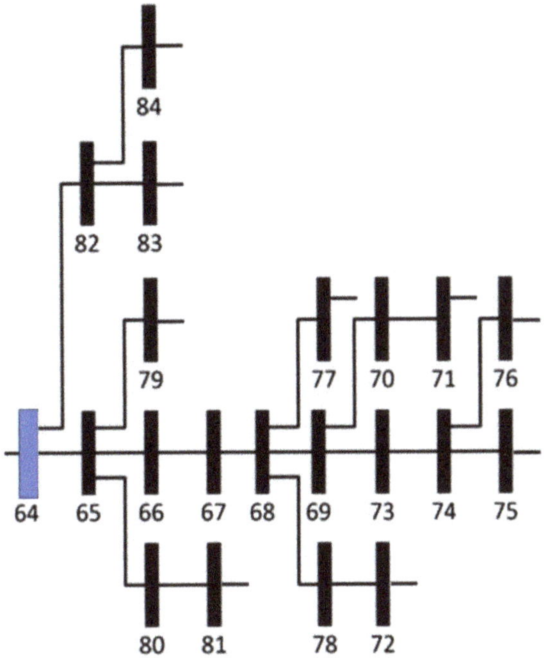

Figure 9. Part of the single diagram of an existing network in Figure 7.

In the first example, suppose the agent has been assembled in the recloser (switch 64). This switch is responsible for observing the switches numbered between 65 to 81.

In this situation, the program is executed, providing a random load for each branch, including the branches after the last switches (as 84, 83, 79, and so on). These branches are also included in the list of branches with single short-circuits.

A single operation of the algorithm randomly generates the values presented in these examples, and they are shown to offer a complete illustrative example.

In the initial step, the network structure after switch 64 is introduced to the program. After that, the program randomly creates loads (I_0, in pu) for each branch, as shown below in the second column in Table 3.

Table 3. Loads of the example: Initial Load (I_0), Pre-fault Load (I_{PF}), and Up-date Initial Load.

Bus	Registrated Load (I_{0SWx})	Load (I_{PF})	Up-Date Registrated Load ($I_{SWx\text{-}new}$)
64	5.067	5.168	5.062
65	1.259	1.239	1.258
66	2.836	2.738	2.833
67	2.522	2.637	2.519
68	5.786	5.691	5.780
69	5.421	5.199	5.415
70	6.796	6.521	6.789

Table 3. *Cont.*

Bus	Registrated Load (I_{0SWx})	Load (I_{PF})	Up-Date Registrated Load ($I_{SWx\text{-}new}$)
71	1.675	1.733	1.674
72	3.798	3.837	3.794
73	1.312	1.352	1.310
74	2.515	2.573	2.512
75	4.342	4.358	4.338
76	1.292	1.353	1.291
77	2.379	2.350	2.377
78	5.242	5.269	5.237
79	4.511	4.660	4.506
80	2.483	2.512	2.480
81	4.792	4.966	4.787
82	6.132	6.179	6.125
83	1.138	1.161	1.135
84	6.114	5.836	6.098

Using the network structure and the branch loads, the program computes the agent's initial current (I_0), 77.414 pu.

In this example, the value (ε) was defined as 5%. This value is related to the load forecasting error, and this one is classified as enormous. Usually, this value is below 1% in one-hour forecasting. Next, the branch loads are randomly modified with values between −5% and +5%.

Then, using the network structure and these new branch loads, the program computes the pre-fault current (I_{PF}) in the agent, 77.332 pu. Then, the initial load value (I_{0SWx}) is updated using the value 0.9989 ($=I_{PF}/I_0$ = 77.332/77.414), generating a new set of load branch values ($I_{SWx\text{-}new}$), as shown in the fourth column in Table 3. In this case, this load update computation is not relevant, but sometimes it is.

Following that, the switches for all branches start to be tested individually. The actual current (I) is defined by the new structure of the network and the modified loads. The loss of a load block difference is obtained from $I_{PF} - I$. The loss of load block is compared with the current of each modified initial load to compose the list of possible operated switches. Table 4 shows the results for short-circuits applied in all studied branches. It presents the operated switch, the list of possible operated switches, and the true operated switch's rank.

Table 4. Examples of short-circuits in all branches, with one agent located in switch 64.

Operated Switch	Complete Ordered List	Rank of Correct Switch
64	[64, 65, 66, 67, 68, 69, 82, 73, 78, 70, 74, 80, 84, 81, 79, 75, 72, 77, 71, 76, 83]	1
65	[65, 66, 67, 64, 68, 69, 82, 73, 78, 70, 74, 80, 84, 81, 79, 75, 72, 77, 71, 76, 83]	1
66	[66, 67, 68, 65, 69, 64, 82, 73, 78, 70, 74, 80, 84, 81, 79, 75, 72, 77, 71, 76, 83]	1
67	[67, 68, 66, 65, 69, 82, 73, 78, 64, 70, 74, 80, 84, 81, 79, 75, 72, 77, 71, 76, 83]	1
68	[68, 67, 66, 69, 65, 82, 73, 78, 70, 74, 80, 84, 81, 79, 75, 72, 64, 77, 71, 76, 83]	1
69	[69, 82, 73, 78, 70, 74, 80, 68, 84, 81, 79, 75, 72, 67, 77, 71, 76, 83, 66, 65, 64]	1
70	[74, 70, 78, 80, 73, 84, 81, 79, 75, 72, 82, 77, 71, 76, 83, 69, 68, 67, 66, 65, 64]	2
71	[71, 76, 83, 77, 72, 75, 79, 81, 84, 80, 74, 70, 78, 73, 82, 69, 68, 67, 66, 65, 64]	1
72	[72, 75, 79, 81, 77, 84, 71, 76, 83, 80, 74, 70, 78, 73, 82, 69, 68, 67, 66, 65, 64]	1

Table 4. *Cont.*

Operated Switch	Complete Ordered List	Rank of Correct Switch
73	[73, 78, 70, 74, 80, 82, 84, 81, 79, 75, 72, 77, 71, 76, 83, 69, 68, 67, 66, 65, 64]	1
74	[74, 70, 78, 80, 73, 84, 81, 79, 75, 72, 82, 77, 71, 76, 83, 69, 68, 67, 66, 65, 64]	1
75	[75, 79, 81, 72, 84, 77, 71, 80, 76, 83, 74, 70, 78, 73, 82, 69, 68, 67, 66, 65, 64]	1
76	[76, 83, 71, 77, 72, 75, 79, 81, 84, 80, 74, 70, 78, 73, 82, 69, 68, 67, 66, 65, 64]	1
77	[77, 71, 76, 83, 72, 75, 79, 81, 84, 80, 74, 70, 78, 73, 82, 69, 68, 67, 66, 65, 64]	1
78	[78, 73, 70, 74, 80, 84, 82, 81, 79, 75, 72, 77, 71, 76, 83, 69, 68, 67, 66, 65, 64]	1
79	[81, 79, 75, 72, 84, 77, 80, 71, 76, 83, 74, 70, 78, 73, 82, 69, 68, 67, 66, 65, 64]	2
80	[80, 74, 70, 84, 78, 73, 81, 79, 75, 72, 77, 71, 82, 76, 83, 69, 68, 67, 66, 65, 64]	1
81	[81, 79, 75, 84, 72, 80, 77, 71, 74, 70, 76, 83, 78, 73, 82, 69, 68, 67, 66, 65, 64]	1
82	[82, 73, 78, 70, 74, 80, 84, 81, 79, 75, 72, 69, 77, 71, 76, 83, 68, 67, 66, 65, 64]	1
83	[83, 76, 71, 77, 72, 75, 79, 81, 84, 80, 74, 70, 78, 73, 82, 69, 68, 67, 66, 65, 64]	1
84	[84, 81, 79, 75, 80, 72, 74, 70, 78, 77, 73, 71, 76, 83, 82, 69, 68, 67, 66, 65, 64]	1

By observing the results of Table 4, it is possible to verify that the proposed approach hit the target in most tests. For instance, in the first example, the short-circuit was applied in branch 64–65 with the operation of switch 64; notice that this switch is in the first place on the possible operating switch list. The correct solution was found in second place when switches 70 and 79 were operated.

In the second example, two agents are located in switches 64 and 68 of Figure 9, dividing the observability of this part of the network. The same data in Table 3 were used, and the results are shown in Table 5.

Table 5. Examples of short-circuits in all branches, with two agents located in switches 64 and 68.

Operated Switch	Complete Ordered List	Rank of Correct Switch
64	[64, 69, 65, 73, 74, 70, 66, 75, 67, 78, 76, 77, 68, 72, 71, 82, 79, 80, 83, 84, 81]	1
65	[65, 69, 66, 67, 68, 64, 73, 74, 70, 75, 78, 76, 77, 72, 71, 82, 79, 80, 83, 84, 81]	1
66	[66, 67, 68, 69, 65, 73, 74, 82, 70, 64, 75, 78, 76, 77, 72, 71, 79, 80, 83, 84, 81]	1
67	[67, 66, 68, 69, 65, 73, 74, 82, 70, 75, 64, 78, 76, 77, 72, 71, 79, 80, 83, 84, 81]	1
68	[68, 67, 66, 69, 65, 73, 82, 74, 70, 75, 78, 79, 80, 76, 77, 83, 84, 64, 72, 71, 81]	1
69	[69, 73, 82, 68, 74, 67, 66, 70, 75, 78, 79, 80, 76, 77, 83, 84, 72, 71, 81, 65, 64]	1
70	[70, 75, 78, 79, 80, 76, 77, 83, 84, 72, 71, 81, 74, 82, 73, 69, 68, 67, 66, 65, 64]	1
71	[81, 71, 72, 84, 77, 83, 76, 80, 79, 78, 75, 70, 74, 82, 73, 69, 68, 67, 66, 65, 64]	2
72	[72, 71, 81, 84, 77, 83, 76, 80, 79, 78, 75, 70, 74, 82, 73, 69, 68, 67, 66, 65, 64]	1
73	[73, 82, 74, 70, 75, 69, 78, 79, 80, 76, 77, 83, 84, 72, 71, 81, 68, 67, 66, 65, 64]	1
74	[74, 82, 73, 70, 75, 78, 79, 80, 76, 77, 83, 84, 72, 71, 81, 69, 68, 67, 66, 65, 64]	1
75	[75, 70, 78, 79, 80, 76, 77, 83, 84, 72, 71, 81, 74, 82, 73, 69, 68, 67, 66, 65, 64]	1
76	[76, 83, 77, 80, 79, 78, 84, 72, 71, 81, 75, 70, 74, 82, 73, 69, 68, 67, 66, 65, 64]	1
77	[76, 83, 77, 80, 79, 78, 84, 72, 71, 81, 75, 70, 74, 82, 73, 69, 68, 67, 66, 65, 64]	3
78	[78, 79, 80, 76, 77, 83, 75, 84, 72, 71, 81, 70, 74, 82, 73, 69, 68, 67, 66, 65, 64]	1
79	[80, 79, 83, 84, 81, 71, 72, 77, 76, 78, 75, 70, 82, 74, 73, 69, 68, 67, 66, 65, 64]	2
80	[80, 79, 83, 84, 81, 71, 72, 77, 76, 78, 75, 70, 82, 74, 73, 69, 68, 67, 66, 65, 64]	1

Table 5. *Cont.*

Operated Switch	Complete Ordered List	Rank of Correct Switch
81	[81, 84, 83, 80, 79, 71, 72, 77, 76, 78, 75, 70, 82, 74, 73, 69, 68, 67, 66, 65, 64]	1
82	[82, 71, 72, 77, 76, 78, 75, 70, 79, 80, 83, 84, 81, 74, 73, 68, 67, 66, 69, 65, 64]	1
83	[83, 80, 84, 79, 81, 71, 72, 77, 76, 78, 75, 70, 82, 74, 73, 69, 68, 67, 66, 65, 64]	1
84	[84, 83, 80, 79, 81, 71, 72, 77, 76, 78, 75, 70, 82, 74, 73, 69, 68, 67, 66, 65, 64]	1

Most of these attempts also hit the correct target. Only faults 71 and 79 are ranked in second place, and 77 in the third place.

Note that the program uses random load values, and each time the program is executed, one result appears. The program must be run hundreds of times to validate the proposed approach and create substantiated statistical results.

This computational program can be run at the repository in [50].

6.3. Validating Test of the Proposed Approach

The computational program uses drawn and random processes, so it is necessary to validate the proposed approach with a thousand tests. It includes different branch loads and different values of ε for the same network structure.

In the first validation test, the structure shown in Figure 9 was used, and the agent was also located in recloser 64. One thousand executions of the program were run for each value of ε, and the values of ε chosen were 1%, 2%, 5%, and 10%, so 4000 executions were performed. Table 6 shows these results.

Table 6. Results of the correct switches in the list (in percentages).

	Position in the List				Error	Hits	Error	Hits
ε	1st	2nd	3rd	4th	Total	Total	1st	1st + 2nd
1%	96.19%	3.81%	0.00%	0.00%	0.00%	100.00%	3.81%	0.00%
2%	95.24%	4.29%	0.48%	0.00%	0.00%	100.00%	4.76%	0.48%
5%	92.00%	4.67%	3.33%	0.00%	0.00%	100.00%	8.00%	3.33%
10%	74.29%	18.57%	6.19%	0.48%	0.48%	99.52%	25.71%	7.14%

By observing the results above table, it is possible to verify that the number of hits which are first place in the list is higher when the value of ε is small. This fact is expected, because the difference between the modified loads and the modified initial loads increases with the value of ε. The worst outcome occurred when ε was equal to 10%. This value in practice does not exist because, as mentioned before, the value of ε is related to load forecasting. This size of load change is not common in a distribution operation. This example was included to verify the quality of detection of the proposed approach even in extreme situations.

Other values of Table 6 show the expected error level when the list sent to the operator contains only one element (one switch) or two elements (two switches). For instance, when the value of ε is 2%, the maximum value hit in most situations is under 8% in the one-element list and 3% in the two-element list. These values are 4.76% and 0.48% in the table, respectively.

Now two examples can be presented from this validation process, using intelligent agents located in switches 64 and 68. In this situation, this part of the circuit is divided into two parts: one observed by switch 64 (monitoring switches 65–67 and 79–84) and another observed by switch 68 (monitoring switches 69–78). The same tests as described above were performed for these switches, and the results are shown in Table 7.

Table 7. Results of the correct switches in the list (in percentages)—tests using switches 64 and 68.

	Position in the List					
ε	1st	2nd	3rd	4th	Error	Hits
1%	99.52%	0.48%	0.00%	0.00%	0.00%	100.00%
2%	95.71%	3.81%	0.48%	0.00%	0.00%	100.00%
5%	92.86%	6.67%	0.48%	0.00%	0.00%	100.00%
10%	79.05%	14.29%	3.33%	2.38%	0.95%	99.05%

6.4. Discussions

Other real and simulated distribution circuits have been tested similarly, and the results have been similar. The proposed approach indeed merges smart grid devices and traditional protection elements, increasing the network's observability.

Regarding the results of the two previous tables, we verified that the level of hit degree is increased. This was also expected because the test was for a network of almost the same size, and two observed agents divided the monitoring of the switches. After running the program thousands of times, the error was continuously below 3%, which validates the proposed approach in the worst cases. This computational program can be run at the repository in [51].

An error above 10% would be accepted because the system operator has no observability in traditional distribution networks. Thus, any contribution is very welcome during the operation, even partly unreliable ones. The proposed approach can hit this level of error (around 10%) when an intelligent agent monitors the protection of hundreds of branches. However, this case occurs only when the number of intelligent agents is few compared with the number of supervised protection branches.

7. Conclusions

This study aimed to make smart grid agents compatible with traditional protection devices of urban distribution electrical networks. This study of compatibilization is necessary because protection devices are being changed step-by-step. For a reasonable period of time, they will act together, mainly in concessionaires without significant financial resources. Thus, if agents can, from the beginning of their operations in networks, interact with the protection devices (even indirectly), increasing the observability of the network, it would be helpful.

Unlike traditional articles, this article discussed a more practical implementation of installing a smart grid. This topic had not yet been addressed in scientific papers, but it is of fundamental importance for companies in this phase of technological development. Many of them will still be at this stage of their smart grids for many years, and with the techniques presented in this article, they will be able to use the benefits of some (really) intelligent devices in their networks.

It is noticed that most articles deal with more advanced smart grids by proposing operational and troubleshooting techniques. Thus, a critical review of the main articles was not performed because they do not present strategies involving traditional protection. In addition, and for an analogous reason, no comparisons were made with other methods.

The proposed agent had precisely these purposes: to increase the observability of the network for the operation centers, reduce the response time for interventions in the network, and assist in lowering the reliability indexes (CAIDI and SAIDI). The operation team can focus on other essential tasks, such as sending maintenance teams to the site to repair problems in the faulty section of the feeder, optimizing resources, and reducing service outage time.

The first function analyzed provides the interactions with fuse switches: verifying which fuse supervised by the agent had its link broken by a short-circuit current. It uses the comparative load proportionality metric. When the agent supervises fewer fuses, its accuracy is greater. The proposed method is not always foolproof because it acts with typical proportional loads. However, it is preferable to present a false positive to the operating center than wait for a consumer complaint to be taken.

The second function, for interacting with reclosers and sectionalizers, was then analyzed. Those devices are monitored analogously to the fuses. Despite having completely different roles, they are treated similarly in their integration to intelligent agents; and the comparative metric of load proportionality is used in the same way.

Thus, in the end, the urban protection devices continue to act traditionally. Different treatment only occurs in the analysis of loss of a load block.

This paper also presented the possibility of integrating intelligent agents into the reclosers and sectionalizers existing in the distribution networks. It can be done without any additional action, because the reclosers open short-circuit currents like the circuit breakers, and the sectionalizers, which do not open short-circuit currents, resemble the disconnectors. Both circuit breakers and disconnectors are switches receiving the intelligent agent in the papers available in the literature. Thus, this embedding was done without the inclusion of any extra elements.

The integration proposal in this paper was tested in an existing distribution system, using random data and producing a list of the operated switches. The degree of correct hits in thousands of tests was higher than 97% in various situations, proving the quality of the proposed approach.

Finally, the proposed approach opens many possibilities for new analyses, such as studies about the optimal locations for smart devices, interactive network observability, and innovative protection schemes using smart grid devices and traditional devices. Even in a transitory situation between the traditional protection and smart grid world, this initial phase of smart grid device implementation creates many possibilities with the proposed interaction of the two philosophies.

Author Contributions: B.S.T., C.P.S. and C.H.V.d.M. conceived and designed the experiments; B.S.T., C.H.V.d.M. and C.P.S. performed the experiments; B.S.T., C.P.S. and L.E.B.d.S. analyzed the case studies; B.S.T. and L.E.B.d.S. developed the methodology; and B.S.T., C.P.S. and C.H.V.d.M. analyzed the results of the proposed approach; and B.S.T., C.P.S. and L.E.B.d.S. wrote the paper. All authors have read and agreed to the published version of the manuscript.

Funding: This research received no external funding.

Institutional Review Board Statement: Not applicable.

Informed Consent Statement: Not applicable.

Data Availability Statement: Not applicable.

Acknowledgments: The authors would like to thank the National Council for Scientific and Technological Development (CNPq), the Coordination for the Improvement of Higher Education Personnel (CAPES), and the Brazilian Electricity Regulatory Agency Research and Development (ANEEL R&D) for supporting this project.

Conflicts of Interest: The authors declare no conflict of interest.

References

1. Siebert, L.C.; Aoki, A.R.; Lambert-Torres, G.; Lambert-de-Andrade, N.; Paterakis, N.G. An Agent-Based Approach for the Planning of Distribution Grids as a Socio-Technical System. *Energies* **2020**, *13*, 4837. [CrossRef]
2. Barreto, N.E.M.; Rodrigues, R.; Schumacher, R.; Aoki, A.R.; Lambert-Torres, G. Artificial Neural Network Approach for Fault Detection and Identification in Power Systems with Wide Area Measurement Systems. *J. Control Autom. Electr. Syst.* **2021**, *32*, 1617–1626. [CrossRef]
3. Mahat, P.; Chen, Z.; Bak-Jensen, B.; Bak, C.L. A Simple Adaptive Overcurrent Protection of Distribution Systems with Distributed Generation. *IEEE Trans. Smart Grid* **2011**, *2*, 428–437. [CrossRef]
4. Fletcher, S.D.A.; Norman, P.J.; Fong, K.; Galloway, S.J.; Burt, G.M. High-Speed Differential Protection for Smart DC Distribution Systems. *IEEE Trans. Smart Grid* **2014**, *5*, 2610–2617. [CrossRef]

5. Zeineldin, H.H.; Sharaf, H.M.; Ibrahim, D.K.; El-Zahab, E.E.-D.A. Optimal Protection Coordination for Meshed Distribution Systems With DG Using Dual Setting Directional Over-Current Relays. *IEEE Trans. Smart Grid* **2015**, *6*, 115–123. [CrossRef]
6. Noudjiep Djiepkop, G.F.; Krishnamurthy, S. Multi-Objective Feeder Reconfiguration Using Discrete Particle Swarm Optimization. *Mathematics* **2022**, *10*, 531. [CrossRef]
7. Kim, M.-S.; Haider, R.; Cho, G.-J.; Kim, C.-H.; Won, C.-Y.; Chai, J.-S. Comprehensive Review of Islanding Detection Methods for Distributed Generation Systems. *Energies* **2019**, *12*, 837. [CrossRef]
8. Ates, Y.; Uzunoglu, M.; Karakas, A.; Boynuegri, A.R. The case study based protection analysis for smart distribution grids including distributed generation units. In Proceedings of the 12th IET International Conference on Developments in Power System Protection (DPSP 2014), Copenhagen, Denmark, 31 March–3 April 2014; pp. 1–5. [CrossRef]
9. Wei, L.; Qi, Y.; Qi, H. Research on design and implementation of relay protection in smart grid. In Proceedings of the 2018 Chinese Control and Decision Conference (CCDC), Shenyang, China, 9–11 June 2018; pp. 1439–1443. [CrossRef]
10. Jimenez, S.; Vázquez, E.; Gonzalez-Longatt, F. Methodology of Adaptive Instantaneous Overcurrent Protection Setting. *Electronics* **2021**, *10*, 2754. [CrossRef]
11. Liu, X.; Zhang, B.; Chen, B.; Aved, A.; Jin, D. Towards Optimal and Executable Distribution Grid Restoration Planning with a Fine-Grained Power-Communication Interdependency Model. *IEEE Trans. Smart Grid* **2022**, *11*, 9973. [CrossRef]
12. Sayed, M.M.; Mahdy, M.Y.; Abdel Aleem, S.H.E.; Youssef, H.K.M.; Boghdady, T.A. Simultaneous Distribution Network Reconfiguration and Optimal Allocation of Renewable-Based Distributed Generators and Shunt Capacitors under Uncertain Conditions. *Energies* **2022**, *15*, 2299. [CrossRef]
13. Erenoğlu, A.K.; Sancar, S.; Terzi, I.S.; Erdinç, O.; Shafie-khah, M.; Catalão, J.P.S. Resiliency-Driven Multi-Step Critical Load Restoration Strategy Integrating On-Call Electric Vehicle Fleet Management Services. *IEEE Trans. Smart Grid* **2022**, *11*, 5438. [CrossRef]
14. Singhal, A.; Vu, T.L.; Du, W. Consensus Control for Coordinating Grid-Forming and Grid-Following Inverters in Microgrids. *IEEE Trans. Smart Grid* **2022**, *9*, 254. [CrossRef]
15. Assad, U.; Hassan, M.A.S.; Farooq, U.; Kabir, A.; Khan, M.Z.; Bukhari, S.S.H.; Jaffri, Z.U.A.; Oláh, J.; Popp, J. Smart Grid, Demand Response and Optimization: A Critical Review of Computational Methods. *Energies* **2022**, *15*, 2003. [CrossRef]
16. ANEEL—Brazilian Electricity Regulatory Agency, "Module 8—Quality of Electric Energy—Electric Power Distribution Procedures in the National Electrical System—PRODIST. 2021. p. 88. Available online: https://www.aneel.gov.br/documents/656827/14866914/Módulo_8-Revis~ao_12/342ff02a-8eab-2480-a135-e31ed2d7db47 (accessed on 8 June 2021). (In Portuguese)
17. Ferreira, L.R.; Aoki, A.R.; Lambert-Torres, G. A Reinforcement Learning Approach to Solve Service Restoration and Load Management Simultaneously for Distribution Networks. *IEEE Access* **2019**, *7*, 145978–145987. [CrossRef]
18. Ferreira, L.R. Hybrid Intelligent Control for Smart Grid Functionalities Integration. Ph.D. Thesis, Federal University of Paraná, Curitiba, Brazil, 2020. Available online: https://hdl.handle.net/1884/70724 (accessed on 10 September 2021).
19. Siebert, L.C. Demand Response Optimization System for Intelligent Power Grids. Ph.D. Thesis, Federal University of Paraná, Curitiba, Brazil, 2013. Available online: http://hdl.handle.net/1884/35731 (accessed on 18 October 2021). (In Portuguese).
20. Sallam, A.A.; Malik, O.P. *Electric Distribution Systems*; Wiley-IEEE Press: Piscataway, NJ, USA, 2019; p. 624. ISBN 9781119509332.
21. Russell, S.J.; Norvig, P. *Artificial Intelligence: A Modern Approach*, 4th ed.; Prentice-Hall: Hoboken, NJ, USA, 2020; p. 2579. ISBN 9780134610993.
22. Ramesh Babu, N. *Smart Grid Systems: Modeling and Control*; Apple Academic Press: Waretown, NJ, USA, 2018; p. 290. ISBN 9781771886253.
23. Belu, R. *Smart Grid Fundamentals: Energy Generation, Transmission and Distribution*; CRC Press: Boca Raton, FL, USA, 2022; p. 568. ISBN 978-1482256673.
24. Tanenbaum, A. *Modern Operating Systems*; Pearson/Prentice Hall: Hoboken, NJ, USA, 2008; p. 160. ISBN 9780136006633.
25. Falk, H. *IEC 61850 Demystified*; Artech House Publishers: Norwood, MA, USA, 2019; p. 332. ISBN 9781630813291.
26. Cleveland, F. IEC TC57 Security Standards for the Power System's Information Infrastructure—Beyond Simple Encryption. In Proceedings of the 2005/2006 IEEE/PES Transmission and Distribution Conference and Exhibition, Chicago, IL, USA, 4–7 May 2006; pp. 1079–1087. [CrossRef]
27. Reda, H.T.; Ray, B.; Peidaee, P.; Anwar, A.; Mahmood, A.; Kalam, A.; Islam, N. Vulnerability and Impact Analysis of the IEC 61850 GOOSE Protocol in the Smart Grid. *Sensors* **2021**, *21*, 1554. [CrossRef]
28. Schneider Electric. Intelligent Loop Automation. 2013. p. 84. Available online: https://www.se.com/library/SCHNEIDER_ELECTRIC/SE_LOCAL/APS/213039_E57A/SEAU26709_Loop_Automation_Manual_WEB.pdf (accessed on 21 December 2021).
29. S&C Electric Company. IntelliTeam®SG Automatic Restoration System. 2022. p. 113. Available online: https://www.sandc.com/globalassets/sac-electric/documents/sharepoint/documents---all-documents/instruction-sheet-1044-570.pdf (accessed on 28 November 2021).
30. Eaton. Yukon Feeder Automation (YFA)I. 2021, p. 16. Available online: https://www.eaton.com/content/dam/eaton/products/utility-and-grid-solutions/grid-automation-systems/yukon-feeder-automation/yukon-feeder-automation-ps818001en.docx (accessed on 28 December 2021).
31. NovaTech. Distribution Automation Controller. 2019. p. 6. Available online: https://back.novatechautomation.com/wp-content/uploads/2020/10/DS_DAMaster_111210.pdf (accessed on 28 October 2021).

32. Schweitzer Engineering Laboratories. Distribution Automation Controller. Available online: https://selinc.com/engineering-services/distribution-network-automation/ (accessed on 16 February 2022).
33. Hitachi Energy. Self-Healing Distribution Grid. Available online: https://www.hitachienergy.com/rtu (accessed on 28 February 2022).
34. Siemens, A.G. *Self-Optimizing Grid—Intelligent Grid Automation*. 2018, p. 12. Available online: https://assets.new.siemens.com/siemens/assets/api/uuid:28ae65dd-359c-4b96-85bd-280facec62ea/self-optimizing-grid.pdf (accessed on 18 November 2021).
35. Railsback, S.; Grimm, V. *Agent-Based and Individual-Based Modeling: A Practical Introduction*, 2nd ed.; Princeton University Press: Princeton, NJ, USA, 2019; 360p, ISBN 9780691190839.
36. McArthur, S.D.J.; Davidson, E.M.; Catterson, V.M.; Dimeas, A.L.; Hatziargyriou, N.D.; Ponci, F.; Funabashi, T. Multi-Agent Systems for Power Engineering Applications—Part I: Concepts, Approaches, and Technical Challenges. *IEEE Trans. Power Syst.* **2007**, *22*, 1743–1752. [CrossRef]
37. Sharma, A.; Srinivasan, D.; Trivedi, A. A Decentralized Multiagent System Approach for Service Restoration using DG Islanding. *IEEE Trans. Smart Grid* **2015**, *6*, 2784–2793. [CrossRef]
38. Torres, B.S.; Ferreira, L.R.; Aoki, A.R. Distributed Intelligent System for Self-Healing in Smart Grids. *IEEE Trans. Power Deliv.* **2018**, *33*, 2394–2403. [CrossRef]
39. Kezunovic, M. Smart Fault Location for Smart Grids. *IEEE Trans. Smart Grid* **2011**, *2*, 11–22. [CrossRef]
40. Zidan, A.; El-Saadany, E.F. A Cooperative Multiagent Framework for Self-healing Mechanisms in Distribution Systems. *IEEE Trans. Smart Grid* **2012**, *3*, 1525–1539. [CrossRef]
41. Ibrahim, A.M.; Yau, K.-L.A.; Chong, Y.-W.; Wu, C. Applications of Multi-Agent Deep Reinforcement Learning: Models and Algorithms. *Appl. Sci.* **2021**, *11*, 10870. [CrossRef]
42. Xu, Y.; Liu, W. Novel Multiagent Based Load Restoration Algorithm for Microgrids. *IEEE Trans. Smart Grid* **2011**, *2*, 152–161. [CrossRef]
43. Nikkhah, S.; Allahham, A.; Bialek, J.W.; Walker, S.L.; Giaouris, D.; Papadopoulou, S. Active Participation of Buildings in the Energy Networks: Dynamic/Operational Models and Control Challenges. *Energies* **2021**, *14*, 7220. [CrossRef]
44. Jamal, S.; Tan, N.M.L.; Pasupuleti, J. A Review of Energy Management and Power Management Systems for Microgrid and Nanogrid Applications. *Sustainability* **2021**, *13*, 10331. [CrossRef]
45. Lim, I.-H.; Sidhu, T.S.; Choi, M.S.; Lee, S.J.; Hong, S.; Lim, S.I.; Lee, S.W. Design and Implementation of Multiagent-Based Distributed Restoration System in DAS. *IEEE Trans. Power Deliv.* **2013**, *28*, 585–593. [CrossRef]
46. Karavas, C.-S.; Kyriakarakos, G.; Arvanitis, K.G.; Papadakis, G. A multiagent decentralized energy management system based on distributed intelligence for the design and control of autonomous polygeneration microgrids. *Energy Convers. Manag.* **2015**, *103*, 166–179. [CrossRef]
47. Siebert, L.C.; Aoki, A.R.; Fernandes, T.S.P.; Lambert-Torres, G. Customer Targeting Optimization System for Price-Based Demand Response Programs. *Int. Trans. Electr. Energy Syst.* **2018**, *29*, e2709. [CrossRef]
48. Frederiks, E.R.; Stenner, K.; Hobman, E.V. The Socio-Demographic and Psychological Predictors of Residential Energy Consumption: A Comprehensive Review. *Energies* **2015**, *8*, 573–609. [CrossRef]
49. Siebert, L.C.; Sbicca, A.; Aoki, A.R.; Lambert-Torres, G. A Behavioral Economics Approach to Residential Electricity Consumption. *Energies* **2017**, *10*, 768. [CrossRef]
50. Integrating Smart Grid Devices into the Traditional Protection of Distribution Networks. Available online: https://colab.research.google.com/github/carloshvmoraes/ISGDTPDN/blob/main/ISGD_basic.ipynb (accessed on 18 November 2021).
51. Integrating Smart Grid Devices into the Traditional Protection of Distribution Networks. Available online: https://colab.research.google.com/github/carloshvmoraes/ISGDTPDN/blob/main/ISGD_benchmark.ipynb (accessed on 18 November 2021).

energies

MDPI

Article

Intelligent Power Distribution Restoration Based on a Multi-Objective Bacterial Foraging Optimization Algorithm

Carlos Henrique Valério de Moraes [1,*], Jonas Lopes de Vilas Boas [1], Germano Lambert-Torres [2], Gilberto Capistrano Cunha de Andrade [2] and Claudio Inácio de Almeida Costa [2]

1 Institute of Systems Engineering and Information Technology, Itajuba Federal University, Itajuba 37500-903, Brazil; jonasmassume@gmail.com
2 R&D Department, Gnarus Institute, Itajuba 37500-052, Brazil; germanoltorres@gmail.com (G.L.-T.); gilbertocapistrano@gmail.com (G.C.C.d.A.); claudiocosta@yahoo.com (C.I.d.A.C.)
* Correspondence: valerio@unifei.edu.br

Abstract: The importance of power in society is indisputable. Virtually all economic activities depend on electricity. The electric power systems are complex, and move studies in different areas are motivated to make them more efficient and solve their operational problems. The smart grids emerged from this approach and aimed to improve the current systems and integrate electric power using alternative and renewable sources. Restoration techniques of these networks are being developed to reduce the impacts caused by the usual power supply interruptions due to failures in the distribution networks. This paper presents the development and evaluation of the performance of a multi-objective version of the Bacterial Foraging Optimization Algorithm for finding the minor handling switches that maximize the number of buses served, keeping the configuration radial system and within the limits of current in the conductors and bus voltage. An electrical system model was created, and routines were implemented for the network verification, which was used as a function of the Multi-Objective Bacterial Foraging Optimization Hybrid Algorithm. The proposed method has been applied in two distribution systems with 70 buses and 201 buses, respectively, and the algorithm's effectiveness to solve the restoration problem is discussed.

Keywords: bio-inspired algorithms; multi-objective optimization; network reconfiguration; distribution system; smart grids

Citation: Moraes, C.H.V.d.; Vilas Boas, J.L.d.; Lambert-Torres, G.; Andrade, G.C.C.d.; Costa, C.I.d.A. Intelligent Power Distribution Restoration Based on a Multi-Objective Bacterial Foraging Optimization Algorithm. *Energies* **2022**, *15*, 1445. https://doi.org/10.3390/en15041445

Academic Editor: Elyas Rakhshani

Received: 18 January 2022
Accepted: 6 February 2022
Published: 16 February 2022

1. Introduction

Electrical power systems exist to produce and deliver electricity to various consumers. These systems move studies in different areas to make them more efficient and solve their operational problems. Telecommunications technologies and computational tools have been implemented in other applications, such as smart power grids, known as smart grids, which emerged from the approximation of these areas and aimed to improve current electrical systems [1]. Smart grids are equipped with monitoring systems and remote operation of your equipment. These networks allow the supply of energy in a controlled and intelligent way, analyzing, among other factors, the consumption patterns of users, and can integrate ways of generating electricity through alternative, renewable sources [2] and electrical and gas storage [3].

Damage to the infrastructure of electrical networks, caused by storms, traffic accidents, vandalism, among other factors, is widespread and can cause interruptions in the electricity supply. In 2012, about 8 million people were left without electricity during Hurricane Sandy's passage in the United States. The country's economy is estimated to lose more than $150 billion a year due to interruptions in the electricity supply [4].

Techniques have been developed to reduce the impacts caused by these interruptions. These techniques try to recover the power distribution system, aiming to minimize the time required to restore supply and reduce the number of people affected [5]. Among them, computational intelligence optimization techniques, such as bio-inspired techniques [6], have been widely applied to design optimal plants, find the best configuration of restore switches and select the optimal sequence of operation of these switches.

With the expansion of new smart grids technologies for monitoring and remote performance in networks, efficient computational techniques will be increasingly important. The process known as self-healing has been studied to make intelligent power grids capable of recovering from faults in a fully autonomous way.

Current equipment effectively detects failures, and some initiatives can already suggest the actions needed to restore networks. However, the role of operators is still required to implement the solutions [7]. In addition, restoration plants need to be constantly updated. Due to the growth of cities and increased energy consumption, substations become overloaded and quickly become obsolete.

The restoration problem in the distribution network was initially treated by numerical methods, using mathematical programmings like mixed-integer linear optimization [8] or second-order optimization [9], or mathematical morphology [10]. However, the quality of the solutions found by these methods depends on the search space. In multi-modal search spaces or nonconvex search spaces, these methods can not have a good solution because it is impossible to produce tunneling.

With the advent of intelligent meta-heuristics techniques, some solutions have been developed to optimize restoring these networks. Optimization techniques such as PSO (particle swarm optimization) [11] and GA (genetic algorithm) [12] have been widely applied to design optimal plants, find the best positioning of restore switches and select the optimal sequence of operation of these switches. References [13–15] present examples of applications of PSO to provide restoration solutions for distribution systems, while references [16–18] deliver these solutions using GA techniques.

All meta-heuristic techniques are stochastic algorithms, meaning they involve random processes. In these algorithms, each initial condition and set of initial parameters produce a given solution. For this reason, a suitable adjustment of the initial parameters is necessary to avoid problems, like local minimums or premature convergence. And more, the algorithms must be executed several times, comparing their responses and selecting the best solution.

This paper proposes another intelligent meta-heuristic approach using the Multi-Objective Bacteria Culture Optimization Hybrid Algorithm (MOBCOHA) to solve the restoration problem. This new algorithm merges ideas shown in two references [19,20]. Then, given a system that failed, the proposed program performs power flow calculations on the network to find a minimum amount of switch opening and closing operations that result in a combination that maximizes the number of loads served.

This paper is structured as follows. The proposed MOBCOHA is presented with their most essential concepts in the second section. After, in the third section, the details of the development of the proposed approach are given. This section also includes information about the techniques and algorithms used. In Section 4, the results of some experiments compared to other algorithms are shown in an 11 kV radial distribution network used in many published studies and a 201-buses distribution network. Finally, Section 5 is presented the conclusion about the analysis of the experiments.

2. The Multi-Objective Bacteria Culture Optimization Algorithm

Algorithms that model the behavior of living beings or some bio-logical phenomenon, called bio-inspired algorithms, assume that nature's various processes always find optimal strategies. They are part of computational intelligence studies, more precisely in natural computing. Using initial conditions and relatively simple rules, they are widely applied for machine learning, element classification, and optimization issues, among others.

Most of these methods use populations of individuals who spread through the search space and can simultaneously search for the best solution. Each individual represents a feasible solution. The way these individuals move in the search space and are selected are specific to each method.

Some models use Darwin's evolutionary method of living beings as the basis. Evolution algorithms model evolution-related processes such as gene transfer between generations, crossover, mutation, and natural selection. Genetic algorithm (GA), proposed in [9], is an example of a technique that uses these concepts.

Another example of a technique inspired by biology is the intelligence of swarms. These techniques try to model the social behavior of some species of living beings while searching for food or protecting themselves from predators. The method proposed in [13], called particle swarm optimization (PSO), is based on the movement of birds flying in flocks. In contrast, the proposal in [21] uses as a basis the strategy of searching for food for ants, called ant colony.

The search for living beings for food, or foraging, has inspired the development of various meta-heuristic optimization techniques. Among them is the BFOA (Bacterial Foraging Optimization Algorithm), which serves as the basis for developing the proposal of this paper.

2.1. General Overview of the BFOA

The algorithm represents the search of bacteria for nutrients and the distancing of toxins, the interaction between bacteria and their reproduction. While changing position in the search space, which can have several dimensions, each individual of a population finds nutrients and toxins. In the optimization procedure, this combination is the value of the objective function intended to maximize or minimize when the state variables are the coordinates of the position of the bacterium in each iteration.

Four processes are implemented to model the behavior of the bacteria population while performing the search. They are chemotaxis, swarming (the interaction between individuals in a swarm), reproduction, and elimination and dispersion. Figure 1 shows a simplified flowchart of the method. Before performing these steps, a population is created with several individuals positioned randomly by the decision space. By default, the calculations shown below seek to minimize the objective function.

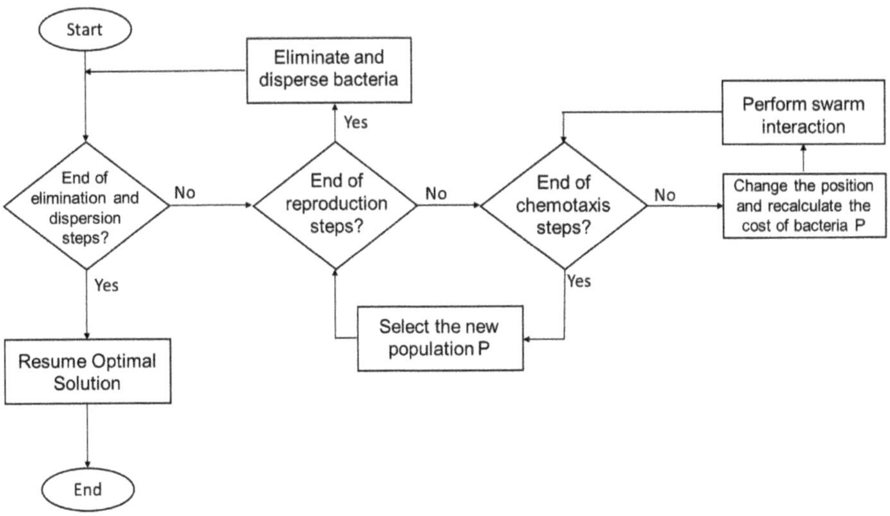

Figure 1. Flowchart of BFOA steps.

2.2. Foraging

Some of the techniques of natural computing are inspired by the strategies of animals in the food search. According to researchers in the field of foraging theory, animals and other heterotroph beings, such as bacteria, always tend to take actions that maximize the energy obtained during the time spent searching for food. This search is made respecting their limitations (for example, cognitive and sensory) and the limitations of the environment (for example, physical characteristics of the region, risk of predators, and the presence of food) [22].

Better strategies in the search and obtaining of food increase the chances of survival and reproduction. These strategies may be related to the distance traveled by the individual by time interval. Some animals use the "cruise" strategy and always look for some prey through the environment. Others use the "ambush" strategy, spending a lot of time waiting for the target to approach. Most animals have strategies classified between these two extremes, called "jumping" strategies, which seek small concentrations of food and remain in the location for a while before "jumping" into another environment [23].

Social behavior among individuals of a species is also significant in obtaining food. Through communication between the individuals, they can increase their chance of finding food, hunting larger prey, and protecting themselves from predators. The association between individuals is widespread in the animal kingdom. The intelligence of these collectives is widely studied and used as models in various applications, such as flocks of birds, schools of fish, ant colonies, and wolf packs, for example.

Based on the concepts of foraging theory, which include some of the concepts of reproduction and natural selection addressed by evolutionary algorithms and the concepts proposed by swarm intelligence algorithms, in [24] is presented an optimization method.

2.3. Bacterial Foraging Optimization Algorithm

The BFOA, proposed by [25], is another swarm optimization algorithm where the behavior of the bacterium *Escherichia coli* (also called *E. coli*) is modeled. This bacterium is found in the human intestine [26]. This bacterium uses flagella to move and is probably the microorganism best understood by science [27].

The algorithm tries to model the search for these bacteria for nutrients or the distancing of toxins and the interaction between bacteria and their reproduction. While changing position in the search space, which can have several dimensions, each individual of a population of bacteria finds a specific combination that merges nutrients and toxins. In the optimization procedure, the combination of nutrients and toxins provides the value of the objective function intended to maximize or minimize when state variables are the coordinates of the position of the bacterium in each iteration.

In the BFOA processes, four main steps are implemented to model the behavior of the bacteria population while performing the search. They are chemotaxis, swarming (the interaction between individuals in a swarm), reproduction, and elimination and dispersion.

2.3.1. Chemotaxis

The strategy of searching for bacteria's food is of the jumping type and is di-straightly related to its mobility. Among the types of bacteria, some are independent of scourge to move. Some bacteria move through flagella, such as *Escherichia coli*. And more, biologists discovered that the movement of bacteria is not random but rather attracted to a direction or repelled by stimulus and called this behavior chemotaxis.

E. coli can move in two ways: advancing (swimming) or changing direction (tumble). It alternates between these two states during their lifetime, randomly changing direction, allowing space scanning in search of nutrients. The central idea of the algorithm is that when the bacteria detect a nutrient gradient, it advances toward the highest concentration and tends to spend more time moving than changing direction.

The change in the position of bacteria is made by equation (1), where Θ^i is the position of the i-th bacterium of the population, $\phi(j)$ is the random direction, $C(i)$ is the size of the forward movement taken in the direction $\phi(j)$, j is the stage of the chemotaxis process, k is the stage of the reproduction process, l is the stage of the process of elimination and dispersion.

$$\Theta^i(j+1,k,l) = \Theta^i(j,k,l) + C(i) \times \varnothing(j)\,, \quad (1)$$

The cost J is calculated, given by the objective function for the bacterium i in the new position, and the swarm interaction with the other bacteria is performed. If the cost of the position $\Theta^i(j + 1, k, l)$ is better than the cost of the position is $\Theta^i(j, k, l)$, the bacterium performs one more advance of size $C(i)$ in that same direction $\phi(j)$. This process defines the scanning of the bacterium in a gradient. It is repeated as long as the cost of the new position is better than the previous one or if the amount of steps taken is less than N_s, adjusted initially.

The chemotaxis process is done for each N_{pop} bacteria in the population and is repeated N_{ch} times.

2.3.2. Swarm

Several types of bacteria exhibit swarm behavior. In the case of *E. coli* that releases a signal in the direction of other bacteria when subjected to a stress condition. According to the situation in which it is subjected, this signal can attract or repel others. In the optimization process, this interaction causes the value of the objective function for a bacterium, given by cost J, to receive a decrease (considering minimization as the default) if it is attracting the others or an increase if it is repelling.

Equation (2) shows the change caused in the cost value, where J_{cc} is the attraction/repelling factor that the bacterium i causes in the population P. A new interaction between individuals is made with each change of position.

$$J(i,j+1,k,l) = J(i,j+1,k,l) + J_{cc}\left(\Theta^i(j+1,k,l,), P(j+1,k,l)\right), \quad (2)$$

This change in the values of the objective function creates a region in the search space around each bacterium. Other bacteria are attracted or reread according to this region. The value J_{cc} is given by Equation (3), where p is the number of dimensions of the decision space, d_{atrac} is the depth and w_{atrac} is the width of the attractiveness region, h_{repel} is the height (h_{repel} is equal to d_{atrac}), and w_{repel} is the width of the region of repulsiveness.

$$\begin{aligned} J_{cc}(\Theta, P(j,k,l)) &= \textstyle\sum_{i=1}^{N_{pop}} J_{CC}^{i}\left(\Theta, \Theta^i(j,k,l,)\right) \\ = \textstyle\sum_{i=1}^{N_{pop}} \left(-d_{atrac} e^{\left(-w_{atrac}\sum_{m=1}^{p}(\Theta_m - \Theta_m^i)^2\right)}\right) &+ \textstyle\sum_{i=1}^{N_{pop}} \left(-h_{repel} e^{\left(-w_{repel}\sum_{m=1}^{p}(\Theta_m - \Theta_m^i)^2\right)}\right) \end{aligned} \quad (3)$$

The health value J_{health} of bacterium i, given by Equation (4), indicates the amount of nutrient it accumulates during the chemotaxis process. This health value is used to select bacteria in the reproduction step.

$$J_{health}^{i} = \sum_{j=1}^{Nch+1} J(i,j,k,l)\,, \quad (4)$$

2.3.3. Reproduction

Healthier bacteria that have accumulated more nutrients are more likely to reproduce. *E. coli* bacteria are prokaryote bacteria and reproduce asexually, copying the chromosome and dividing the cell into two without mutations. On the contrary, bacteria that have absorbed fewer nutrients or failed to prevent toxins are eliminated.

To model this process, after all stages of chemotaxis, the health values of all bacteria, given by the value J_{health}, are increasingly ordered. The set is divided into two and a half, with the best health values surviving, while the better half's replica replaces the other half. The amount of reproduction iterations is defined by the constant N_{rep}, adjusted initially.

2.3.4. Elimination and Dispersion

The environment in which bacteria are searching for food can change gradually or instantly, causing some individuals to disperse to other regions during this change.

In the iterations of this process, each bacterium is eliminated with one probability p_{ed}, and another is randomly positioned in the search space. This process is performed by N_{ed} iterations. A high value of N_{ed} increases the complexity of the solution and allows a better sweep of the search space, decreasing the chance of the solution being found to be a great location. The p_{ed} value should not be great, as it would lead to an exhaustive search.

2.4. MOOP and Bio-Inspired Algorithms

Multi-objective optimization problems (MOOP) involve optimizing two or more objectives simultaneously, which can be conflicting, i.e., improving one of the objectives causes the degradation of another. In addition, these problems may have restrictions of equality and inequality that must be satisfied for a feasible solution.

Each objective function can be maximized or minimized independently in a real application. However, optimization algorithms require that functions are only maximized or minimized. For this, the denial of the function can be made, changing its objective to suit the method. A technique widely used in optimization techniques is the definition of a normalization value to facilitate the comparison of results. Thus, one can easily standardize objective functions such as maximization or minimization. The objectives in a MOOP usually have a conflicted relationship with each other. If the value of an objective function f_1 is improved, the value of another function, f_2 worsens; it can be said that f_1 is conflicting with f_2.

Bio-inspired algorithms have several advantages over classical optimization methods. It is worth highlighting the ability to avoid optimal locations and the possibility of these algorithms working with multiple objectives. Although bio-inspired implementations are initially proposed to optimize a single objective function, the adaptations needed to solve a MOOP are simple and presented in several articles [6,28–30].

Some developments have also proposed BFOA versions with multi-objective optimization. The main change in the original technique is the issue of the reproduction of bacteria. In [19], a method of selecting bacteria that reproduce based on the health and dominance of individuals' Pareto is developed. In [31], the same selection strategy is applied but with a repository of better solutions and an adaptation of the size of the jump of bacteria to each generation. A similar approach is proposed in [32].

Using the concepts of NSGA-II (Non-Dominated Sorting Genetic Algorithm-II) [33], it is proposed the selection of individuals for reproduction through non-dominated sorting, followed by a step of ordering by crowding distance and an ordering force of adapted Pareto. This approach considers each dominated bacterium and dominant solutions [20]. This last method also excludes the interaction of bacteria swarm and the repetition of chemotaxis, making each position change and evaluation a new selection and reproduction.

3. Proposed Methodology

The proposed methodology involves modeling the system to be restored, the isolation of the regions involved in the faults, the calculation of the power flow for a given system configuration, and the search for combinations of feasible and optimal switching that restore most of the system loads.

Figure 2 shows a flowchart of the proposed method. All prototype routines have been implemented in the C language. Initially, input data is collected from the observed electrical network, such as power demanded by consumers and conductor impedances. Then, the routines are executed for the search and isolation of failures and for the power flow calculation, which goes through the model checking the state of the switches. These routines are used as objective functions of the MOBFOA, which generates a random combination of system switch states and stores each bacterium to be evaluated. The following stage of development is described in more detail.

The proposed methodology can be classified as "branch exchange" and is used to reduce load losses after a contingency, like a short-circuit. The technique starts with a switches configuration that maintains the radial system. Then a set of switches is chosen, according to the algorithm proposed for reducing load losses, and is operated. Finally, another group of switches is chosen to be operated to keep the system in the radial configuration.

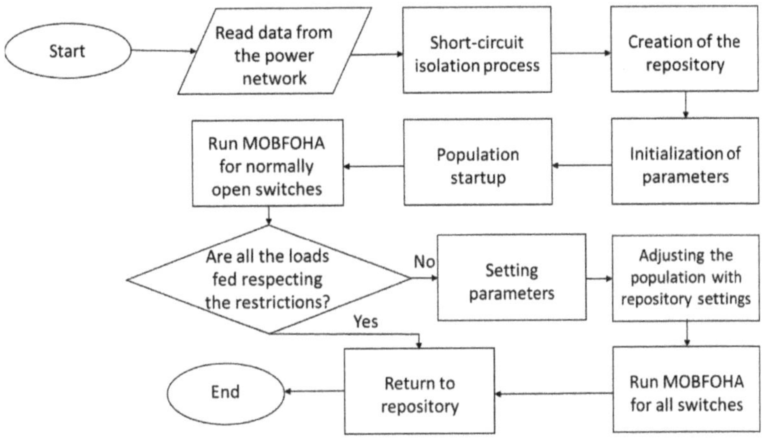

Figure 2. Flowchart of the proposed method.

Initially, the algorithm is applied to the NO (normally open) switches to make the execution faster. If any configuration of these switches is sufficient to meet all loads respecting all restrictions, the program finalizes returning which NO switches should be closed. If disconnected loads or constraints are violated, the algorithm runs again for all system switches.

On the first run, the population is initialized with the value of the initial configuration of the switches. If necessary, the population starts with the best combinations found in the previous run in the second run. These initializations increase the chance of optimal solutions being found.

In the following sections, the electrical network model is presented based on the input data, as power demanded by consumers and impedance of the conductors. After, the routines for searching and isolating the short-circuit and calculating the power flow in the model are detailed. Then, the operational restrictions are checked, and, finally, the proposed version for the MOBFOHA algorithm and its application for the resolution of the problem is discussed in detail.

3.1. Network Representation

The distribution network is modeled as an acyclic, weighted, and non-related graph, where the vertices are the buses, and the feeders are the conductors that connect one bus to another. The topology of the distribution system is always radial. This feature creates a network where the buses are connected only to one substation, and there is no formation of loops. Thus, the formed graph is a set of trees, called a forest, where the root nodes are the substations that provide power to the buses.

Some of the connections between buses may be normally open or closed switches. The open switches enable network recovery in the event of failures. These switches can have their state changed to close according to the necessity of the network operation. The combination of open and closed switches must respect system constraints discussed below.

The graph that models the system is represented by an adjacency matrix $N_{bus} \times N_{bus}$, where N_{bus} is the number of buses. Each element of the matrix stores information about the connection between the bus i and the bus j, being $i, j = 0, 1, \ldots, N_{bus} - 1$. A position vector $N_{switches}$ is also implemented to store the information of each of the switches.

The system information that is relevant to the recovery procedure is arranged in an input file. The data is formatted so that the program can read and organize it in the data structure. The first information is the number of buses (N_{bus}) and the number of switches of the network (N_{switch}). The following information refers to the lower and upper-voltage limits for the buses, $V^{(min)}$ and $V^{(max)}$, respectively. The base power value, the base voltage value, and the threshold for convergence analysis during the power flow study are also reported.

The following information is related to the feeders: (a) connection between two buses i and j, expressed by a pair (i, j); (b) identification *id* of the switch; (c) state of the switch; (d) impedance of the connection (Z_{ij}), expressed in resistance and reactance; (e) power S of the bus j, expressed in active and reactive power; and (f) current limit on the conductor.

3.2. Fault Isolation

Whenever a short-circuit is detected in the system, it must be isolated; that is, the part of the tree that has been affected is disconnected from the system, with its buses and connections. This process is important because, at first, it is not known for sure what caused the short-circuit. Therefore, failures can hinder the process of restoring other system loads.

The routine for isolating faults is based on an in-depth network search, starting from all roots (substations) until the short-circuit location is found. Then all adjacent close switches of the short-circuit open. These switches are flagged not to change in the restore process.

An auxiliary routine has also been implemented to perform some comparative tests. As in [13], a method is proposed that simplifies the insulation process. By deleting the feeders with system failures, considering this feeder as an open connection, this strategy makes the isolation process faster. However, this implies that the short-circuit has necessarily occurred in one of the conductors and that it is not short-circuited.

3.3. Power Flow Computation

The power flow computation is essential to know if all loads are fed and all network and operational restrictions are respected. The proposed solution is only applied in radial networks, then a simple method to perform this calculation is the sum of currents.

The current sum method is implemented as an iterative routine that ends when the difference between the voltages value calculated in iteration n and the value calculated in the $n + 1$ interaction is less than the convergence threshold initially defined.

The algorithm procedure is as follows:

1. The following values must be normalized: impedance (Z_{ij}) of each conductor, power bus consumed (S), the voltage of the power supply (V_{base}).
2. The normalized reference voltage (of the power supply) is initially assumed, and its normalized value is denoted in 1 pu. Each consumer is then assigned the stress value equal to that of the reference.
3. A scan is made from the end nodes determining the values of the system currents according to the Equation (5):

$$I_{ij} = \left(\frac{S_{ij}}{V_{ij}}\right)^* + \sum Ik, \tag{5}$$

4. Starting from the substation to the final bars, the values of the system voltages are determined according to Equation (6):

$$V_j = V_i - Z_{ij} I_{ij}\,, \tag{6}$$

5. After updating the stress values, an absolute error test is performed between the previous stress value and $V_i{}^0$ the current value for a previously $V_i{}^1$ specified ϵthreshold, as described in the Equation (7):

$$\left|V_i^0 - V_i^1\right| < \epsilon\,, \tag{7}$$

6. If the test fails for some stress value, the procedure repeats from step 3 until convergence occurs.

3.4. Objective Functions and Constraints

The main restrictions that must be respected during the verification of the optimal configuration of a power transmission system proposed by this approach are maintaining the network structure in radial topology, keeping the bus's voltage limits, and maintaining the current limits in the connections between the buses.

The optimal solution of the proposed restore system is composed of the following two objective functions:

1. Maximize the number of loads restored: After calculating the power flow, if all constraints have been respected, all powered buses are added, considering the priorities of each bus.
2. Minimize the number of switch operations: A comparison is made between the switch state solution vector with the initial vector to identify the number of changed switches.

3.5. The Multi-Objective Bacteria Culture Optimization Hybrid Algorithm

The version of MOBFOHA proposed in this paper merges the methodologies proposed in [19,20]. The implementation focuses on changing the stages of selection and reproduction of bacteria so that they do not depend on the health of individuals, making it possible to use the method adapted to a multi-objective more easily. In addition, how bacteria walk through the research space has been changed to a more adaptive version and focused on interaction with the rest of the population, as discussed in more detail in this section. Figure 3 shows the adjustments made in the method concerning the traditional algorithm.

As previously presented, the size of the jump defines the movement of bacteria while scanning the search space. Adjusting this jump size can significantly impact method convergence. A fixed and significant jump size value causes convergence to be achieved faster but can cause the individuals to jump "over" the optimal points, destabilizing the search. Otherwise, a fixed and small value improves the stability of the search but has a slower convergence and can cause the method to end before finding an optimal solution or can cause the solution to come to a great location. Adjusting through generations makes it possible to take advantage of the best benefits during scanning, maintaining stability, improving convergence speed, and avoiding optimal locations.

The proposed MOBFOHA belongs to the class of stochastic algorithms involving random processes. Then each time the algorithm is run, it can give a different answer. Therefore, a suitable adjustment of the initial parameters is essential to obtain reasonable solutions. Another problem is the efficiency of the proposed algorithm. This aspect can be divided into two topics: the time for convergence and the time to provide a restoration solution for a failure in the distribution network. For the first topic, the time for convergence of the algorithm in a large distribution network is not a problem because it does not contain recursive computations. Its order of complexity is O(n). Providing a solution in real-time is not a problem in the second topic. Even though the algorithm is being executed many times, the total time is inferior to the time of the operator's analysis.

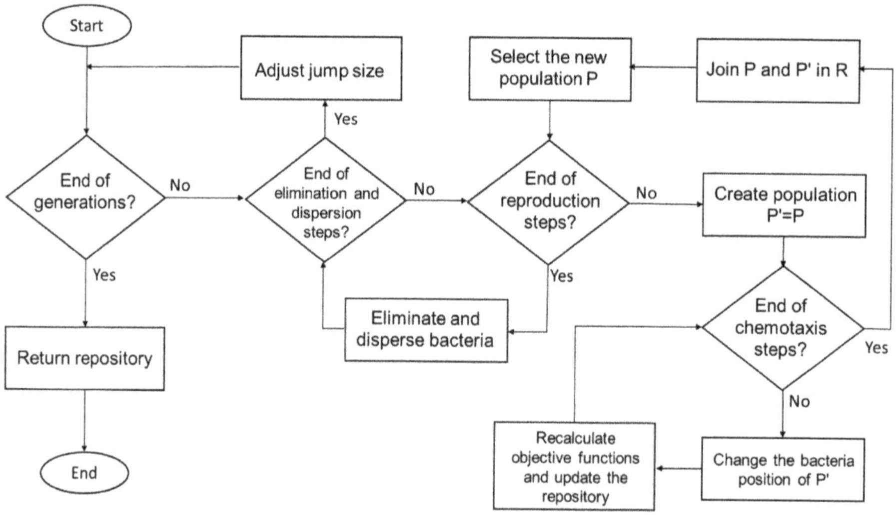

Figure 3. Flowchart of the MOBFOHA.

The same problem discussed in implementing the Multiple-Objective Particle Swarm Optimization (MOPSO) version proposed in [32] is considered in this proposed approach. Bacteria should not be attracted to a single point but the Pareto-optimal region. Thus, swarm operations must be adapted so that attraction and repelling interactions respect this characteristic. The version proposed in this paper is simplified and integrated into the chemotaxis stage, as presented in [34]. Each bacterium chooses the direction to move attracted by another dominant bacterium.

Then, given a MOOP with objectives, the algorithm steps and the routines mentioned above are presented in detail below:

1. Create N population with S-size and randomly spread individuals through the search space.
2. Evaluate each individual for each one of the M objectives.
3. Individuals who present solutions not dominated by other individuals are inserted in file P, that is, solutions that are not worse for each of the objectives found and better in at least one goal. These solutions are considered the approximate Pareto border [35].
4. The solutions in file A are mapped in the parallel cell coordinate system (PCCS) [32]. That is, each solution receives a label that is an integer between 1 and $|A|$ (size of A) according to the Equation (8), where n is the solution indicator $(1, 2, \ldots, |A|)$, m is the objective indicator, $f_{k,m}$ is the value of solution n, for objective m, f_m^{max} is the highest value found for objective m and f_m^{min} is the lowest value found for objective m. The value is rounded up, and when $f_{k,m} = f_m^{min}$, the value is changed to 1 to avoid division by zero.

$$L_{k,m} = K \frac{f_{k,m} - f_m^{min}}{f_m^{max} - f_m^{min}} \quad (8)$$

5. If the file size exceeds the maximum size S, the densities of all solutions must be calculated, and the highest density is eliminated. The density calculation of a solution is done based on the distances between pairs of solutions, as shown in the following equations:

$$DCP(P_i, P_j) = \begin{cases} \sum_{m=1}^{M} \left|L_{i,m} - L_{j,m}\right|, & if\ \exists m \mid L_{i,m} \neq L_{j,m} \\ 0.5, & otherwise \end{cases}, \quad (9)$$

$$D(P_i) = \sum_{j=1}^{K} \frac{1}{DCP(P_i, P_j)^2}\ . \tag{10}$$

6. For each generation, with the values of the labels, the entropy of the population is calculated, which measures the uniformity of the approximation of the Pareto border. Equation (11) shows the entropy calculation for generation t, where K is $|A|$, M is the number of objectives and $Cell_{k,m}(t)$ is the number of solutions with the label $L_{k,m}$ in the k-th row and m-th column of the PCCS.

$$E(t) = -\sum_{k=1}^{K}\sum_{m=1}^{M} \frac{Cell_{k,m}(t)}{KM} \log \frac{Cell_{k,m}(t)}{KM} \tag{11}$$

7. The adjustment in the jump size of bacteria is made with the differential $\Delta E(t) = E(t) - E(t-1)$ according to the evolutionary state of the population, which can be a state of convergence, state of diversity, and state of stagnation. The threshold for the convergence state is given by $\delta_C = \frac{2}{K} log2$ and the threshold for the stagnant state is given by $\delta_E = \frac{2}{KS} log2$ where S is the maximum file size.
8. Given the thresholds, the adjustment of the jump size C is given by the following equation, where λ and μ are the adjustments and $K(t)$ is the number of solutions in A the generation t.

$$\begin{array}{c} \text{Convergence}: |\Delta E(t)| > \delta_C \text{ or } |K(t) - K(t-1)| > 0 \\ \text{Diversity}: \delta_C < |\Delta E(t)| < \delta_E \text{ and } K(t) = K(t-1) = S \\ \text{Stagnation}: |\Delta E(t)| > \delta_E \text{ and } K(t) = K(t-1) \end{array} \tag{12}$$

$$C_t(i) = \begin{cases} C_{t-1}(i) - \lambda(1 + |\Delta E(t)|), & if\ convergence \\ C_{t-1}(i) + \mu\,|\Delta E(t)|, & if\ diversity \\ C_{t-1}(i), & if\ stagnation \end{cases} \tag{13}$$

9. For each one of the N_{ed} steps of elimination and dispersion is made the reproduction process.
10. For each of the N_{rep} steps of reproduction is carried out the process of chemotaxis. A new population is generated P′ copying the values of the population P, and for each step of chemotaxis N_{ch}, the population P′ is updated recalculating the position of each individual according to the equation (14), where is $\theta^i(ch, r,\ ed)$ is the position of the i-th bacterium in the ch-th step of chemotaxis, r- th stage of reproduction and ed-th elimination and dispersion step, $\Delta(i)$ is a random vector that represents the direction in which the bacterium jump and $C_t(i)$ is the size of the jump.

$$\theta^i(ch+1, r,\ ed) = \theta^i(ch, r, ed) + C_t(i)\ \frac{\Delta(i)}{\sqrt{\Delta^T(i)\Delta(i)}} \tag{14}$$

11. All bacteria are evaluated for all purposes, and file P is updated with the previous procedure.
12. Once the chemotaxis process is finished, a new population is generated through non-dominance ordering [20]. The N population is merged with the N' population in a set R. All solutions from the Pareto border are selected. When the number of solutions found is less than S, they are added in the new population N and removed from R. A new Pareto border, that is, the rank 2 border, is selected from R. If the number of members of that border is less than S − $|N|$, they are also added in N and removed from R, and so on.

If any of the borders cannot be added in N, that is, it is larger than the space available for N to be size S, it is necessary to perform crown distance ordering of the elements of the border [36]. The crowd distance is the average distance between its two neighboring solutions. Sorting elements based on each of the objectives is required. Infinite distance is allocated to solutions with value f_m^{min} and f_m^{max} and the other elements of the border in question have the distance calculated by $F[k]_{distance} = F[k]_{distance} + (f_{k+1,m} - f_{k-1,m})/(f_m^{max} - f_m^{min})$. Solutions with the highest values are added in N until $|N|$ = S to ensure diversity on the Pareto border. Figure 4 shows how the construction of the new population is made.

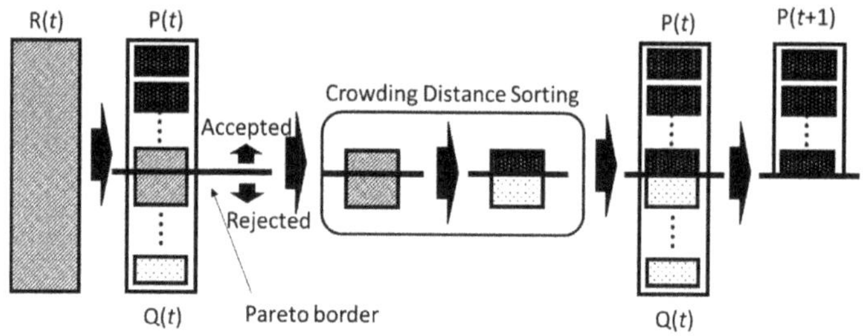

Figure 4. Construction of the new population.

13. Once the reproduction steps are completed, individuals are eliminated and dispersed in the search space with a probability P_{ed}.
14. After the elimination and dispersion steps, a new adjustment in the jump size is made and finished the generations. File P contains the best solutions found.

4. Results and Comparisons in Distribution Networks

The proposed methodology's effectiveness for restoring an electrical distribution network is evaluated by applying the method in two systems with 70 and 201 buses. Some contingency cases with one and more short-circuit are tested for each system.

4.1. Applying the Proposed Methodology

The method is initially applied to the NO (Normally Open) switches. If any configuration of these switches is sufficient to meet all loads respecting all constraints, the program terminates by returning which or which NO switches should be closed.

After the first run, it is checked if any optimal solutions violate any conductor's current limit. If this happens, there may be an NC (Normally Closed) switch that eliminates the overhead that occurred. In addition, there may be another NO switch that can be closed to meet the rest of the loads after opening this NC switch. In this way, a second execution of the algorithm is done considering all system switches.

Even if overloads have not occurred during the first run, it is also checked whether any solution reestablishes all the buses outside the isolation area. If this does not happen, the new execution is also performed for the search to continue. It is noteworthy that there may not be a solution that restores 100% of the required load, and the second execution is only performed to ensure that this possible optimal solution was not found in the first.

For each execution, individuals are initialized to optimize the search for the best settings. On the first run, the population is initialized with the value of the initial configuration of the switches after the isolation of the faults. If necessary, the population starts with the best combinations found in the previous run in the second run.

4.2. Parameter Adjustment

The parameters used for the tests were chosen based on studies present in the literature and by observing the convergence of the method. As discussed above, the algorithm runs twice. Each run has an initial configuration of different parameters because they involve other variables. Table 1 shows the value chosen for each parameter in the two runs.

Table 1. MOBFOHA Parameters.

Parameters	Execution 1	Execution 2
N_{ed}	2	2
p_{ed}	0.25	0.25
N_{rep}	4	4
N_{ch}	60	120
C	50	50
Λ	2	2
M	10	10
$maxgen$	5	10

4.3. Test System I: 70-Buses Network

The first network used for the tests was presented in [37], and it has been used in many published studies. This network comprises 70 buses, 2 substations, 9 NO switches, and 11 NC switches. The base voltage values are set to 11, and the base power value is 1. The voltage values on the buses must be in the range of 0.8 to 1.05. Current limits are given for each conductor. Figure 5 shows the initial network configuration.

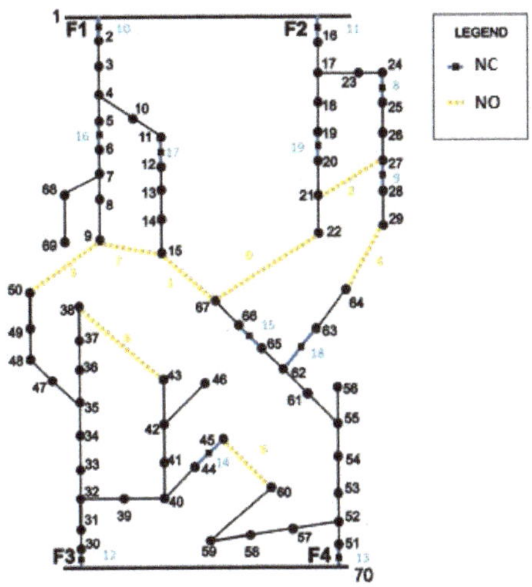

Figure 5. Test system I: 70-buses network [37].

4.3.1. Single Short-Circuit

This section presents two single short-circuits in the feeders 25–26 and 3–4.

Then, for the first experiment, a short-circuit is considered in the feeder between buses 25 and 26. For failure isolation, switches (NC) 8 and 9 are opened and marked, so they are not used in the restoration step and the NO switch 2.

The two Pareto-optimal solutions, present in Table 2, are found by performing a brute force analysis of all possible switching combinations. As shown in the table, the first solution meets all the demand by performing maneuvers on 5% of the switches, closing the NO switch 4 to meet buses 28 and 29. The second solution does not operate in the available switches and no longer meets about 5% of the demand power. Loads of buses 28 and 29 are disconnected from the system after isolation. It is noteworthy that the demand of the fault region is subtracted from the total demand during isolation.

Table 2. Pareto-optimal solutions for the 70-buses system with short-circuit in feeder 25–26.

Faulty Feeders	NO Switches	NC Switches	P(%)
25–26	29–64	None	100%
	None	None	94.83%

The solutions found by the algorithm studied are shown in Figure 6 using the Pareto curve. The method found all optimal solutions in 35 s, using 2400 iterations.

Figure 6. Pareto border for 70 buses system with short-circuit in feeder 25–26.

In the second experiment, a short-circuit is considered in the feeder between buses 3 and 4. NC switches 10, 16, and 17 are used in isolation and are therefore marked so that they are not typed in the optimization step.

Again, the optimal solutions in Table 3 are found when performing a brute force analysis. In the first solution, all demand is applied by closing the NO switches 67–15 and 9–50, 10% of the available switches. Only the NO switch 67–15 is closed in the second solution. 5% of the switches and only 5% of the load are not applied. The final solution is the configuration after isolation. About 89.34% of the demand is supplied without any switch changes. Approximately 10.66% of the required load is disconnected from the system. It is important to remember that all optimal settings respect all operating restrictions of the system.

Table 3. Pareto-optimal solutions for the 70-buses system with short-circuit in feeder 3–4.

Faulty Feeders	NO Switches	NC Switches	P(%)
3–4	67–15 9–50	None	100%
	67–15	None	95.00%
	None	None	89.34%

The proposed algorithm also finds all the optimal solutions for this case, taking about 75 s and 2400 iterations. The solutions are shown at the Pareto border in Figure 7.

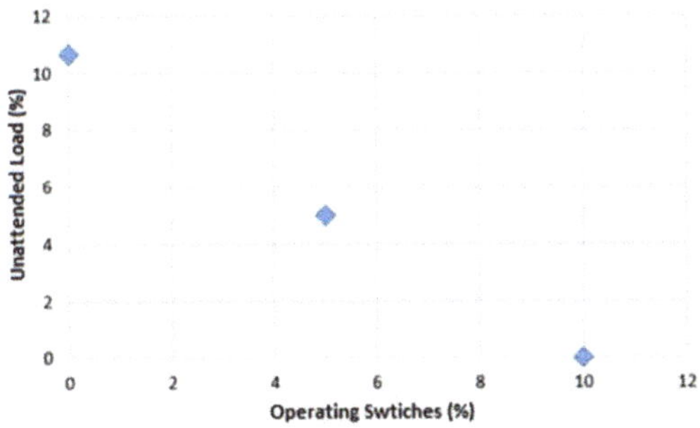

Figure 7. Pareto border for 70 buses system with short-circuit in feeder 3–4.

4.3.2. Double Short-Circuit

This section shows an example where two short-circuit happen at the same time. In this case, the short-circuits occur in the same two feeders of the previous example: between buses 3 and 4 and between buses 25 and 26. The fault isolation process locked the same NC switches: 10, 16, and 17 due to fault in 3–4; and 8 and 9, and NO switch 2, due to 25–26.

After brute force analysis, the solutions presented in Table 4 are found. In this case, performing only the closing of NO switches could not restore all the buses without violating the current limit on some conductors. Therefore, in two solutions, the NC switch 14 must be opened, and in two others, the NC switch 15 is changed.

Table 4. Pareto-optimal solutions for the 70-buses system with short-circuit in feeders 3–4 and 25–26.

Faulty Feeders	NO Switches	NC Switches	P(%)
3–4 25–26	9–50 29–64 45–60 9–15	44–45	100%
	22–67 67–15 9–50 29–64	65–66	100%
	22–67 67–15 29–64 9–15	65–66	100%
	9–15 29–64 9–50	44–45	97.56%
	67–15 9–50	None	94.22%
	67–15	None	88.86%
	None	None	89.34%

In Figure 8, the Pareto curve of the solutions found by the proposed algorithm is displayed with the best possible solutions. The solutions are found in 633 s with 12,000 iterations. Figure 9 shows the second of the 5 solutions found. This solution restores all the required load.

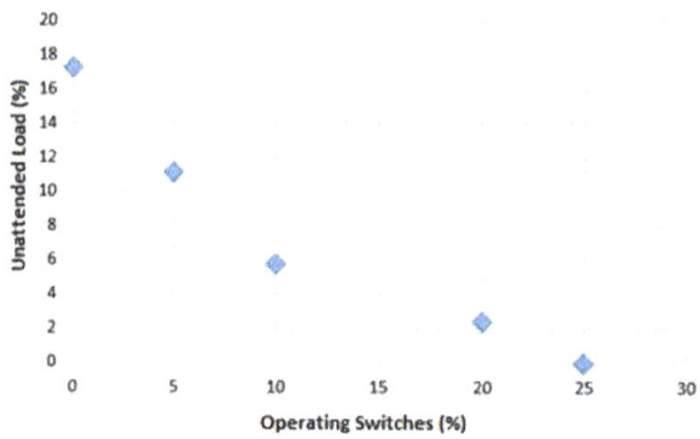

Figure 8. Pareto border for 70 buses system with short-circuit in feeders 3–4 and 25–26.

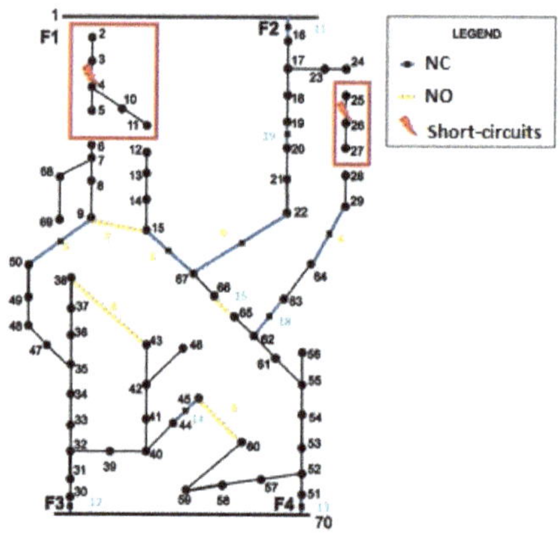

Figure 9. Optimal configuration after the restoration of the system of 70 bus with faults in 3–4 and 25–26.

4.4. Test System II: 210-Buses Network

The second system used for the experiments is adapted from [13]. It has 201 buses, one substation with three feeders, 39 NO switches, and 37 NC switches. The base voltage value is 10 kV, and the base power value is 100 kVA. This system does not present the impedance values for the conductors, so the voltage drop calculation is not made, and, consequently, the voltage value for each bus is set at 1 pu. The maximum power of 7983 kW gives the flow limit on the conductors. Thus, the current limit of 79.83 A for each conductor is considered to adapt to the method. Figure 10 shows the initial system configuration.

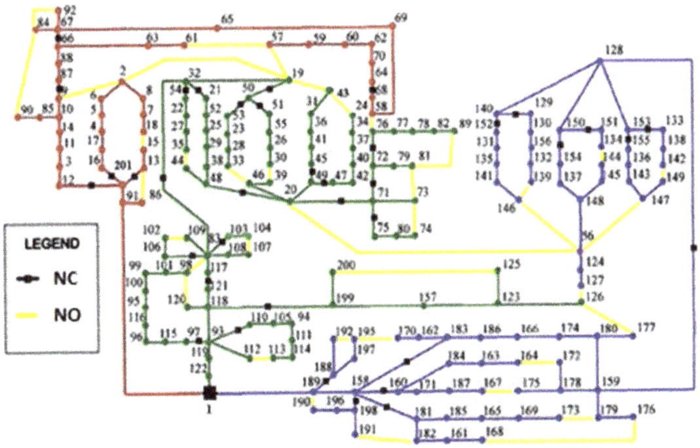

Figure 10. Test system II: 210-buses network [13].

4.4.1. Single Short-Circuit

This section presents an example considering a short circuit between buss 72 and 79. After isolation, the feeders between buses 76 and 89 are unsupplied.

After the execution of the method, the solutions in Table 5 are found, shown at the Pareto border in Figure 11. The proposed method used 2400 iterations and took 90 s to deliver all responses.

Table 5. Pareto-optimal solutions for the 210-buses system with short-circuit in feeder 72–79.

Faulty Feeders	NO Switches	NC Switches	P(%)
72–79	58–64	None	100%
	None	None	97.77%

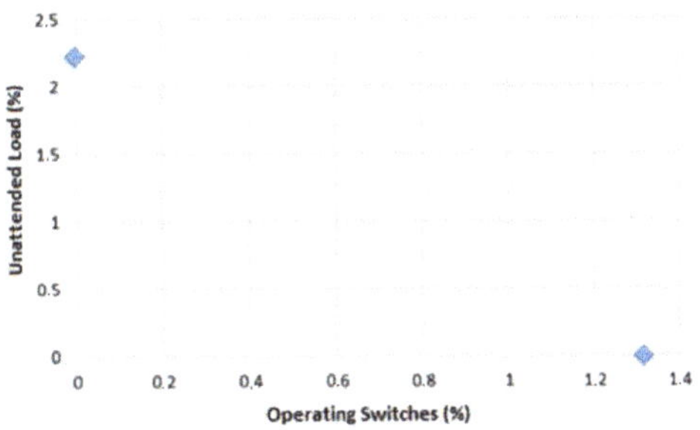

Figure 11. Pareto border for 210 buses system with short-circuit in feeder 72–79.

4.4.2. Double Short-Circuit

A more complex experiment is now performed, where short-circuit happens simultaneously between buses 9 and 10 and between buses 20 and 49. Joint failures disconnect about 30% of network load demand. Figure 12 shows the network configuration after the two faults are isolated.

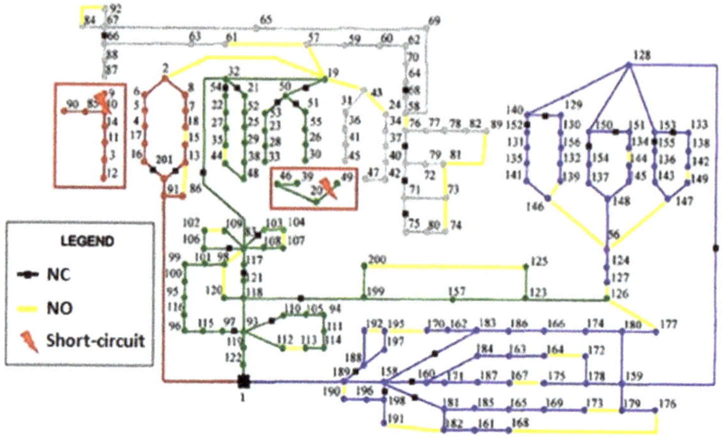

Figure 12. Test system II: 210-buses network after the isolation process with short-circuits in the feeders 9–10 and 20–49.

The first execution of the proposed algorithm failed to restore all the load, although the best solutions found do not violate any operational limit. Therefore, a new execution of the algorithm is made, and, in this case, a solution that meets all the demand can be found. Table 6 shows all the solutions found, and Pareto's boundary for these solutions is shown in Figure 13.

Table 6. Pareto-optimal solutions for the 210-buses system with short-circuits in feeders 9–10 and 20–49.

Faulty Feeders	NO Switches	NC Switches	P(%)
9–10 20–49	2–19 19–43 19–57 24–43 58–76	117–121	100%
	19–43 19–57 24–43 58–76	83–106	99.02%
	19–43 19–57 58–76	None	95.83%
	19–57 58–76	None	93.14%
	19–57	None	85.48%
	None	None	75.91%

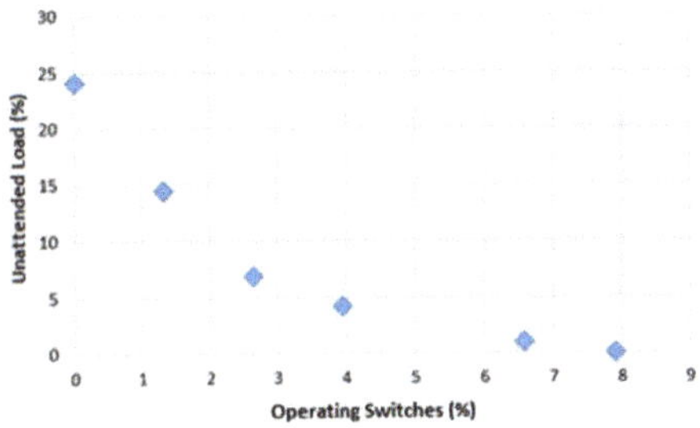

Figure 13. Pareto border for 210 buses system with short-circuits in feeders 9–10 and 20–49.

The first configuration, which restores all load demand for this case, is shown in Figure 14.

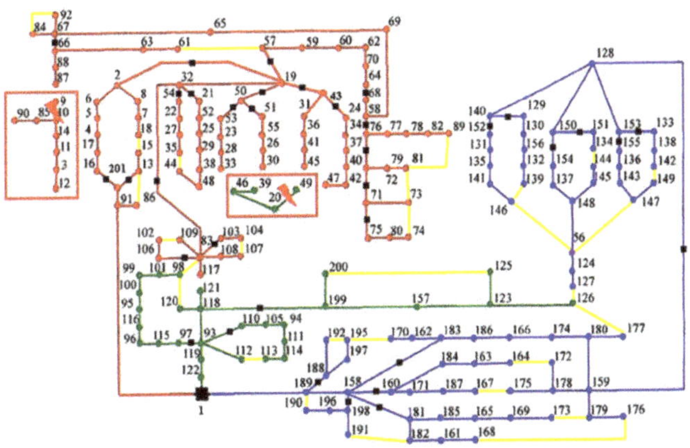

Figure 14. Test system II: 210-buses network after the restoration process with short-circuits in the feeders 9–10 and 20–49.

4.4.3. Comparison between Methods

The quality of the found solutions by the proposed approach to the 201-buses system is made by comparing it with the results obtained using the PSO technique proposed in [13] and the GA technique proposed in [38]. Both techniques use only one objective function and consider the other objectives through weighting. However, in different runs, the method finds equally optimal different solutions. Thus, it is possible to compare with a multi-objective method proposed in this work.

First, a set of short-circuits is compared in which the PSO and the GA presented a single solution. Table 7 shows each of these cases and the results found by the two methods in parallel. Observing the solutions on this table, MOBFOHA presents more alternatives to the operators, and new objectives can be applied to find the best solution among the better solutions.

Table 7. Comparison of MOBFOHA, PSO, and GA methods results - Single solution.

Faulty Feeders	MOBFOHA NO	MOBFOHA NC	MOBFOHA P(%)	PSO NO	PSO NC	PSO P(%)	GA NO	GA NC	GA P(%)
1–91	2–19 20–56	20–48	100%	2–19 20–56	20–48	100%	2–9 19–57 76–58	70–62 67–65	82.62%
	19–57 20–56	20–48	100%	-	-	-	-	-	-
	2–19	12–201	86.75%	-	-	-	-	-	-
	None	None	82.62%	-	-	-	-	-	-
76–77	81–89	None	100%	81–89	None	100%	81–89	None	100%
	None	None	98.16%	-	-	-	-	-	-
93–110	112–113	None	100%	112–113	None	100%	112–113	None	100%
	None	None	95.68	-	-	-	-	-	-
128–140	56–146	None	100%	56–146	None	100%	56–146	None	100%
	None	None	93.62	-	-	-	-	-	-

Also, the results in which the PSO and the GA have more than one solution been compared. Table 8 shows the solutions found. It is possible observing that MOBFOHA finds a more significant number of solutions. Some objective functions considered in the mono-objective method have greater weight than others. Another considerable remark is the solutions found by MOBFOHA are the result of a single random execution of the method. In the case of the PSO and the GA, for each solution found, one execution is performed.

Table 8. Comparison of MOBFOHA, PSO, and GA methods results - Multiple solutions.

Faulty Feeders	MOBFOHA NO	MOBFOHA NC	MOBFOHA P(%)	PSO NO	PSO NC	PSO P(%)	GA NO	GA NC	GA P(%)
1–91	58–76 126–127	117–121	100%	-	-	-	-	-	-
	58–76 126–177	117–121	100%	58–76 126–177	117–121	100%	-	-	-
	19–57 20–56	20–48	100%	19–57 20–56	20–48	100%	-	-	-
	19–57 126–127	117–121	100%	19–57 126–127	117–121	100%	-	-	-
	19–57 126–177	117–121	100%	19–57 126–177	117–121	100%	-	-	-
	2–19 20–56	20–48	100%	-	-	-	-	-	-
	2–19 126–177	117–121	100%	2–19 126–177	117–121	100%	-	-	-
	19–57	117–121	87.68%	-	-	-	-	-	-
	58–76	117–121	87.68%	-	-	-	2–9 19–57	58–68	82.62
	2–19	117–121	87.68%	-	-	-	2–9 19–57 76–58	69–58 19–57	82.62
	None	None	54.51%	-	-	-	-	-	-

Table 8. *Cont.*

Faulty Feeders	MOBFOHA NO	MOBFOHA NC	MOBFOHA P(%)	PSO NO	PSO NC	PSO P(%)	GA NO	GA NC	GA P(%)
1–189	2–19 126–127	117–121	100%	-	-	-	-	-	-
	58–76 126–127	117–121	100%	58–76 126–127	117–121	100%	126–127	-	62.87%
	19–57 126–177	117–121	100%	19–57 126–177	117–121	100%	126–127 126–177	124–127	62.87%
	2–19 126–177	117–121	100%	2–19 126–177	117–121	100%	-	-	-
	58–76 126–177	117–121	100%	58–76 126–177	117–121	100%	58–76 126–177	117–121	100%
	None	None	62.87%	-	-	-	-	-	-
19–32	33–46	None	100%	33–46	None	100%	33–46	None	100%
	2–19	None	100%	2–19	None	100%	2–19	None	100%
	19–43	None	100%	19–43	None	100%	19–43	None	100%
	30–39	None	100%	30–39	None	100%	30–39	None	100%
	None	None	93.91	-	-	-	-	-	-

Observing the solutions shown in the previous tables, notice that MOBFOHA finds more solutions. It occurs because some objective functions considered in the mono-objective method have greater weight than others. It is worth noting that the solutions found by MOBFOHA are the result of a single random execution of the method. In the cases of PSO and GA, for each solution found, an execution was performed.

Unlike the PSO and the GA approach, MOBFOHA also considers the solutions that do not restore all the load demanded due to Pareto dominance rules. In this way, the concessionaire can weigh according to the need, choosing the best solution for each specific situation. Previously, weighing the functions makes this type of analysis impossible and should ensure that the best weights are chosen.

5. Conclusions

Through the proposed model, the performance of a version of the bio-inspired technique in bacteria's foraging strategy can be evaluated. The adaptations made to the multi-objective version proved effective in searching for Pareto-optimal solutions. Through the comparations with the results obtained with the brute force routine in the case of the 70-buses system and with the results obtained by the PSO method for the 201-buses system, it can be noted that the method can deliver all the optimal solutions expected for the cases tested, considering the two proposed objectives.

Multi-objective optimization is very important for the operation of systems. With this approach, solutions that maximize the number of loads served are placed in parallel with others that minimize the number of switchings, even if not serving many loads. Thus, the distribution companies can make the best decision, considering the impact caused for each of the decisions, concerning the difficulty and cost generated to implement the proposed configuration and the possible fine for lack of supply, among other factors not addressed by the solution. In addition, when all other objectives can be considered, such as those suggested as future work, a much more assertive decision can be made, especially in the context of smart grids.

Author Contributions: C.H.V.d.M. and J.L.d.V.B. conceived and designed the experiments; C.H.V.d.M., J.L.d.V.B. and G.L.-T. performed the experiments; C.H.V.d.M., J.L.d.V.B. and G.C.C.d.A. analyzed the case studies; C.H.V.d.M., J.L.d.V.B. and G.L.-T. developed the methodology; and C.H.V.d.M., J.L.d.V.B., G.C.C.d.A. and C.I.d.A.C. analyzed the results of the proposed approach and C.H.V.d.M., J.L.d.V.B., G.L.-T. and C.I.d.A.C. wrote the paper. All authors have read and agreed to the published version of the manuscript.

Funding: This research received no external funding.

Institutional Review Board Statement: Not applicable.

Informed Consent Statement: Not applicable.

Data Availability Statement: Not applicable.

Acknowledgments: The authors would like to thank the National Council for Scientific and Technological Development (CNPq), Coordination for the Improvement of Higher Education Personnel (CAPES), and Brazilian Electricity Regulatory Agency Research and Development (ANEEL R&D) for supporting this project.

Conflicts of Interest: The authors declare no conflict of interest.

References

1. Del Real, A.J.; Pastor, A.; Durán, J. Generic Framework for the Optimal Implementation of Flexibility Mechanisms in Large-Scale Smart Grids. *Energies* **2021**, *14*, 8063. [CrossRef]
2. Akil, M.; Dokur, E.; Bayindir, R. The SOC Based Dynamic Charging Coordination of EVs in the PV-Penetrated Distribution Network Using Real-World Data. *Energies* **2021**, *14*, 8508. [CrossRef]
3. Gu, C.; Zhang, Y.; Wang, J.; Li, Q. Joint planning of electrical storage and gas storage in power-gas distribution network considering high-penetration electric vehicle and gas vehicle. *Appl. Energy* **2021**, *301*, 117447. [CrossRef]
4. Golshani, A.; Sun, W.; Zhou, Q.; Zheng, Q.P.; Tong, J. Two-Stage Adaptive Restoration Decision Support System for a Self-Healing Power Grid. *IEEE Trans. Ind. Inform.* **2017**, *13*, 2802–2812. [CrossRef]
5. Siebert, L.C.; Sbicca, A.; Aoki, A.R.; Lambert-Torres, G. A Behavioral Economics Approach to Residential Electricity Consumption. *Energies* **2017**, *10*, 768. [CrossRef]
6. Esmin, A.A.A.; Lambert-Torres, G.; de Souza, A.C.Z. A hybrid particle swarm optimization applied to loss power minimization. *IEEE Trans. Power Syst.* **2005**, *20*, 859–866. [CrossRef]
7. Torres, B.S.; Ferreira, L.R.; Aoki, A.R. Distributed Intelligent System for Self-Healing in Smart Grids. *IEEE Trans. Power Deliv.* **2018**, *33*, 2394–2403. [CrossRef]
8. Poudel, S.; Dubey, A. Critical Load Restoration Using Distributed Energy Resources for Resilient Power Distribution System. *IEEE Trans. Power Syst.* **2019**, *34*, 52–63. [CrossRef]
9. Li, Y.; Xiao, J.; Chen, C.; Tan, Y.; Cao, Y. Service Restoration Model With Mixed-Integer Second-Order Cone Programming for Distribution Network With Distributed Generations. *IEEE Trans. Smart Grid* **2019**, *10*, 4138–4150. [CrossRef]
10. Gush, T.; Bukhari, S.B.A.; Haider, R.; Admasie, S.; Oh, Y.U.; Cho, G.Y.; Kim, C.H. Fault detection and location in a microgrid using mathematical morphology and recursive least square methods. *Int. J. Electr. Power Energy Syst.* **2018**, *102*, 324–331. [CrossRef]
11. Eberhart, R.; Kennedy, J. A new optimizer using particle swarm theory. In Proceedings of the Sixth International Symposium on Micro Machine and Human Science, Nagoya, Japan, 4–6 October 1995; pp. 39–43. [CrossRef]
12. Holland, J.H. Genetic Algorithms and the Optimal Allocation of Trials. *SIAM J. Comput.* **1973**, *2*, 88–105. [CrossRef]
13. Lambert-Torres, G.; Martins, H.G.; Coutinho, M.P.; Salomon, C.P.; Vieira, F.C. Particle Swarm Optimization applied to system restoration. In Proceedings of the 2009 IEEE Bucharest PowerTech, Bucharest, Romania, 28 June–2 July 2009; pp. 1–6. [CrossRef]
14. Lin, W.-C.; Huang, W.-T.; Yao, K.-C.; Chen, H.-T.; Ma, C.-C. Fault Location and Restoration of Microgrids via Particle Swarm Optimization. *Appl. Sci.* **2021**, *11*, 7036. [CrossRef]
15. Zou, K.; Mohy-ud-din, G.; Agalgaonkar, A.P.; Muttaqi, K.M.; Perera, S. Distribution System Restoration with Renewable Resources for Reliability Improvement Under System Uncertainties. *IEEE Trans. Ind. Electron.* **2020**, *67*, 8438–8449. [CrossRef]
16. Mendoza, J.; Lopez, R.; Morales, D.; Lopez, E.; Dessante, P.; Moraga, R. Minimal loss reconfiguration using genetic algorithms with restricted population and addressed operators: Real application. *IEEE Trans. Power Deliv.* **2006**, *21*, 948–954. [CrossRef]
17. Tomoiagă, B.; Chindriş, M.; Sumper, A.; Sudria-Andreu, A.; Villafafila-Robles, R. Pareto Optimal Reconfiguration of Power Distribution Systems Using a Genetic Algorithm Based on NSGA-II. *Energies* **2013**, *6*, 1439–1455. [CrossRef]
18. Čađenović, R.; Jakus, D.; Sarajčev, P.; Vasilj, J. Optimal Distribution Network Reconfiguration through Integration of Cycle-Break and Genetic Algorithms. *Energies* **2018**, *11*, 1278. [CrossRef]
19. Niu, B.; Wang, H.; Wang, J.; Tan, L. Multi-objective bacterial foraging optimization. *Neurocomputing* **2013**, *116*, 336–345. [CrossRef]
20. Kaur, M.; Kadam, S. A novel multi-objective bacteria foraging optimization algorithm (MOBFOA) for multi-objective scheduling. *Appl. Soft Comput.* **2018**, *66*, 183–195. [CrossRef]
21. Dorigo, M.; Maniezzo, V.; Colorni, A. Ant system: Optimization by a colony of cooperating agents. *IEEE Trans. Syst. Man Cybern. Part B* **1996**, *26*, 29–41. [CrossRef]

22. Selva Rani, B.; Aswani Kumar, C. A Comprehensive Review on Bacteria Foraging Optimization Technique. In *Multi-Objective Swarm Intelligence. Studies in Computational Intelligence*; Dehuri, S., Jagadev, A., Panda, M., Eds.; Springer: Berlin/Heidelberg, Germany, 2015; Volume 592. [CrossRef]
23. Panda, A.; Pani, S. A Symbiotic Organisms Search algorithm with adaptive penalty function to solve multi-objective constrained optimization problems. *Appl. Soft Comput.* **2016**, *46*, 344–360. [CrossRef]
24. Ustun, D.; Çarbas, S.; Toktas, A. A symbiotic organisms search algorithm-based design optimization of constrained multi-objective engineering design problems. *Eng. Comput.* **2020**, *38*, 632–658. [CrossRef]
25. Passino, K.M. Biomimicry of bacterial foraging for distributed optimization and control. *IEEE Control. Syst. Mag.* **2002**, *22*, 52–67. [CrossRef]
26. Berg, H.; Brown, D. Chemotaxis in Escherichia coli analysed by Three-dimensional Tracking. *Nature* **1972**, *239*, 500–504. [CrossRef] [PubMed]
27. Zhang, Z.; Long, K.; Wang, J.; Dressler, F. On Swarm Intelligence Inspired Self-Organized Networking: Its Bionic Mechanisms, Designing Principles and Optimization Approaches. *IEEE Commun. Surv. Tutor.* **2014**, *16*, 513–537. [CrossRef]
28. Li, Z.; Tian, M.; Zhao, Y.; Zhang, Z.; Ying, Y. Development of an Integrated Performance Design Platform for Residential Buildings Based on Climate Adaptability. *Energies* **2021**, *14*, 8223. [CrossRef]
29. Nasrullah, A.I.H.; Puji Santosa, S.; Widagdo, D.; Arifurrahman, F. Structural Lattice Topology and Material Optimization for Battery Protection in Electric Vehicles Subjected to Ground Impact Using Artificial Neural Networks and Genetic Algorithms. *Materials* **2021**, *14*, 7618. [CrossRef]
30. Qi, J.; Wu, D. Green Energy Management of the Energy Internet Based on Service Composition Quality. *IEEE Access* **2018**, *6*, 15723–15732. [CrossRef]
31. Yi, J.; Huang, D.; Fu, S.; He, H.; Li, T. Multi-Objective Bacterial Foraging Optimization Algorithm Based on Parallel Cell Entropy for Aluminum Electrolysis Production Process. *IEEE Trans. Ind. Electron.* **2016**, *63*, 2488–2500. [CrossRef]
32. Hu, W.; Yen, G.G. Adaptive Multiobjective Particle Swarm Optimization Based on Parallel Cell Coordinate System. *IEEE Trans. Evol. Comput.* **2015**, *19*, 1–18. [CrossRef]
33. Mansour, M.R.; Santos, A.C.; London, J.B.; Delbem, A.C.B.; Bretas, N.G. Energy restoration in distribution systems using multi-objective evolutionary algorithm and an efficient data structure. In Proceedings of the 2009 IEEE Bucharest PowerTech, Bucharest, Romania, 28 June–2 July 2009; pp. 1–7. [CrossRef]
34. Pandi, V.R.; Panigrahi, B.K.; Hong, W.C.; Sharma, R. A multi-objective bacterial foraging algorithm to solve the environmental economic dispatch problem. *Energy Sources Part B Econ. Plan. Policy* **2014**, *9*, 236–247. [CrossRef]
35. Zaenudin, E.; Kistijantoro, A.I. pSPEA2: Optimization fitness and distance calculations for improving Strength Pareto Evolutionary Algorithm 2 (SPEA2). In Proceedings of the 2016 International Conference on Information Technology Systems and Innovation (ICITSI), Bandung, Indonesia, 24–27 October 2016; pp. 1–5. [CrossRef]
36. Deb, K.; Pratap, A.; Agarwal, S.; Meyarivan, T. A fast and elitist multiobjective genetic algorithm: NSGA-II. *IEEE Trans. Evol. Comput.* **2002**, *6*, 182–197. [CrossRef]
37. Das, D. A fuzzy multiobjective approach for network reconfiguration of distribution systems. *IEEE Trans. Power Deliv.* **2006**, *21*, 202–209. [CrossRef]
38. Salomon, C.P.; Coutinho, M.P.; de Moraes, C.H.V.; Borges da Silva, L.E.; Lambert-Torres, G.; Aoki, A.R. Applications of Genetic Algorithm in Power System Control Centers. In *Genetic Algorithms in Applications*; Popa, R., Ed.; InTech: London, UK, 2012; pp. 201–222. [CrossRef]

energies MDPI

Article

Power Scheduling Scheme for DSM in Smart Homes with Photovoltaic and Energy Storage

Sławomir Zator

Department of Production Engineering and Logistics, Opole University of Technology, Sosnkowskiego 31, 45-272 Opole, Poland; s.zator@po.edu.pl; Tel.: +48-77-449-8746

Abstract: This article presents a case study of a single-family house with several photovoltaic micro-installations oriented in different directions, in which the energy electricity storage systems have been operating for several months. In the house, the heat source is the air–water heat pump cooperating with heat buffers. The first photovoltaic installation was installed in 2016 and, in the subsequent five years, was expanded using microinverters. The final amount of energy from photovoltaics covers 50% of the energy demand of the building. The procedure for dealing with technical and economic aspects was presented, allowing us to determine whether it is profitable to install energy storage in the given conditions of energy prices, equipment efficiency, and prices, as well as government support. This paper presents the effects of the designed and built home energy management system that supervises energy storage in heat and batteries, mainly through its impact on the self-consumption of energy from the photovoltaic system and on final costs. Comparative calculations were performed with the demand-side management, which dictated the instantaneous energy costs. Attention was paid to the possibility of obtaining a high self-consumption, but the economic calculations showed that it was not always beneficial. An annual self-consumption increased by approximately one-sixth upon installation of the electrical energy storage system and by one-third from the start of use of the home energy management system. Concurrently, by utilising energy storage in heat and batteries, almost 95% of energy was consumed in the cheapest multi-zone tariff. The impact of inverters and battery charging systems on the power grid is also presented. Often, when the active energy was nearing zero, the capacitive reactive energy was significant.

Keywords: energy storage; self-consumption; reactive energy

Citation: Zator, S. Power Scheduling Scheme for DSM in Smart Homes with Photovoltaic and Energy Storage. *Energies* **2021**, *14*, 8571. https://doi.org/10.3390/en14248571

Academic Editor: Germano Lambert-Torres

Received: 26 November 2021
Accepted: 17 December 2021
Published: 19 December 2021

1. Introduction

One of the main challenges world economies and scientific research focus on is the reduction of fossil fuel dependency due to their fluctuating market prices, which compromise the stability of the world's economy and the harmful emissions they cause. In order to reduce the share of energy originating from fossil fuels and simultaneously increase the share of renewable energy sources and an energy storage of both a small and large scale is required. A variety of accumulation solutions are currently being subjected to numerous tests in order to improve storage efficiency. The least costly method of accumulating energy storage is heat storage. However, it is limited in versatility and its efficiency peaks at low outside temperatures, making it dependent on the climate and geography. Analysis showed that the cost of investment in heat storage is a quarter of its equivalent in electricity storage [1]. Heat storage in domestic hot water buffer (DHWB) and heating water buffer is economically beneficial but does not allow for the use of energy in form other than heat. In buildings with a photovoltaic system (PVS), the main goal is to ensure as high a self-consumption coefficient (SC) as possible. Accumulation of energy in batteries could allow for reduction of the peak energy exported to the grid by creating a buffer that separates the production from its consumption. A substantial drawback of energy storage is that it is not yet profitable for prosumers with the absence of public support system [2]. The Energy

Transformation Programme in Europe promotes renewable energy and shifting from coal to gas. A shortage of natural gas as a result of the pandemic and the lack of wind increased electricity prices [3]. In 2021, electricity prices in the European Union have soared. In July of that year, most countries experienced a 1.5-year-long record in wholesale prices [4].

PVS costs have been falling in line with forecasts until the pandemic, but those of the energy electricity storage systems (EESS) are still high, over 200 EUR/kWh [5]. The cost projection shows an increase from 28% to even 58% of capital cost reductions by 2030. High energy prices and a drop in the EESS prices contribute to their increasing popularity.

Distribution system operators (DSOs) have a significant influence on energy prices. Their method portfolio consists of instruments through which they can manipulate the demand for electricity: demand-side management (DSM) and demand-side request (DSR). In Poland, the DSR system is still not available for individual consumers [6]. The only mechanism that households can use is DSM in the form of time-based tariffs. Potentially, it is possible to use DRS if an energy cluster is established in an energy community. An example of which in Poland is an energy cluster in which energy self-sufficiency can almost be achieved. It is much easier to implement the concept of a smart grid within it.

2. Review of the Literature

In article [7], the new model considering the Real-Time Demand Response Pricing scenario in the smart grid was tested. The Pricing Suggestion Unit was proposed based on a real-time pricing algorithm by considering users' preferences using stochastic optimisation techniques, better than real-time pricing. This system suggests to the end-users when they should turn on their load, but the final decision belongs to the people, not the machine.

Article [8] proposes a realistic scheduling mechanism to reduce user frustration and enhance appliance utility by classifying appliances with respective constraints and their time-of-use effectively. Algorithms are proposed regarding the functioning of home appliances. A 24 h time slot was divided into four logical sub-time slots, each composed of 360 min, or 6 h.

For the end-user, and for the DSO, when they make the decisions, there are a few evaluation criteria. They are almost identical for both sides. However, the importance is different. These criteria are: energy security, power reliability, and secondly investment and operating costs. There are several methods for decision making, from the simple weighted sum method [9,10], to multicriteria decision analysis optimisation approaches [11,12].

Paper [13] proposed the model which aims to optimise three different criteria: minimising electricity costs, reducing the probability of power loss, and maximising the use of locally available renewable energy. It was shown that lithium iron phosphate and lithium nickel cobalt alumina were the leaders among the five battery technologies. For the calculations used the Metaheuristic Canonical Differential Evolutionary Particle Swarm optimisation, proposed previously in article [14], which merges distinct principles of evolutionary computation and swarm intelligence.

There are three main goals for designing hybrid microgrid systems in the literature. Firstly, to minimise the cost of electricity per unit generated [15]; secondly, to minimise the loss of load probability or breakdowns [16]; and thirdly, to maximise the share of renewable energy sources in electricity generation [17].

Systems that will have optimal scheduling are possible to be designed and built, which is a requirement for current and future microgrids [18], but end-user optimisation would be the best option. The desired goal of the system's operation is to reduce the peaks of both the consumed energy and the energy fed into the network [19–21]. These are the most requested features by DSOs. All the proposed solutions are not simple enough in terms of user-friendliness and are yet to be implemented on a large scale in households.

There are publications that propose models that coordinate the benefits of households and utility operators. The results revealed a 38% reduction in the electricity consumption cost and an 18% reduction in peak demand of the distribution transformer [22]. In article [21], the authors made appliance classification as: non-deferrable (home lightings, washing machine), interruptible (water heater, iron), and must-run loads (TV, PC). The interruptible appliances have sequences of operations that can be interrupted. Must-run loads must be run immediately at any time, and these appliances must be run at any cost. This division assumes that the devices will be used as designed by the producer. For some devices, it is possible to temporarily interrupt the energy supply, too, e.g., heaters, which would not be of significant importance to the user. Ovens, washing machines, dishwashers and kettles are such devices. Had they been equipped with the internet of things (IoT), it would have been much easier to manage energy and reduce the peak-to-average (PAR) ratio. Another way to allow for uninterrupted functioning of devices while temporarily depriving them of external power supply is to equip them with an energy buffer, i.e., a battery or a supercapacitor. This solution could be used by almost all low-power devices that are powered by a direct current of several volts. We use cell phones, laptops, and battery-powered tools that allow for a number of hours of such work on a daily basis. Theoretically, any electrical equipment could have an energy buffer. Here it is merely an idea. It is conceivable that instead of building large EESS, it would be easier and cheaper to manage the energy of small loads using IoT.

3. Motivations

Unfortunately, contrary to their declared goal, most DSM mechanisms do not encourage consumers to actively manage energy [23]. There exists a prosumer support system in the form of feed-in tariffs in European countries. The increase in energy production from renewable energy sources resulted in a reduction in the feed-in tariffs [24], which prompted prosumers to increase auto-consumption [25]. In Poland, there is a system called rebates, which is based on net metering [1]. The prosumer can feed energy into the DSO grid and receive this energy during the year. The cost to the prosumer is 30% of the energy for a photovoltaic system (PVS) over 10 kWp but not exceeding 50 kWp, or 20% of energy for PVS under 10 kWp.

If the distribution grid acts as an energy storage, then prosumers who have concluded 15-year contracts with the DSO will not be interested in investing in the EESS at all. Most prosumers in Poland, whose installations allow for the annual balance of energy generated in PVS and used in the building, use the G11 tariff. However, according to the new governmental proposals, the new contracts concluded with DSOs will be less beneficial to prosumers. Surplus energy will be resold to the grid at wholesale prices, almost twice lower than the costs of its purchase. This is expected to increase SC, also through the accumulation of energy in the EESS. Calculations will be performed considering the new conditions.

The authors of [1] showed the impact of heat accumulation on the COP and SC. The additional effect of reducing electricity costs (CE) was achieved through load shifting and the multi-zone tariff. In the building, the SC reached a high level of 46% during the heating season. However, in the summer season, due to limited possibilities of energy accumulation, it was low and amounted to 32%. This was one of the reasons for the decision to invest in an accumulation system based on batteries. Before this happened, however, a preliminary economic analysis was made.

Usually, no attention is paid to the kind of energy, active and reactive, consumed by home equipment, especially in an intelligent home. In the presented house, there are over 30 control devices, such as temperature controllers, roller shutters, garage gate, intercom, intelligent lighting and others. Their power is only 3–5 W each, but they draw about 160 W active power and 280 VAr reactive power together. If powered from the EESS, the DC load of batteries is 250 W. These devices consume 1.4 MWh energy per year when powered from the grid, but if we would like to power them only by night from the EESS, they would have taken about 1 MWh instead of 0.7 MWh. The most common proposal is to use the energy stored in the batteries during the night hours. It turned out not to be cost-effective, as will be shown later in the article.

Often during analyses, catalogue values of the efficiency of energy converting devices (inverters, chargers, converters, batteries) are adopted [26,27], although these are usually the highest values that the device can achieve. However, during operation, these devices very rarely operate at the nominal (maximum) powers. This leads to a reduction in their actual effectiveness. When calculating economic indicators, real efficiencies are necessary to be able to make rational decisions. Therefore, the efficiencies of the energy accumulation system were determined based on measurements.

The next three chapters describe the steps to achieve the described end results. Section 7 presents forecasts of potential benefits, and Section 8 the actual results. Section 9 presents some reflections related to overvoltage and reactive power that have arisen during the research. This problem is merely highlighted, which is a contribution to further research and publications.

4. Home Energy Management System

Prosumers can import and export energy at the same time when the building has a three-phase electrical installation. In the studied building, the first energy meter installed by the OSD performed vector summation mode (Ferraris mode) and balanced both energy streams before counting them in registers. After the expansion of the PVS in November 2018, the OSD replaced the energy meter with an implemented algebraic summation mode, which does not balance energy between phases before it is counted [1]. This has a financial effect. Each unbalanced energy unit is a loss of 20% of its value. To reduce these negative effects, the HEMS was designed and implemented. Depending on the instantaneous power balance, it switched the installation circuits under load in such a way as to maximise auto-consumption. A new approach is active load management by switching the load circuits and controlling the battery charging power. The HEMS was intended to lower electricity bills by increasing self-consumption. Prior to the launch of the HEMS, there was a system that controlled the following: the heat accumulation in water buffers, heat distribution, and the air–water heat pump (AWHP) work [1]. Although the PVS is three-phase, it consists of one main inverter and four single-phase microinverters. Photovoltaic (PV) panels are oriented in different directions and at different slopes. One of them has a simple one-axis tracker. They were originally connected to different phases. The AWHP and the main PV inverter are connected to the same power phase. An example of the power generated in June 2020 is shown in Figure 1.

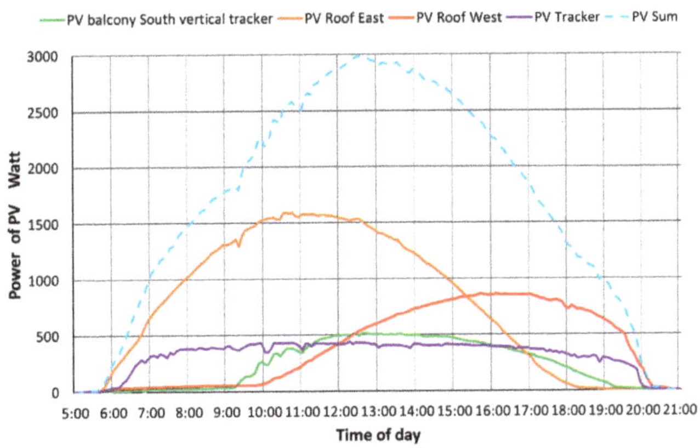

Figure 1. Power generated by PVS on sunny day on 6 June 2020.

In HEMS Version 0, which started in April 2019, only the microinverters could have been connected to different phases, and it is possible to connect them all to the same phase as the AWHP. However, switching occurs only when the AWHP starts and the power consumed by it exceeds the power generated by the main inverter. The switching of microinverters caused only small energy losses as they started up relatively quickly (in under 1 min). The current HEMS (Version 1), which has been working since November 2019, additionally controls five load circuits. One of them contains the AWHP, the second the battery charger, and the remaining three are the building installations. An example of energy flows during the cooperating HEMS and EESS is shown in Figure 2.

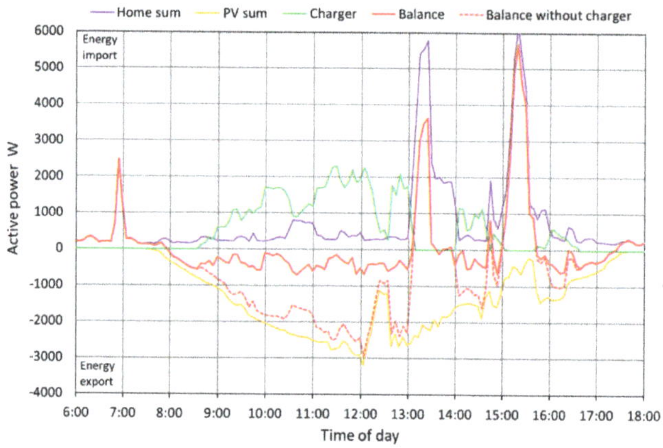

Figure 2. Energy flows during sun operation on 24 October 2021.

On this day, the energy generated by PVS was 15.0 kWh, the energy taken from the grid, 13.9 kWh, the energy exported to the grid, 2.8 kWh, and the daily SC was 81%. The annual effects of HEMS' working are presented in Figure 3. It shows the cumulative SC in 2019, 2020, and 2021. The timeline shows the moments of power changes in PVS, the HEMS and the EESS commissioning. In the year following the launch of the HEMS, the SC grew by 1/6 from 34.7% to 40.5%. There is also a visible increase in the SC caused by the EESS.

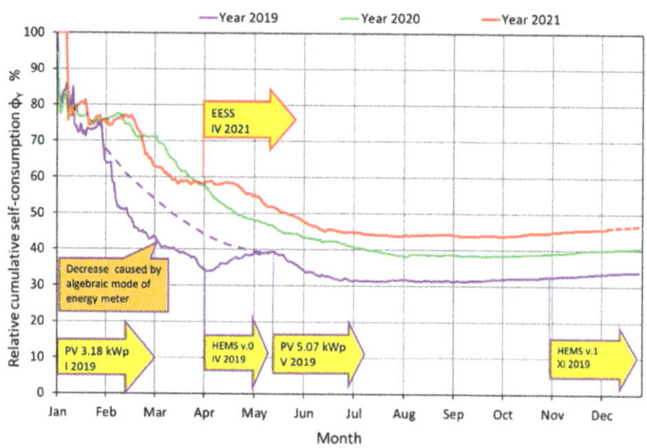

Figure 3. The cumulative SC average in a period of one year.

The cost of introducing the HEMS was omitted as load control systems already existed. It was only necessary to add a load switching module to the software. Measurement data, i.e., voltages, currents, active and apparent powers of all separated circuits, including inverters, are made in 10 s intervals and saved as five-minute averages. Therefore, it will not be included in the economic analysis of its implementation. If three-phase balancing had taken place during the settlements with DSO, there would have been no need to build it. However, this system realises a real increase in self-consumption. The SC based on vector-mode meters does not consider real energy flows.

5. Efficiency of the Electric Energy Storage Systems

The EESS is dedicated to off-grid or hybrid systems. Off-grid systems do not collaborate with the grid, and they shall not be analysed in this article. Nowadays, there are two types of hybrid systems available on the market. The first type works with low-voltage batteries, and the second collaborates with high-voltage batteries, but the differences are more significant. The main difference is in the location of the batteries in the electricity conversion chain and the number of DC/AC converters. Placing a high-voltage battery downstream of the MPPT and before the DC/AC converter simplifies the construction and eliminates the need for voltage reduction to 24 or 48 V. This results in greater efficiency of the inverter. Alas, the inverters for the studied house were purchased long before hybrid inverters with high-voltage batteries were available. Paramount importance was given to the investment costs. A hybrid inverter was used as a battery charger and DC/AC converter, which enabled the control of electrical parameters, such as the charge and discharge current state, battery voltage, and battery charge status. It turned out that these measurements were deficient in accuracy. Therefore, additional direct voltage and alternating voltage energy meters were installed to accurately determine the efficiency of the EESS.

If the efficiency was calculated from Equation (1) assuming the catalogue data, where the charger efficiency, EF_{CHG}, is 95% and the AC/DC inverter efficiency, EF_{INV}, is 93%, the total efficiency, EF_{BAT}, would be 88%. In the period of a few days, the efficiency of battery storage, EF_{ACC}, can be assumed as 100%.

$$EF_{BAT} = EF_{CHG} \times EF_{ACC} \times EF_{INV}, \tag{1}$$

Figure 4a shows the EESS efficiency calculated as the amount of output energy versus input energy, both measured on the AC side. After eight months of operation, the efficiency of the EESS stabilised at 72%.

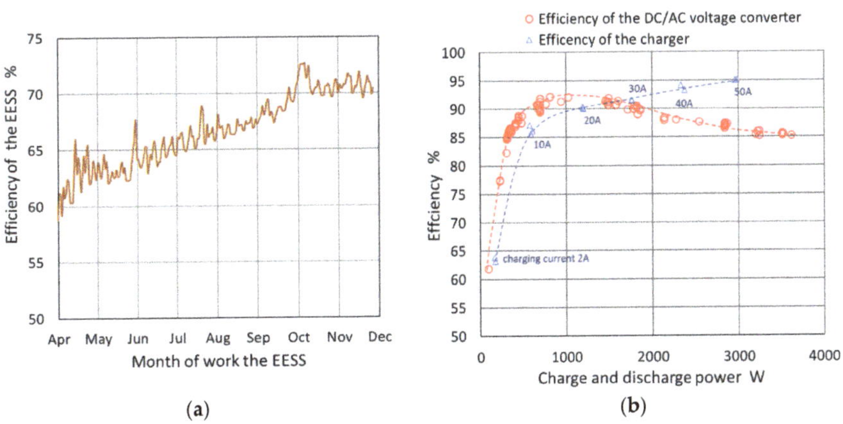

Figure 4. The efficiency of: (**a**) EESS (**b**) charger and DC/AC voltage converter.

The visible peaks attribute to cloudy days when the batteries are only discharging. The average efficiency of the battery charger and the DC/AC voltage converter was also measured (Figure 4b). Their average values are below the catalogue maximum values, respectively EF_{CHG} = 85.9% and EF_{INV} = 83.8%. This is due to the fact that most of the time, the charging current does not exceed 20 A, and the load fluctuates around 500 W.

6. Energy Costs as the Basis for Decisions

One might ask if battery storage is profitable. In order to answer this question, several calculations have been made. The conducted analysis takes current increases in energy prices into consideration. A discount rate $r = 0$ was also assumed. The result of the calculations is given as the mean 20-year cost and as a percentage change in the cumulative costs.

Firstly, the EESS expenditure was calculated. The EESS was built with ready components: batteries LiFePo4 (200 Ah, 52 V), battery management system (BMS) 125 A, a battery charger and DC/AC inverter (a hybrid invert 5 kVA was used). With additional equipment, the total cost was EUR 2480, or approximately 240 EUR/kWh. At the beginning, the maximum amount of energy that could be accumulated in the batteries in the period of twenty years was estimated. The number of full battery cycles is 4000, which corresponds to eleven years. The batteries could be operating for twenty years, with an average load of 60% in such a cycle. During this time, the battery capacity will decrease by 20%, so the annual capacity reduction ratio is 0.9883. During this period, the batteries could have stored 41.6 MWh. Considering the investment costs and the EESS efficiency, the average unit energy cost will be 93.5 EUR/MWh.

Batteries can be charged with both PVS and utility power. Solar energy itself is free of charge. However, current methods of its collection can be costly. In the studied building, EUR 3300 was invested in PVS (accounting for the tax relief). The installation's capacity is 5.07 kWp, and the investment cost was around 650 EUR/kWp. Considering the measurements made so far and the reduction in PVS efficiency, collected solar energy in twenty years will be 69.6 MWh. The average unit cost of obtaining energy will be 47.5 EUR/MWh.

Finally, an average twenty-year cost of purchased energy should be determined. In Poland, there various scenarios of electricity prices are forecasted, depending on the costs of the energy transition. One such realistic scenario is presented in the study [28]. It presents four circumstances for the development of coal-based generation capacity, the diversification of sources with and without nuclear energy, and the dynamic development of renewable energy sources. Wholesale price forecasts in the form of graphs are shown in Figure 5. Retail prices are derived from average wholesale prices, and it has been assumed that their annual changes will be the same.

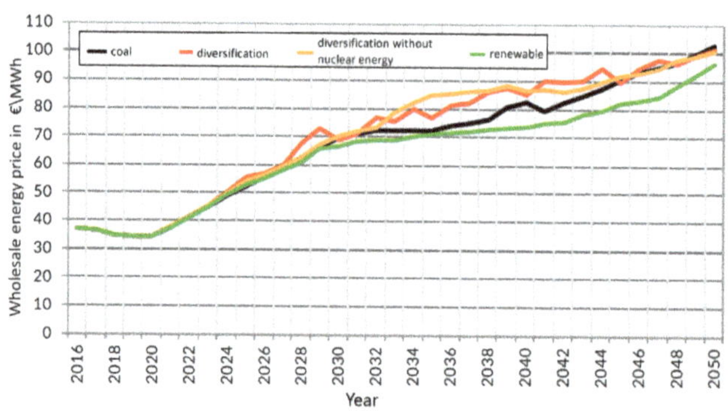

Figure 5. Energy price forecast to 2050 [28].

Wholesale energy prices are projected to increase by 90% to 130% over a twenty-year period. Achieving this price level means an average annual increase in electricity prices of 6.4% to 8%. The calculations also considered the very optimistic average price increase over the last four years, amounting to 4.8%. The calculations were made for the used multi-zone G13 tariff. Table 1 presents time zones in G tariffs for OSD Tauron. Only for multi-zone tariffs may it be profitable to recharge batteries not solely with solar energy.

Table 1. Time zones T1, T2, T3 in G tariffs for OSD Tauron.

Hour	G11 Mo-Su All Year	G12 Mo-Su All Year	G12w Mo-Fr All Year	G12w Sa-Su * All Year	G13 Mo-Fr I-III X-XII	G13 Mo-Fr IV-IX	G13 Sa-Su * All Year
1	T1	T2	T2	T2	T3	T3	T3
...	T1	T2	T2	T2	T3	T3	T3
6	T1	T2	T2	T2	T3	T3	T3
7	T1	T1	T1	T2	T3	T3	T3
8	T1	T1	T1	T2	T1	T1	T3
...	T1	T1	T1	T2	T1	T1	T3
13	T1	T1	T2	T2	T1	T1	T3
14	T1	T2	T2	T2	T3	T3	T3
15	T1	T2	T1	T2	T3	T3	T3
16	T1	T1	T1	T2	T3	T3	T3
17	T1	T1	T1	T2	T2	T3	T3
...	T1	T1	T1	T2	T2	T3	T3
20	T1	T1	T1	T2	T2	T2	T3
21	T1	T1	T1	T2	T2	T2	T3
22	T1	T1	T1	T2	T3	T2	T3
23	T1	T2	T2	T2	T3	T3	T3
24	T1	T2	T2	T2	T3	T3	T3

* also on public holidays.

Table 2 shows the prices of consumed energy and the percentage use of energy in each time zone over the last four years. The share of consumed energy in the cheapest T3 zone increases yearly. This is the result of energy accumulation. The calculated average energy prices over the period of twenty years for three variants of its average annual growth: 4.8%, 6.4% and 8% are presented in Table 3. The energy share of particular time zones was assumed as in 2021.

Table 2. Prices and structure of energy imported in G13 tariff zones in four year period.

	Zone				Zone		
	T1	**T2**	**T3**	**Avg.**	**T1**	**T2**	**T3**
	Unit Energy Cost				Participation		
Year	**EUR/MWh**	**EUR/MWh**	**EUR/MWh**	**EUR/MWh**	%	%	%
2018	119.1	189.1	63.6	78.6	4.6	9.9	85.5
2019	119.1	189.1	63.6	74.6	3.1	7.4	89.5
2020	137.1	217.2	74.9	86.6	3.3	6.8	89.9
2021	138.3	221.3	76.1	82.5	2.8	32	94.0

This table also presents the wholesale energy price because, from 2022, prosumers will sell surplus energy to DSOs at wholesale prices. In the bill project, it is called net-billing. In line with the current net-metering billing system, a prosumer can recover 80% of the energy fed into the grid. The unit cost of this recovered energy corresponds to the average price of purchased energy. Accounting for the efficiency of the EESS, it seems the energy storage in the net-metering billing system will be unprofitable. Using the calculations made, the cost-effective use of the EESS can be determined. Table 4 shows the energy prices for various combinations of charging sources in comparison to utility energy prices.

Table 3. Calculated 20-year average price of energy for 3 variants of price increase.

	Zone				Wholesale Price
	T1	T2	T3	Avg.	
	Unit Energy Cost				
Price Increase Variant	EUR/MWh				
Low (4.8%)	223.8	358.3	123.2	132.9	75.7
Medium (6.4%)	265.5	425.0	146.1	158.5	89.8
High (8.0%)	316.3	506.4	174.1	188.9	106.9

Table 4. Energy prices for various combinations of charging sources in comparison to energy price from grid.

		Unit Energy Cost Form the EESS				Unit Energy Cost in G13 Tariff			
	Price Increase Variant	2021 Price	Low (4.8%)	Medium (6.4%)	High (8.0%)	2021 Price	Low (4.8%)	Medium (6.4%)	High (8.0%)
		EUR/MWh							
Charging source of the EESS	PVS	138.4	138.4	138.4	138.4				
	Zone T1	229.2	317.8	356.4	407.3	138.3	223.8	265.5	316.3
	Zone T2	312.3	449.2	515.9	597.3	221.3	358.3	425.0	506.4
	Zone T3	173.5	214.1	237.1	265.0	82.5	123.2	146.1	174.1

The efficiency of the EESS was accounted for in the calculation. Considering current prices, the most profitable scenario is charging batteries from PVS and discharging them in the T2 zone tariff. What is also profitable is charging batteries from the grid in the T3 zone and discharging them in the T2 zone tariff. Charging the batteries with solar energy and using them in zones T1 and T3 is not profitable.

Considering the average twenty-year price of energy (with assumption, that energy costs in the zones will not be flattened), it will be profitable to use solar energy stored in batteries in the T2 tariff zone. It will be profitable to charge the batteries from the grid in the cheapest T3 zone and use it in the more expensive zones T1 and T2 and also charge in the T1 zone and discharge in the T2 zone. The conclusions from the price analysis are more visible in Table 5. Profitable combinations have a negative percentage change, which means that the energy stored in the battery from the indicated source is cheaper than the energy from the grid in the given tariff zone. Columns for which neither variant is profitable are not shown.

Table 5. Profitable combination of battery source charging and time zone their use.

Electricity Price Forecast		Energy Cost Changes in Comparison to Year without the EESS								
		2021 Price	Low (4.8%)		Medium (6.4%)			High (8.0%)		
		Discharge in G13 Tariff Zone								
		T2	T1	T2	T1	T2	T3	T1	T2	T3
		%								
Source of battery charging	PVS	−37.4	0.1	−61.4	−47.9	−67.4	−5.3	−56.2	−72.7	−20.5
	Grid T1 Zone	3.6	40.6	−12.1	34.2	−16.2	144.0	28.7	−19.6	134.0
	Grid T3 Zone	−21.6	-4.3	−40.2	−10.7	−44.2	62.2	−16.2	−47.7	52.2

7. Potential Profits

Considering the previous analysis, the calculations were made for two scenarios and, within them, two variants. The first scenario is the current net-metering billing system, and the second is the future net-billing system when surplus energy would be sold at wholesale prices. The first variant assumes that the electricity is not taken from the grid in the T2 zone tariff for the entire year. The latter variant means that the same was assumed for both T1 and T2 zones. This means that energy will be taken from the grid only in the T3 tariff, both for current needs and for charging the battery. Data collections used for calculations, i.e., imported and exported energies, energy from PVS and the SC are presented in the graphical form on Figures 6–8. Monthly summary of the data is presented in Table 6.

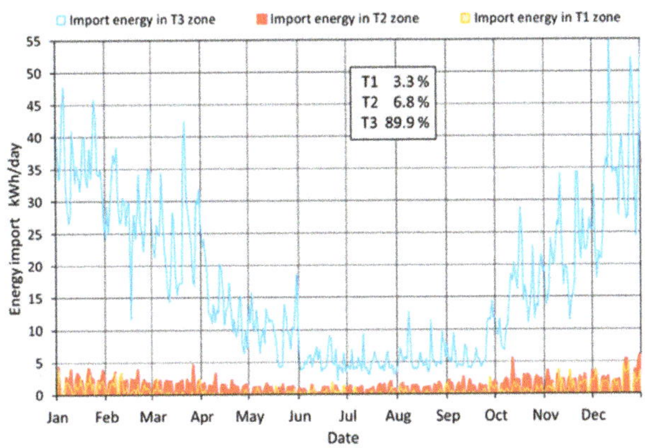

Figure 6. Energy imported in 2020.

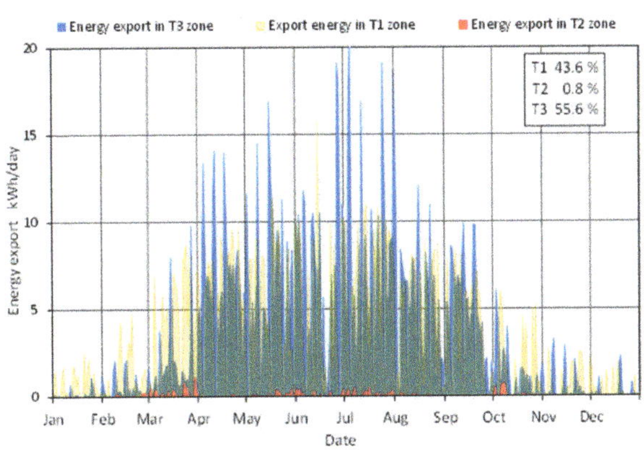

Figure 7. Energy exported in 2020.

In the winter months, the amount of energy exported to the grid is lower than that taken from the grid in each of the T1 and T2 time zones. In order not to consume energy in zones T1 and T2, the missing energy will have to be replenished from the grid in the T3 zone, charging the batteries there.

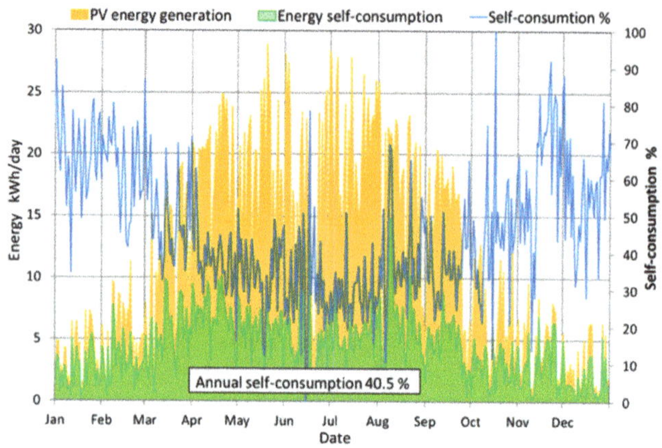

Figure 8. PV generation and the SC in 2020.

Table 6. Monthly summary of imported, exported, produced PVS and SC.

Months	Jan	Feb	Mar	Apr	May	Jun	Jul	Aug	Sep	Nov	Dec
					Energy kWh						
PVS generation	98.2	134.7	328.8	558.8	509.9	449.7	594.8	501.2	384.0	168.2	61.6
PVS SC	74.6	90.7	163.9	212.4	181.2	135.5	186.7	198.1	139.8	81.9	26.6
Energy export	23.6	43.9	164.9	346.4	328.6	314.1	408.1	303.1	244.2	86.3	35.0
Energy import T1	17.8	9.2	6.1	5.0	6.3	10.8	4.4	8.2	11.3	19.6	32.6
Energy import T2	71.5	46.2	45.5	32.2	20.6	18.5	21.7	26.2	31.3	52.0	45.0
Energy import T3	1083.4	801.2	754.9	425.0	272.8	148.9	133.5	180.4	192.0	473.0	659.6

In the first scenario, the energy cost, *EC*, was calculated in accordance with Equation (2) for the year 2020. These are not only the current costs incurred but also ones related to the amortisation of investments in the PVS and the EESS.

$$EC = \left(E_{grid\ T1} + E_{grid\ T2} + E_{grid\ T3}\right) \times Uc_{grid\ avg} + E_{PV\ selfcons} \times Uc_{PV} + \\ -E_{grid\ Exp} \times k \times Uc_{grid\ avg}, \tag{2}$$

The annual energy cost, *EC*, was calculated according to Equation (3) in the first variant and in accordance with Equation (4) in the second one. In the second scenario, the annual energy cost was calculated according to almost the same equations; the only difference is in the last part of the expressions, where instead of income from energy sent to the network as a rebate, is income for electricity supplied to the grid at wholesale price. Instead of the expression $k \times Uc_{grid\ avg}$, it is $Uc_{\ whp}$.

$$EC = E_{grid\ T1} \times Uc_{grid\ T1} + E_{grid\ T3} \times Uc_{grid\ T3} + E_{PV\ selfcon} \times Uc_{PV} + \\ +p \times \frac{E_{grid\ T2}}{\eta_{BAT}} \times (Uc_{PV} + Uc_{BAT}) + (1-p) \times \frac{E_{grid\ T2}}{\eta_{BAT}} \times (Uc_{grid\ T3} + Uc_{BAT}) + \\ -(E_{grid\ Exp} - \frac{E_{grid\ T2}}{\eta_{BAT}}) \times k \times Uc_{grid\ avg}, \tag{3}$$

$$EC = E_{grid\ T3} \times Uc_{grid\ T3} + E_{PV\ selfcon} \times Uc_{PV} + \\ +p \times \frac{E_{grid\ T1} + E_{grid\ T2}}{\eta_{BAT}} \times (Uc_{PV} + Uc_{BAT}) \\ +(1-p) \times \frac{E_{grid\ T1} + E_{grid\ T2}}{\eta_{BAT}} \times (Uc_{grid\ T3} + Uc_{BAT}) + \\ -(E_{grid\ Exp} - \frac{E_{grid\ T1} + E_{grid\ T2}}{\eta_{BAT}}) \times k \times Uc_{grid\ avg} \tag{4}$$

where:

E_i—energy component i
Uc_i—unit cost energy of component i
k—factor rebate 0.8
p—share of energy obtained from PVS in winter months (in other $p = 1$)

Indexes *grid Ti, grid avg, PV, BAT, wph* describe energy component and unit cost.

When determining the unit costs, Uc, the energy conversion chain was included. For example, the unit energy cost from the EESS is the sum of the costs depending on the charging source, grid $Uc_{grid\ T3}$ or PV energy Uc_{PV} and storage Uc_{BAT}. The impact of different options for accumulation on the annual energy costs in the described building is shown in Table 7. Profitable options have a negative percentage change in the cost of energy.

Table 7. Impact of different options of accumulation on the annual energy costs.

		Electricity Price Forecast Variants							
		2021 Price	Low (4.8%)	Medium (6.4%)	High (8.0%)	2021 Price	Low (4.8%)	Medium (6.4%)	High (8.0%)
		Annual Energy Price				Cost Changes Compare to 2020			
		EUR	EUR	EUR	EUR	%	%	%	%
Current support system	Base	556	894	1047	1233				
	Variant 1	604	943	1069	1234	8.6	2.2	0.4	−1.2
	Variant 2	640	822	945	1094	15.2	5.8	3.2	0.9
Net billing	Base	622	894	1047	1233				
	Variant 1	698	934	1069	1243	12.3	4.5	2.1	0.1
	Variant 2	666	883	1008	1160	7.2	−1.3	−3.7	−5.9

It would seem that, accounting for the data in Table 5, it should be profitable to use energy from PVS and stored batteries in zones T1 and T2. However, this is not the case as the EESS accumulation efficiency is less than 0.8, which is less than the rebate value. Unfortunately, lowering the average energy cost reduces the benefits of transferring excess electricity to the grid. The Equations (1)–(3) can be used to control the EESS, deciding whether to charge the battery from two sources simultaneously, meaning from PVS and grid.

Using the same data collections, it is also possible to calculate the maximum SC coefficient, Φ, based on Equations (5) and (6).

$$\Phi_{variant1} = \frac{E_{PV\ selfcon} + E_{grid\ T2}}{E_{PV generate}}, \tag{5}$$

$$\Phi_{variant2} = \frac{E_{PV\ selfcon} + E_{grid\ T1} + E_{grid\ T2}}{E_{PV generate}}, \tag{6}$$

The conclusion of these calculations is that limiting the export of energy in the zones T1 and T2 by charging the batteries can maximally increase the SC for the first variant to 57.6% and for the second variant to 64.4%. Only for prosumers in the new net-billing system would the second variant be profitable while investing in the EESS.

8. Profits from the Operation of the HEMS and the EESS

The HEMS system has been operating since November 2019. From the beginning of April 2021, the EESS was installed and launched. The research was conducted under the first scenario, in which the surplus of non-consumed PV energy was stored in batteries and then used in the T2 time zone, in which energy is the most expensive. There are some limitations associated with charging and discharging of batteries, which were implemented at the HEMS, which controlled the EESS:

- Currently, it is not possible to simultaneously charge the battery and to power the building from the battery (this is why only scenario first was realised).
- The battery charging current can be set as stepwise with the minimum current of 2 A, the next values being 10 A and its multiples up to 80 A.
- The charger is only turned on when the excess PV power over the instantaneous consumption exceeds 200 W. The batteries are then charged with a current of 2 A. The charging efficiency for this current is exceptionally low (65%).
- Batteries are charged to 94–95% of their capacity (after reaching the absorption voltage), which will extend their life. In addition, the overall efficiency of the EESS also increases due to the reduction of the charging time with a low current.
- Energy measurements and flow control are taken every 10 s.

The algorithm of operation of this system is based on the state machine. Its inputs are:

- measured instantaneous powers consumed by five load circuits;
- measured powers generated by five PV inverters;
- 24 h weather forecast in one-hour steps;
- date and time;
- system operation mode (manual/battery/auto)
- electricity price forecast (current price/low/medium/high)

Parameters of the state machine are:

- determined unit energy cost for calculating price increase variants and current price

Outputs of the three-state type control they switch:

- four circuits of load;
- four inverters circuits.

The results of three-year measurements of the SC are presented in Figure 9. Each of the actions (introduction of HEMS and EESS) taken increased the SC. The cumulative yearly SC is already presented in Figure 3.

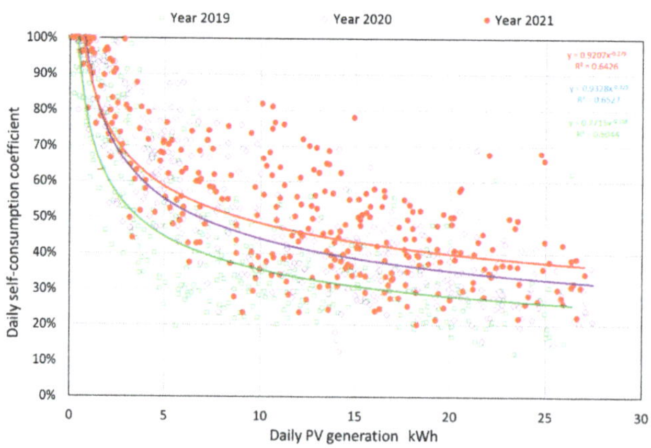

Figure 9. Influence of the daily PV generation on the SC.

In Table 8, the SC for 2019, 2020 and 2021 are presented, divided into working and non-working days as well as the heating season and summer season. The same data is graphically shown in Figure 10 to better show the relationship between them.

Table 8. The self-consumption over year 2019, 2020 and 2021.

Calculation Period		2019	2020	2021
		Self-Consumption Coefficient		
		%		
All year	All days	34.7	40.5	45.9
	Free days	41.2	46.7	50.2
	Work days	31.7	38.0	43.8
Heating season	All days	46.3	51.4	56.4
	Free days	55.7	59.2	59.6
	Work days	42.2	48.1	54.9
Summer season	All days	31.8	33.8	40.3
	Free days	37.7	38.7	45.5
	Work days	29.0	31.9	37.9

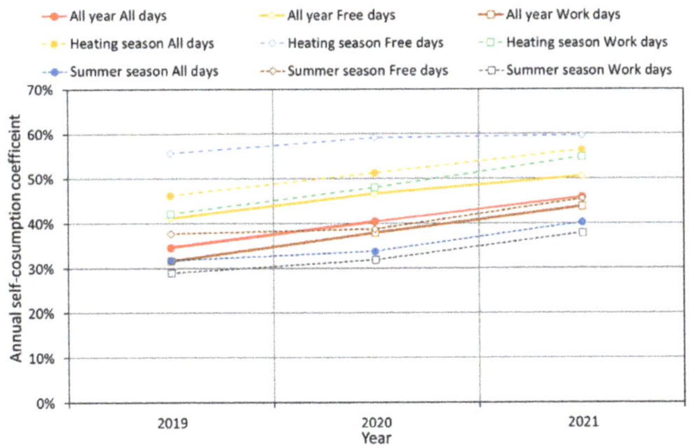

Figure 10. The annual self-consumption in 2019, 2020 and 2021.

In most cases, both of the introduced systems (HEMS and EESS) brought about similar effects, i.e., increased the SC. There were annual increases of 5–6% on working days. The increase in auto-consumption in the summer due to the EESS was more than twice as high as for the HEMS.

According to the estimates shown in the previous section, the investment into the EESS would not be profitable in the first scenario, except for the forecast with the highest increase in energy prices. To compare the effects of the HEMS and EESS in the net-metering and net-billing billing systems, the cost of energy was calculated assuming the same amount of energy consumed and its price in 2019–2021. Results are shown in Table 9.

Table 9. Energy cost in tariff G13 before and after introducing the HEMS in 2019 and the ESSS in 2021.

	Energy Consump.	Energy Import				Energy Export	PV Generation		Net Metering	Net Billing
		Sum	T1	T2	T3	Sum	Sum	SC	Energy Cost	
Year	kWh	kWh	%	%	%	kWh	kWh	%	EUR	EUR
2019	8579	7113	3.1	7.4	89.5	2454	3920	34.7	675	749
2020	8579	6991	3.3	6.8	89.9	2332	3920	40.5	667	733
2021	8579	6815	2.8	3.3	94.0	2156	3920	46.9 *	657	719

* estimated value at the end of the year.

It can be seen that despite the increase in SC, energy costs have not decreased in the current net-metering system. The reason is, among others, that the increase in SC was indirectly reduced the average unit costs of exported and imported energy due to the reduction of energy consumed in zone T2. A slight drop in cost would take place if the future prosumer billing system were to apply. The energy cost of a future system would be 13% higher. However, when looking at the trends in Table 7, it can be concluded that the increase in energy prices will increase the profitability of energy accumulation. Table 10 shows the yearly actual energy flows and the purchase costs of energy without depreciation. Considering an increase in energy consumption, the energy costs decreased, particularly in 2021, by 5%.

Table 10. Real energy flows and energy bills in years 2019–2021.

	Energy Import	Energy Export	PV Generation	Energy Consumption	Energy Bill
Year	kWh	kWh	kWh	kWh	EUR
2019	7163	2198	3366	8331	511
2020	6991	2332	3920	8571	481
2021	7252	2074	3905	9083	487

One might ask why the G11 tariff was not used. For comparison, the cost that would be incurred if it were used with the same SC as in the G13 tariff is presented in Table 11. Additionally, for the year 2021, energy costs were calculated assuming the maximum possible SC with the use of the EESS, assuming that surplus energy will be accumulated, but only as much as can be consumed.

Table 11. Hypothetical energy in tariff G11 cost before and after introducing the HEMS and the ESSS.

	Energy Consump.	Energy Import	Energy Export	PV Generation		Net-Metering	Net-Biling	Remarks
				Sum	SC.	Energy Cost		
Year	kWh	kWh	kWh	kWh	%	EUR	EUR	
2019	8579	7218	2560	3920	34.7	919	1089	
2020	8579	6991	2332	3920	40.5	912	1068	SC like in G13
2021	8579	6740	2082	3920	46.9	927	1066	
	8579	5178	520	3920	86.7	976	1011	Max. possible SC

The reasons for choosing the G13 tariff are explained in article [1], but here the DSM benefits can also be seen as time-based multi-zone tariffs. The energy cost in the G13 tariff is about 40% lower than it would be in the G11 tariff.

9. Overvoltage and Reactive Power in Building

A growing number of the PVS, especially in the suburban areas, contributes to certain problems to the current grid system, such as shutdown due to overvoltage. Figure 11 shows the daily voltage changes collected in the summer season in a small housing estate in which 25% of buildings have photovoltaic installations. Each point represents a five-minute window in which the maximum value is determined. As can be seen, the permissible voltage of 253 V is frequently exceeded. Each such voltage excess results in the inverters being switched off. The problem, however, is not a recent phenomenon. The main cause is the insufficient capacity of distribution networks in suburban areas, which were designed as energy supply networks with low simultaneity factors. The energy provided by the sun is at its highest at the same time that the self-consumption is low. These hurdles might be overcome by modernising the network, which is expensive. Therefore, often, a cheaper solution is shaving off the peaks. One of the easiest to implement and most effective solutions is to turn on receivers with high power consumption, such as heaters in the

DHWB or the AWHP. This solution, however, has a finite capacity to absorb excess energy. A good yet costly solution is the EESS. Another one is reactive power compensation, which can help in some cases [29,30].

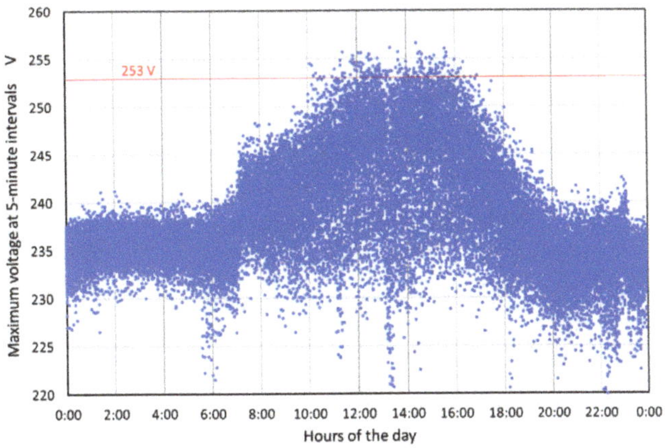

Figure 11. Measured voltage variability in the electrical socked in summer season.

Almost all publications pay attention to inductive power compensation, even today [31]. Here, the problem is the capacitive reactive power. A large number of electronic devices can lead to the situation described here. The PVS and the EESS have, in many cases, a positive impact on the power grid, but there is also the negative impact related to the reactive power.

In prosumers' households, energy is usually recorded in four quadrants energy meters, in registers described by the QBIS code, as shown in Figure 12. The energy meter installed in the described building, unfortunately, does not record the exported reactive energy. Registers 7.8.0 and 6.8.0 are not available.

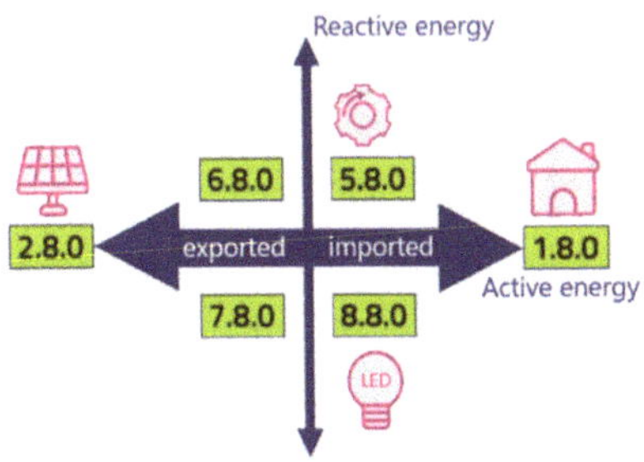

Figure 12. Registers in four quadrants energy meter.

For the illustration as big is the reactive power, Table 12 shows energies imported and exported. A better illustration can be obtained by calculation of the power factor cos(φ) and tan(φ). In zone T2, taken from the grid, active energy is exceptionally low because the amount of consumed energy is equal to the energy generated from the PVS, but capacitive reactive energy is noticeably high. The power factor is substantially low, equal to 0.013.

Table 12. Structure of energy imported and exported and calculated average powers factors in 3 months summer period.

	Registers				Power Factors			
	1.8.0	2.8.0	5.8.0	8.8.0	Inductive		Capacitive	
Period	kWh	kWh	kVAr	kVAr	cos(φj)	tan(φ)	cos(φ)	tan(φ)
All	522	636	3	468	1.000	0.006	0.744	0.898
T1	21	259	0	31	1.000	0.019	0.558	1.488
T2	12	4	0	964	1.000	0.000	0.013	79.000
T3	489	373	3	396	1.000	0.005	0.777	0.811

Within contracts for the supply of energy to institutional consumers, it is allowed to consume the inductive, reactive energy with a tan(φ) not greater than 0.4. For a greater value, this will result in additional fees [32]. Charges are made for the consumption of the capacitive reactive energy, regardless of the power factor, but its cost is much higher than that of the active energy. Both consumers and prosumers pay for energy in municipal tariffs (flats, houses, and small enterprises can use tariffs G in Poland). They pay only for the active energy, and they do not know the effect of reactive power consumption. They are only interested in energy quality in an area important to them. The main parameter is the voltage range (without overvoltage) and with an uninterrupted power supply. Prosumers also want to generate and sell energy from their installations continuously.

The simple (but not free of cost) way to reduce the capacitive reactive power is to use low-power inductors. Simple calculations show that this way allows a reduction of the voltage drop on the building connection by 1–2 V and at the same time reduce the voltage drop in the power line by another 2–3 V.

A simple experiment was carried out with reactive power compensation in the building, known as shunt compensation. Only the capacitive reactive power was compensated with three compensation inductors of the powers 100, 250, and 500 VAr, allowing for an eight-step regulation. As there are no ready-made compensators with such low power, a prototype was built using the existing meter with an RS485 interface, modules for switching coils and a software module implemented in the HEMS. How the power factor varies without and with compensation is shown in Figure 13. The maximum voltages obtained in the period of 5 min are shown in Figure 14. In many cases, it is enough to keep the grid inverter from turning off.

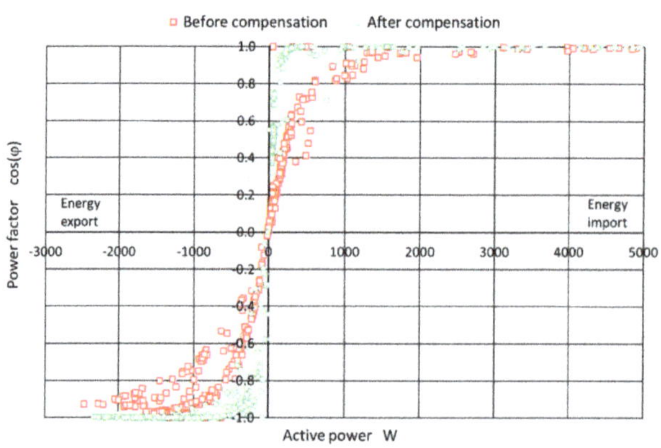

Figure 13. Power factor for imported and exported energy before and after capacitive reactive power compensation.

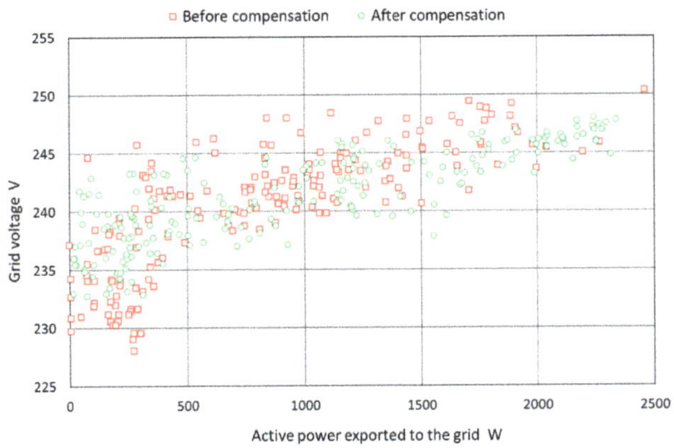

Figure 14. Voltage at energy export before and after capacitive reactive power compensation.

Nowadays, all inverters have the option to set a fixed power factor value from −0.8 (cap.) to 0.8 (ind.). There are also inverters in which the power factor may vary (according to the segmental approximation) depending on the amount of generated power. The newest inverters can compensate for the reactive power during the operation. This functionality is, unfortunately, rarely used. Obviously, they are a bit more expensive as they require a measurement of the power factor or the reactive energy.

10. Summary and Conclusions

Presented calculations were made to determine whether and in what case the investment in the EESS could be profitable. The current prosumer support system (by net-metering) seems to be so attractive that it is difficult to convince them to invest in the EESS without significant additional support. It has been shown that in some cases, it is worthwhile to invest in an EESS when actively using DSM.

This article discusses impacts on the SC of two activities: the introduction of the HEMS and secondly the installation of the EESS. It has been shown that the algorithm implemented in the HEMS can significantly increase the SC. In the analysed building, for over 2019, the increase was 34.7%. Following the launch of the HEMS, the yearly SC increased to 40.5% in 2020. Using the EESS in 2021 resulted in a further increase in the yearly SC to 45%. This is a very good result as it amounts to 25% on average. For the case at hand, both the HEMS and the EESS have a positive impact on SC, but they will not reduce the energy cost consumed. The profitability of the investment depends largely on the efficiency of energy accumulation. All in all, this is beneficial given the forecast that the increase in energy prices will make energy accumulation profitable for already installed batteries. It must be remembered that both systems were experimentally built at a very low cost. In the year 2021 alone, a few reasonably priced commercial systems with hybrid energy storage appeared on the market, which signals a positive trend in the development of this industry

An increasing nuisance for prosumers in Poland is the increasingly frequent shutdowns of inverters. Overvoltage caused by the PVS, which was still incidental in 2018, is already becoming a serious problem in the year 2021. Currently, Polish DSOs do not control micro-installations both below and up to 10 kWp, but new inverters that meet the requirements of the grid code are already prepared for this. However, there is no infrastructure that would allow for remote control of the generated power. For now, one of the best ways to increase grid stability is to shave energy peaks. The article also highlights the growing capacitive reactive power and presents its effects.

Energy storage is a solution that can be implemented, but, currently, it is not the most common practice in Poland. From the DSO point of view, the EESS is the more advantageous form, but as it has been shown, it is on the verge of profitability. Perhaps in the next announced support program for existing prosumers, the subsidies will be large enough to encourage investments into the accumulation of electric energy.

Funding: This research received no external funding.

Institutional Review Board Statement: Not applicable.

Informed Consent Statement: Not applicable.

Data Availability Statement: Not applicable.

Acknowledgments: I would like to thank my daughter, Katarzyna Zator, for proofreading.

Conflicts of Interest: The authors declare no conflict of interest.

Abbreviations

AWHP	air–water heat pump
BMS	battery management system
COP	coefficient of performance
DHWB	domestic hot water buffer
DSM	demand-side management
DSO	distribution system operator
DSR	demand-side request
ESS	energy storage system
EESS	electrical energy storage system
FiT	feed-in tariff
HEMS	home energy management system
IoT	internet of things
MPPT	maximum power point tracking
PAR	peak-to-average ratio
PV	photovoltaic
PVS	photovoltaic system
SC	self-consumption coefficient

References

1. Zator, S.; Skomudek, W. Impact of DSM on Energy Management in a Single-Family House with a Heat Pump and Photovoltaic Installation. *Energies* **2020**, *13*, 5476. [CrossRef]
2. Hoppmann, J.; Volland, J.; Schmidt, T.S.; Hoffmann, V.H. The economic viability of battery storage for residential solarv photovoltaic systems—A review and a simulation model. *Renew. Sustain. Energy Rev.* **2014**, *39*, 1101–1118. [CrossRef]
3. Why Has the Price of electricity in Europe Reached Record Highs. Available online: https://www.economist.com/the-economist-explains/2021/09/15/why-has-the-price-of-electricity-in-europe-reached-record-highs (accessed on 17 October 2021).
4. Average Monthly Electricity Wholesale Prices in Selected Countries in the European Union (EU) from January 2020 to July 2021. Available online: https://www.statista.com/statistics/1267500/eu-monthly-wholesale-electricity-price-country/ (accessed on 10 October 2021).
5. Cole, W.; Frazier, A.W.; Augustine, C. Cost Projections for Utility-Scale Battery Storage: 2021 Update, Technical Report NREL/TP-6A20-79236, June 2021. Available online: https://www.nrel.gov/docs/fy21osti/79236.pdf (accessed on 10 September 2021).
6. Polish Power Grids—DSR Programs. Available online: https://dsr.pse.pl (accessed on 10 September 2021).
7. Mahmud, A.S.M.A.; Sant, P. Real-time price savings through price suggestions for the smart grid demand response model. In Proceedings of the 5th International Istanbul Smart Grid and Cities Congress and Fair (ICSG), Istanbul, Turkey, 19–21 April 2017; pp. 65–69. [CrossRef]
8. Mahmood, D.; Javaid, N.; Alrajeh, N.; Khan, Z.A.; Qasim, U.; Ahmed, I.; Ilahi, M. Realistic Scheduling Mechanism for Smart Homes. *Energies* **2016**, *9*, 202. [CrossRef]
9. Wang, J.-J.; Jing, Y.-Y.; Zhang, C.-F.; Zhao, J.-Z. Review on multi-criteria decision analysis aid in sustainable energy decision-making. *Renew. Energy Rev.* **2009**, *13*, 2263–2278. [CrossRef]
10. Shaaban, M.; Scheffran, J.; Böhner, J.; Elsobki, M.S. Sustainability Assessment of Electricity Generation Technologies in Egypt Using Multi-Criteria Decision Analysis. *Energies* **2018**, *11*, 1117. [CrossRef]
11. Rigo, P.D.; Rediske, G.; Rosa, C.B.; Gastaldo, N.G.; Michels, L.; Neuenfeldt Júnior, A.L.; Siluk, J.C.M. Renewable Energy Problems: Exploring the Methods to Support the Decision-Making Process. *Sustainability* **2020**, *12*, 10195. [CrossRef]

12. Ridha, H.M.; Gomes, C.; Hizam, H.; Ahmadipour, M.; Heidari, A.A.; Chen, H. Multi-objective optimization and multi-criteria decision-making methods for optimal design of standalone photovoltaic system: A comprehensive review. *Renew. Sustain. Energy Rev.* **2021**, *135*, 110202. [CrossRef]
13. Marcelino, C.; Baumann, M.; Carvalho, L.; Chibeles-Martins, N.; Weil, M.; Almeida, P.; Wanner, E. A combined optimisation and decision-making approach for battery-supported HMGS. *J. Oper. Res. Soc.* **2020**, *71*, 762–774. [CrossRef]
14. Marcelino, C.; Baumann, M.; Weil, M.; Carvalho, L.; Wanner, E.; Almeida, P.; Miranda, V. Solving security constrained optimal power flow problems: A hybrid evolutionary approach. *Appl. Intell.* **2018**, *48*, 3672–3690. [CrossRef]
15. Barelli, L.; Bidini, G.; Bonucci, F. A microgrid operation analysis for cost-effective battery energy storage and RES plants integration. *Energy* **2016**, *113*, 831–844. [CrossRef]
16. Borhanazad, H.; Mekhilef, S.; Ganapathy, V.; Modiri-Delshad, M.; Mirtaheri, A. Optimization of micro-grid system using MOPSO. *Renew. Energy* **2014**, *71*, 295–306. [CrossRef]
17. Levron, Y.; Guerrero, J.M.; Beck, Y. Optimal Power Flow in Microgrids with Energy Storage. *IEEE Trans. Power Syst.* **2013**, *28*, 3226–3234. [CrossRef]
18. Silva, V.A.; Aoki, A.R.; Lambert-Torres, G. Optimal Day-Ahead Scheduling of Microgrids with Battery Energy Storage System. *Energies* **2020**, *13*, 5188. [CrossRef]
19. Hwang, J.S.; Rosyiana Fitri, I.; Kim, J.-S.; Song, H. Optimal ESS Scheduling for Peak Shaving of Building Energy Using Accuracy-Enhanced Load Forecast. *Energies* **2020**, *13*, 5633. [CrossRef]
20. Sant'Ana, W.C.; Gonzatti, R.B.; Lambert-Torres, G.; Bonaldi, E.L.; Andrade de Oliveira, P.; Torres, B.S.; Foster, J.G.L.; Pereira, R.R.; Borges-da-Silva, L.E.; Mollica, D.; et al. Implementation of Automatic Battery Charging Temperature Compensation on a Peak-Shaving Energy Storage Equipment. In Proceedings of the IEEE 15th Brazilian Power Electronics Conference and 5th IEEE Southern Power Electronics Conference (COBEP/SPEC), Santos, Brazil, 1–4 December 2019. [CrossRef]
21. Iqbal, Z.; Javaid, N.; Iqbal, S.; Aslam, S.; Khan, Z.A.; Abdul, W.; Almogren, A.; Alamri, A. A Domestic Microgrid with Optimized Home Energy Management System. *Energies* **2018**, *11*, 1002. [CrossRef]
22. Amer, A.; Shaban, K.; Gaouda, A.; Massoud, A. Home Energy Management System Embedded with a Multi-Objective Demand Response Optimization Model to Benefit Customers and Operators. *Energies* **2021**, *14*, 257. [CrossRef]
23. Luthander, R.; Widén, J.; Nilsson, D.; Palm, J. Photovoltaic self-consumption in buildings: A review. *Appl. Energy* **2015**, *142*, 80–94. [CrossRef]
24. Campoccia, A.; Dusonchet, L.; Telaretti, E.; Zizzo, G. An analysis of feed'in tariffs for solar PV in six representative countries of the European Union. *Solar Energy* **2014**, *107*, 530–542. [CrossRef]
25. Best Practices on Renewable Energy Self-Consumption. Commission Staff Working Document, Brussels. Available online: https://ec.europa.eu/energy/sites/ener/files/documents/1_EN_autre_document_travail_service_part1_v6.pdf (accessed on 15 January 2020).
26. Pan, X.; Khezri, R.; Mahmoudi, A.; Yazdani, A.; Shafiullah, G.M. Energy Management Systems for Grid-Connected Houses with Solar PV and Battery by Considering Flat and Time-of-Use Electricity Rates. *Energies* **2021**, *14*, 5028. [CrossRef]
27. Knutel, B.; Pierzyńska, A.; Dębowski, M.; Bukowski, P.; Dyjakon, A. Assessment of Energy Storage from Photovoltaic Installations in Poland Using Batteries or Hydrogen. *Energies* **2020**, *13*, 4023. [CrossRef]
28. Ecke, J.; Steinert, T.; Bukowski, M.; Śniegocki, A. Polski Sektor Energetyczny 2050. 4 Scenariusze. Warszawa. Available online: https://forum-energii.eu/public/upload/articles/files/Polski-sektor-energetyczny-2050_druk.pdf (accessed on 10 September 2021).
29. Wang, W.; He, W.; Cheng, J.; Huang, X.; Liu, H. Active and reactive power coordinated control strategy of battery energy storage system in active distribution network. In Proceedings of the 32nd Youth Academic Annual Conference of Chinese Association of Automation (YAC), Hefei, China, 19–21 May 2017; pp. 462–465. [CrossRef]
30. Hashmi, M.U.; Deka, D.; Bušić, A.; Pereira, L.; Backhaus, S. Arbitrage with Power Factor Correction Using Energy Storage. *IEEE Trans. Power Syst.* **2020**, *35*, 2693–2703. [CrossRef]
31. Amiel, I.; Rajput, S.; Averbukh, M. Capacitive reactive power compensation to prevent voltage instabilities in distribution lines. *Int. J. Electr. Power Energy Syst.* **2021**, *131*, 107043. [CrossRef]
32. Rozporządzenie Ministra Energii z Dnia 6 Marca 2019r. w Sprawie Szczegółowych Zasad Kształtowania I Kalkulacji Taryf Oraz Rozliczeń w Obrocie Energią Elektryczną, Dz. U. 2019 r. Poz. 503 and Dz.U. 2020 poz. 2053. Available online: https://www.ure.gov.pl/pl/urzad/prawo/rozporzadzenia/rozporzadzenia-w-spraw/7322,Rozporzadzenie-w-sprawie-taryf-energia-elektryczna.html (accessed on 15 October 2021).

energies

MDPI

Article

A Robust Algorithm for Real-Time Phasor and Frequency Estimation under Diverse System Conditions

Babak Jafarpisheh [1] and Anamitra Pal [2,*]

[1] Depsys SA, Route du Verney 20B, 1070 Puidoux, Switzerland; babak_jafarpisheh@yahoo.com
[2] School of Electrical, Computer and Energy Engineering, Arizona State University, Tempe, AZ 85281, USA
* Correspondence: anamitra.pal@asu.edu; Tel.: +1-(480)-965-2882

Abstract: This paper presents a comprehensive approach for performing phasor and frequency estimation from voltage and/or current signals of the modern power system. Undesirable components, such as decaying DC, if present in the input signal, are first attenuated using a complex-gain filter. The initial estimates of phasor and frequency are obtained next using the discrete Fourier transform and an improved estimation of signal parameters via rotational invariance technique, respectively. Finally, the accuracy of phasor and frequency estimates are increased based on the identified system condition. Simulations performed to evaluate the proposed approach confirm that it can do fast and accurate estimation of phasor and frequency under diverse operating conditions, making it ideal for wide-area monitoring, protection, and control applications in power systems.

Keywords: discrete Fourier transform (DFT); estimation of signal parameters via rotational invariance techniques (ESPRIT); frequency estimation; phasor estimation

Citation: Jafarpisheh, B.; Pal, A. A Robust Algorithm for Real-Time Phasor and Frequency Estimation under Diverse System Conditions. *Energies* **2021**, *14*, 7112. https://doi.org/10.3390/en14217112

Academic Editor: Germano Lambert-Torres

Received: 3 September 2021
Accepted: 15 October 2021
Published: 1 November 2021

1. Introduction and Background

Optimal energy management in smart grids requires an enhanced visualization of its system parameters, e.g., voltage/current phasors and frequency. This heightened visualization is usually provided by devices capable of producing time-synchronized measurements, such as phasor measurement units (PMUs) in the transmission system and micro-PMUs in the distribution system. To ensure the relevance of the measurements obtained from these devices for monitoring, protection, and control applications, it is necessary that the estimation algorithms used in them are accurate, robust against stray components, computationally efficient, and have low response time [1,2]. Hence, digital signal processing techniques such as discrete Fourier transform (DFT) [3–11], least squares (LS) [12–15], maximum likelihood [16], space vector transform [17], artificial neural networks [18], Hilbert transform [19], Stockwell transform [20], matrix pencil method [21], Kalman filters [22,23], subspace-based methods [24,25], and filter-based methods [26,27] have been proposed recently to estimate phasor and/or frequency under different operating conditions. However, many of the techniques mentioned above suffer from long response time during switching transients [9,13,20], high computational complexity [7,21,24], susceptibility to grid disturbances [12,18,22] and noise [19], lengthy observation window [7,10,11,25–27], and performance degradation due to: (i) off-nominal frequency condition [4–6,9,12,14,18], and (ii) presence of undesirable components in the input signals, such as decaying DC (DDC) [3,8,15–17,23,25]. Thus, there is a genuine need to create robust, real-time phasor and frequency algorithms for synchrophasors as it will ensure a more effective management of the smart grid.

The proposed research satisfies this need and goes beyond the state-of-the-art in the following ways. First, stray components (such as DDC), if present in the input signal, are removed/attenuated by a filter that is designed based on the concept of *complex frequencies*. The initial values of phasor and frequency are then estimated in parallel from the filtered signal using DFT and an improved version of estimation of signal parameters via rotational invariance technique (ESPRIT) [28], called iESPRIT that significantly reduces the computational burden. Finally, the initial estimates of phasor and frequency are updated based on the identified system condition (normal/abnormal). Performance evaluation of the proposed approach using different test signals indicates that it makes the following salient contributions:

1. It efficiently suppresses undesirable components present in the input signal, and can therefore be applied to both voltage and current signals.
2. It provides fast, accurate, and robust phasor and frequency estimates under diverse operating conditions (faults, frequency ramps, power swings, harmonics).
3. It detects abnormal system conditions quickly (in less than one-cycle).
4. It satisfies the requirements of relevant IEEE and IEC Standards (this paper focuses on P-class PMUs which have more stringent speed requirements) [29,30].

2. Proposed Filter for DDC Rejection

Considering the most general setting, let the input signal be described by

$$x(t) = A_0 + \sum_{k=1}^{K} A_k \cos(\omega_k t + \theta_k) + D_0 e^{-t/\tau} \tag{1}$$

where A_0 denotes a DC offset; K denotes the total number of sinusoidal components; A_k and θ_k denote the magnitude and phase angle of the k-th sinusoid. The angular frequency of the k-th sinusoid is denoted by ω_k and is equal to $2\pi f_k$, where f_k is the frequency of the k-th sinusoid; the nominal frequency and angular frequency are denoted by f_{nom} and ω_{nom}, respectively. D_0 and τ are the magnitude and time constant of the DDC component and are called DDC parameters. The above signal can be written in discretized form as

$$x[n] = A_0 + \sum_{k=1}^{K} A_k \cos(\omega_k n\Delta t + \theta_k) + D_0 E^n \tag{2}$$

in which Δt is the sampling interval and is equal to $1/f_s$ (where f_s is the sampling frequency), n denotes the sample number, and $E \triangleq e^{-\Delta t/\tau}$, called the *damping factor*, depends on the DDC time constant [4]. It is difficult to predetermine the DDC parameters as they depend on fault characteristics, such as, fault inception time/angle, fault location, and fault resistance [4–6,31]. However, their presence (usually in the current signals) can cause up to 15% error in the estimated phasor [31]. This can result in a maximum overreach of 115% if the first zone of a distance relay is set at 100% of the line impedance. Thus, DDC suppression is important for safe and reliable operation of the smart grid.

The goals of the proposed filter design are: (i) removing/attenuating the DDC for different values of its time constant, (ii) preventing variation in parameters of the fundamental component during the removal process, and (iii) attaining fast response time (or equivalently low-order characteristic). The frequencies present in (1) are shown in Figure 1, where the frequencies of the DC offset, DDC, and sinusoidal components are located at the origin, negative side of the real axis, and imaginary axis of the s-plane, respectively. The DDC time constant (τ) varies due to fault characteristics (such as fault impedance and fault location) between 0.5 and 5 cycles [31,32]; hence, a corresponding lower bound (τ_l) and upper bound (τ_u) are specified. Since the DDC frequency is equal to $-1/\tau$ in the s-plane, the zeros of the proposed filter should lie between $[-1/\tau_l, -1/\tau_u]$, which is called DDC frequency band (see Figure 1), to filter out the DDC. However, the filter's zeros will cause phase shifting of the fundamental component. To compensate for this phase-shift, filter's poles can be added (as was done in [6]). However, doing so will result in: (i) neutralization of the effect of DDC attenuation by the filter's zeros, (ii) additional error in phasor estimation due to the vicinity of the added poles to the fundamental frequency, and (iii) oscillatory behavior in the estimated magnitude and phase angle. To solve this problem, in this paper, we propose the design of a complex gain filter. The transfer function of the proposed filter is given by

$$G(s) = K_F \prod_{m=1}^{R} (s - z_m) \tag{3}$$

where the total number of zeros (or equivalently the filter order) is denoted by R which plays a key role in the response time of the proposed filter. The filter gain is expressed by K_F which is complex (i.e., $K_F \triangleq |K_F|e^{j\gamma_F}$). z_ms are inserted in the DDC frequency band with the aim of reaching minimum multiplication of their distance to any frequency in this band. This can be formulated as the following optimization problem:

$$\min_{z_m} \prod_{m=1}^{R} |x - z_m|^2 \quad \forall x \in [-1/\tau_l, -1/\tau_u]$$
$$\text{subject to:} \ -1/\tau_l < z_m < -1/\tau_u \quad \forall m \in \{1, \cdots, R\} \tag{4}$$

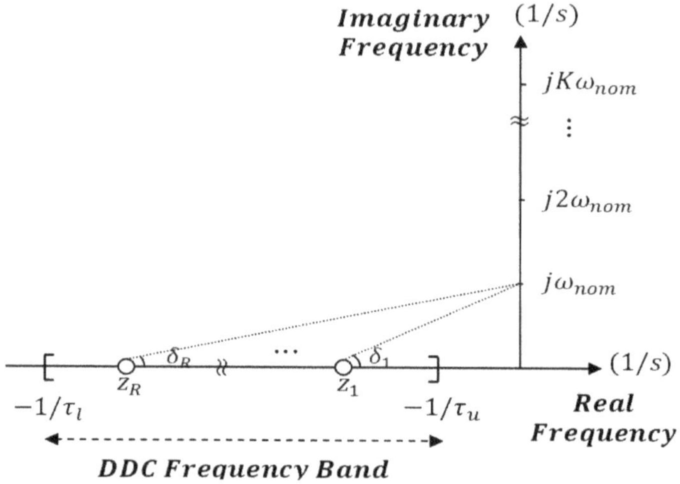

Figure 1. Design of the proposed filter in s-plane.

Note that this optimization problem ensures that the proposed filter suppresses DDC components regardless of their time-constant values, without any re-tuning.

Next, to compensate for the phase shift due to the added zeros, the phase-angle of the filter gain γ_F is adjusted as follows (see Figure 1):

$$\gamma_F + \sum_{m=1}^{R} \delta_m = 0 \tag{5}$$

Finally, the magnitude of the filter gain $|K_F|$ is determined with the aim of reaching unit total filter gain at nominal frequency f_{nom}. This is given by

$$|G(\omega_{nom} = 2\pi f_{nom})| = 1 \tag{6}$$

Since the proposed filter is designed for ω_{nom}, its performance may deteriorate at off-nominal frequencies. Under such circumstances, the filter's parameters can be readjusted as described in Section 6.2. As the number of filter's zeros increases, the filter can suppress DDC components more effectively, at the cost of increased response time. This trade-off between the proposed-filter's capacity for DDC attenuation and its response time is investigated in Section 7.1.

3. Phasor Estimation at Off-Nominal Frequencies

DFT is one of the most widely used techniques for phasor estimation due to its simplicity in implementation, low computational burden, harmonic filtering, and noise alleviation [1]. By applying DFT on the filtered signal, denoted by subscript f, we get

$$\acute{\hat{X}}_1 = \frac{2}{N}\sum_{n=r}^{r+N-1} x_f[n]e^{-j\frac{2\pi}{N}n} = \frac{2}{N}\sum_{n=r}^{r+N-1} x_f[n]e^{-jn\omega_{nom}\Delta t} \quad (7)$$

where r is the number of the first sample in the data string, N denotes the number of samples per cycle, and $\acute{\hat{X}}_1$ refers to estimated fundamental phasor at nominal frequency. Note that the DC offset and harmonic components, present in (2), are filtered out by the DFT, while the DDC has already been attenuated by the proposed filter (see Section 2). However, when the frequency deviates from its nominal value f_{nom}, the phasor estimated using DFT can be erroneous. This requires additional correction, as illustrated below.

Consider an off-nominal frequency condition in which the angular frequency is different from its nominal value ω_{nom}; i.e., let the actual angular frequency be ω_1. The DFT of the filtered input signal in this situation can be written as

$$\hat{X}_1 = \frac{2}{N}\sum_{n=r}^{r+N-1} x_f[n]e^{-jn\omega_1\Delta t} \quad (8)$$

where $\hat{X}_1$ is the estimated phasor by DFT at actual frequency ω_1. After performing trigonometric simplifications on (7) and (8), we get

$$\acute{\hat{X}}_1 = P\hat{X}_1 e^{jr(\omega_1-\omega_{nom})\Delta t} + Q\hat{X}_1^* e^{-jr(\omega_1+\omega_{nom})\Delta t} \quad (9)$$

$$P \triangleq \frac{\sin\left(\frac{N(\omega_1-\omega_{nom})\Delta t}{2}\right)}{N\sin\left(\frac{(\omega_1-\omega_{nom})\Delta t}{2}\right)} e^{j(N-1)\frac{(\omega_1-\omega_{nom})\Delta t}{2}} \quad (10)$$

$$Q \triangleq \frac{\sin\left(\frac{N(\omega_1+\omega_{nom})\Delta t}{2}\right)}{N\sin\left(\frac{(\omega_1+\omega_{nom})\Delta t}{2}\right)} e^{-j(N-1)\frac{(\omega_1+\omega_{nom})\Delta t}{2}} \quad (11)$$

where P and Q are correction coefficients that are independent of 'r' and dependent on N and frequency deviation Δf. Variations of P and Q for different values of N and Δf are shown in Tables A1 and A2 in Appendix A.1. In a typical power system, the variations in the frequency rarely exceed 0.5 Hz in either direction from its nominal value [33]. As shown in Tables A1 and Appendix A.1, Q is negligibly small in comparison to P in the range of -0.5 Hz and $+0.5$ Hz. The impact of Q on the accuracy of the estimated phasor by the proposed algorithm is investigated in more detail in Appendix A.2. From the conducted investigation, it is realized that the impact of Q on the accuracy of the estimated phasor by the proposed algorithm is small (in comparison to the total vector error (TVE) limit of 1% [29,30]). Consequently, (9) is simplified and rewritten as follows:

$$\acute{\hat{X}}_1 \approxeq \underbrace{Pe^{jr(\omega_1-\omega_{nom})\Delta t}}_{\acute{P}} \hat{X}_1 = \acute{P}\hat{X}_1 \quad (12)$$

Since $\acute{P}$ depends on the actual frequency, it is necessary to estimate the frequency precisely and rapidly for correcting the phasor estimate. The frequency estimation algorithm described in Section 4 can not only improve accuracy of the estimated phasor at off-nominal frequency conditions (see Section 6.2), but also estimate the frequency quickly and accurately for other power system applications.

4. Proposed Frequency Estimation Algorithm

ESPRIT is a well-known subspace-based technique for frequency estimation due to its strong immunity against noise [28]. However, due to its high computational complexity [25], ESPRIT is not suitable for power system applications that require fast estimation of frequency. Hence, an improved version of the ESPRIT, called iESPRIT, is proposed in this paper. The input signal for frequency estimation is the output of the filter that was proposed in Section 2, and can be expressed as

$$x_f[n] = A_{0f} + A_1 \cos(\omega_1 n\Delta t + \theta_1) + \sum_{k=2}^{K} A_{kf} \cos(\omega_k n\Delta t + \theta_{kf}) + w_f[n] \tag{13}$$

where $w_f[n]$ denotes an additive noise component with the power of ${\sigma_n}^2$. Note that the filter was designed in such a way that it kept the parameters of the fundamental component unchanged. Equation (13) can be equivalently written in the form of $Y \triangleq 2K+1$ complex exponentials distorted with noise, called the *harmonic model*, as follows:

$$\begin{aligned} x_f[n] &= A_{0f} + \frac{1}{2}A_1\left(e^{j(\omega_1 n\Delta t+\theta_1)} + e^{-j(\omega_1 n\Delta t+\theta_1)}\right) \\ &+ \sum_{k=2}^{K} \frac{1}{2}A_{kf}\left(e^{j(\omega_k n\Delta t+\theta_{kf})} + e^{-j(\omega_k n\Delta t+\theta_{kf})}\right) + w_f[n] = \sum_{k=-K}^{K} \alpha_k e^{j\omega_k n\Delta t} + w_f[n] \end{aligned} \tag{14}$$

in which $\alpha_0 = A_{0f}$ and $\omega_0 = 0$; $\alpha_1 = \alpha^*_{-1} \triangleq |\alpha_1|e^{j\varphi_1} = \frac{1}{2}A_1 e^{j\theta_1}$ and $\omega_1 = -\omega_{-1}$; $\alpha_k = \alpha^*_{-k} \triangleq |\alpha_k|e^{j\varphi_k} = \frac{1}{2}A_{kf}e^{j\theta_{kf}}$ and $\omega_k = -\omega_{-k}$ for $k = 2, \cdots, K$. For simplicity, we denote the Y α_ks and ω_ks for $k = -K, \cdots, 0, \cdots, K$ with $\alpha_1, \alpha_2, \cdots, \alpha_Y$ and $\omega_1, \omega_2, \cdots, \omega_Y$, respectively, and let $\overline{\omega}_k \triangleq (\omega_k/f_s)$, where $\overline{\omega}_k$ is thenormalized angular frequency of the k-th exponential component. Then, by using M consecutive samples of $x_f[n]$, an M-length data vector $\mathbf{x}_f(n)$ is defined as follows:

$$\begin{aligned} \mathbf{x}_f(n) \triangleq \begin{bmatrix} x_f[n] \\ x_f[n+1] \\ \vdots \\ x_f[n+M-1] \end{bmatrix} &= \underbrace{\begin{bmatrix} 1 & \cdots & 1 \\ e^{j\overline{\omega}_1} & \cdots & e^{j\overline{\omega}_Y} \\ \vdots & \ddots & \vdots \\ e^{j(M-1)\overline{\omega}_1} & \cdots & e^{j(M-1)\overline{\omega}_Y} \end{bmatrix}}_{\mathbf{H}_{M\times Y}} \mathbf{\Phi}^n \underbrace{\begin{bmatrix} \alpha_1 \\ \alpha_2 \\ \vdots \\ \alpha_Y \end{bmatrix}}_{\boldsymbol{\alpha}} \\ &+ \underbrace{\begin{bmatrix} w_f[n] \\ w_f[n+1] \\ \vdots \\ w_f[n+M-1] \end{bmatrix}}_{\mathbf{w}_f(n)} = \mathbf{H}\mathbf{\Phi}^n\boldsymbol{\alpha} + \mathbf{w}_f(n) \end{aligned}$$

$$\mathbf{\Phi}_{Y\times Y} = \text{diag}\left[e^{j\overline{\omega}_1}, e^{j\overline{\omega}_2}, \cdots, e^{j\overline{\omega}_Y}\right] \tag{15}$$

where the Y columns of matrix $\mathbf{H}$ are length-M *signal frequency vectors* described by

$$\mathbf{h}(\overline{\omega}_k) \triangleq \begin{bmatrix} 1 & e^{j\overline{\omega}_k} & \cdots & e^{j(M-1)\overline{\omega}_k} \end{bmatrix}^T \tag{16}$$

for $k = 1, \cdots, Y$. The matrix $\mathbf{H}$ is simplified using (16) as

$$\mathbf{H}_{M\times Y} = [\mathbf{h}(\overline{\omega}_1) \quad \mathbf{h}(\overline{\omega}_2) \quad \cdots \quad \mathbf{h}(\overline{\omega}_Y)] \tag{17}$$

In addition, matrix $\mathbf{\Phi}$ is the diagonal matrix of phase shifts and can be expressed as $\mathbf{\Phi} = \text{diag}[\phi_1, \phi_2, \cdots, \phi_Y]$, where $\phi_k = e^{j\overline{\omega}_k}$ for $k = 1, 2, \cdots, Y$. Since the frequencies of the complex exponentials $\overline{\omega}_k$s completely describe the matrix $\mathbf{\Phi}$, the estimated frequencies can be obtained by finding $\mathbf{\Phi}$.

Let us consider two overlapping sub-matrices within the length-M window vector. In this regard, an $(M-1) \times Y$ matrix $\mathbf{H}_{M-1}$, which is the length-$(M-1)$ sub-matrix of $\mathbf{H}$, is defined as

$$\mathbf{H}_{M-1} = [\mathbf{h}_{M-1}(\overline{\omega}_1)\ \mathbf{h}_{M-1}(\overline{\omega}_2)\ \cdots\ \mathbf{h}_{M-1}(\overline{\omega}_Y)] \tag{18}$$

in which $\mathbf{h}_{M-1}(\overline{\omega}_k) \triangleq \begin{bmatrix} 1 & e^{j\overline{\omega}_k} & \cdots & e^{j(M-2)\overline{\omega}_k} \end{bmatrix}^T$ for $k = 1, 2, \cdots, Y$. Then, two new $(M-1) \times Y$ matrices which are called $\mathbf{H}_1$ and $\mathbf{H}_2$ are defined as follows:

$$\mathbf{H}_1 \triangleq \mathbf{H}_{M-1}\mathbf{\Phi}^n \quad \text{and} \quad \mathbf{H}_2 \triangleq \mathbf{H}_{M-1}\mathbf{\Phi}^{n+1} \tag{19}$$

Using (15) and (19), we have:

$$\mathbf{H}\mathbf{\Phi}^n = \begin{bmatrix} & \mathbf{H}_1 & \\ * & * & * \end{bmatrix} = \begin{bmatrix} * & * & * \\ & \mathbf{H}_2 & \end{bmatrix} \tag{20}$$

As seen in (19), the matrices $\mathbf{H}_1$ and $\mathbf{H}_2$ are related as

$$\mathbf{H}_2 = \mathbf{H}_1\mathbf{\Phi} \tag{21}$$

Now, a $G \times M$ matrix, $\mathbf{X}_f$, called the data matrix, can be obtained from an L-length of input data

$$\mathbf{X}_f \triangleq \begin{bmatrix} x_f[1] & x_f[2] & \cdots & x_f[M] \\ x_f[2] & x_f[3] & \cdots & x_f[M+1] \\ \vdots & \vdots & \ddots & \vdots \\ x_f[G] & x_f[G+1] & \cdots & x_f[G+M-1] \end{bmatrix} \tag{22}$$

in which $L \triangleq G + M - 1$ and G is the total number of data vectors used to construct $\mathbf{X}_f$. In conventional ESPRIT, $\mathbf{\Phi}$ is calculated from the singular value decomposition (SVD) of $\mathbf{X}_f$, as shown below [28]

$$\mathbf{X}_f = \mathbf{L}\mathbf{\Sigma}\mathbf{U}^H \tag{23}$$

where $\mathbf{L}_{G\times G}$ and $\mathbf{U}_{M\times M}$ are the left and right singular vectors (both of them are unitary), and $\mathbf{\Sigma}_{G\times M}$ is a diagonal matrix containing singular values ordered in descending magnitude. However, SVD of $\mathbf{X}_f$ will have a complexity of $O(GM^2 + MG^2 + M^3)$ [34], which makes ESPRIT too slow for frequency estimation in P-class PMUs [1,29,30]. To overcome this problem, the proposed iESPRIT estimates an autocorrelation matrix from $\mathbf{X}_f$ as shown below

$$\hat{\mathbf{R}}_x = \frac{1}{G}\mathbf{X}_f^H\mathbf{X}_f \tag{24}$$

where $\hat{\mathbf{R}}_x$ can be expressed in terms of its eigenvalues and eigenvectors using eigendecomposition (ED) as

$$\hat{\mathbf{R}}_x = \mathbf{Q}\mathbf{\Lambda}\mathbf{Q}^H \tag{25}$$

where $\mathbf{\Lambda}$ is an $M \times M$ diagonal matrix containing the eigenvalues of the $\hat{\mathbf{R}}_x$ in descending order. The columns of $\mathbf{Q}$ are also the corresponding eigenvectors. It is proved in Appendix B that the eigenvectors of $\hat{\mathbf{R}}_x$ (columns of $\mathbf{Q}$) are equal to the columns of $\mathbf{U}$ which is used in ESPRIT (see (23)). Note that the complexity of ED of $\hat{\mathbf{R}}_x$ is $O(M^3)$ [34], which is much lower than the complexity of SVD [35]. The Y largest diagonal elements of $\mathbf{\Lambda}$ (i.e., $\lambda_1, \lambda_2, \cdots, \lambda_Y$) are equal to $M|\alpha_k|^2 + {\sigma_n}^2$ (for $k = 1, 2, \cdots, Y$), while the remaining $M - Y + 1$ diagonal elements of $\mathbf{\Lambda}$ (i.e., $\lambda_{Y+1}, \cdots, \lambda_M$) have similar values as ${\sigma_n}^2$ [28]. As a result, the total number of complex exponentials Y can be obtained using the criterion of the relative slope variation between the $\hat{\mathbf{R}}_x$'s eigenvalues as shown in Figure 2. Note that the fundamental frequency corresponds to the largest singular value in $\mathbf{\Lambda}$.

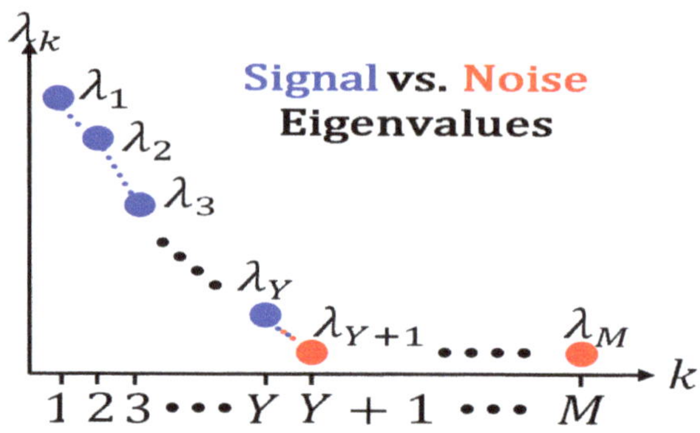

Figure 2. Slope variation between the eigenvalues of the $\hat{\mathbf{R}}_x$.

In order to find $\mathbf{\Phi}$, the matrix $\mathbf{Q}$ is split into two sub-matrices for separation of the signal and noise subspaces

$$\mathbf{Q}_{M\times M} = [\mathbf{Q}_S \mid \mathbf{Q}_n] \tag{26}$$

where $\mathbf{Q}_S$ is an $M \times Y$ sub-matrix that corresponds to the Y largest magnitudes of the eigenvalues (see (25)). Since the elements of the matrix $\mathbf{H}$ defined in (17) completely describe the frequencies of exponential components existing in the input signal, $\mathbf{H}$ similar to $\mathbf{Q}_S$ must also lie in the same subspace, called the *signal subspace*. As a result, an invertible transformation $\mathbf{W}$ exists that maps $\mathbf{Q}_S$ into $\mathbf{H}$ as

$$\mathbf{H} = \mathbf{Q}_S\mathbf{W} \tag{27}$$

Note that it is not necessary to calculate the transformation $\mathbf{W}$, and it is only formulated as a mapping between the matrices $\mathbf{H}$ and $\mathbf{Q}_S$ within the signal subspace. The matrix $\mathbf{Q}_S$ can also be split into two smaller $M-1$-dimensional subspaces as was done with the matrix $\mathbf{H}$

$$\mathbf{Q}_S = \begin{bmatrix} & \mathbf{Q}_1 & \\ * & * & * \end{bmatrix} = \begin{bmatrix} * & * & * \\ & \mathbf{Q}_2 & \end{bmatrix} \tag{28}$$

where both $\mathbf{Q}_1$ and $\mathbf{Q}_2$ are $(M-1) \times Y$ sub-matrices. Since $\mathbf{H}_1$ and $\mathbf{H}_2$ correspond to the same subspaces as $\mathbf{Q}_1$ and $\mathbf{Q}_2$, (27) must also hold for these subspaces, as shown below:

$$\mathbf{H}_1 = \mathbf{Q}_1\mathbf{W} \quad \text{and} \quad \mathbf{H}_2 = \mathbf{Q}_2\mathbf{W} \tag{29}$$

As seen in (21), the rotational matrix $\mathbf{\Phi}$ exists that relates (rotates) $\mathbf{H}_1$ to $\mathbf{H}_2$. In the same way, a rotational matrix (denoted by $\mathbf{\Psi}$ in this paper) must also exist that relates (rotates) $\mathbf{Q}_1$ to $\mathbf{Q}_2$ as

$$\mathbf{Q}_2 = \mathbf{Q}_1\mathbf{\Psi} \tag{30}$$

Note that the matrices $\mathbf{Q}_1$ and $\mathbf{Q}_2$ are known from the ED of $\hat{\mathbf{R}}_x$. Therefore, from (30), the rotational matrix $\mathbf{\Psi}$ can be estimated using LS as

$$\hat{\mathbf{\Psi}} = \left(\mathbf{Q}_1^H\mathbf{Q}_1\right)^{-1}\mathbf{Q}_1^H\mathbf{Q}_2 \tag{31}$$

Finally, using (21), (29), and (30), we get

$$\hat{\mathbf{\Phi}} = \mathbf{W}^{-1}\hat{\mathbf{\Psi}}\mathbf{W} \tag{32}$$

Now, as the diagonal entries of $\hat{\mathbf{\Phi}}$ ($\hat{\phi}_k = e^{j\hat{\overline{\omega}}_k}$ for $k = 1,2,\cdots,Y$) are simply the eigenvalues of $\hat{\mathbf{\Psi}}$, the estimated frequencies can be obtained as

$$\hat{\overline{\omega}}_k = \angle\hat{\phi}_k \qquad \text{for} \quad k = 1,2,\cdots,Y \tag{33}$$

Recall $\overline{\omega}_k = 2\pi f_k/f_s$; therefore, the estimated frequencies are

$$\hat{f}_k = f_s(\angle\hat{\phi}_k)/(2\pi) \qquad \text{for} \quad k = 1,2,\cdots,Y \tag{34}$$

5. Fast Detection of System Operating Conditions

The input-signal parameters (magnitude, phase angle, frequency) may abruptly change/oscillate due to power system events, such as faults. Termed *abnormal* conditions in this paper, these sudden changes can quickly be detected using $\hat{\mathbf{R}}_x$. Let the estimated autocorrelation matrix corresponding to the previous data string be denoted by $\hat{\mathbf{R}}_{x_{old}}$. The new autocorrelation matrix $\hat{\mathbf{R}}_{x_{new}}$ is estimated using (24) when the most recent data string is obtained. We can now create an $M \times M$ matrix $\mathbf{\Delta}\hat{\mathbf{R}}_x$ as shown below:

$$\mathbf{\Delta}\hat{\mathbf{R}}_x \triangleq \hat{\mathbf{R}}_{x_{new}} - \hat{\mathbf{R}}_{x_{old}} \tag{35}$$

The Frobenius norm of $\mathbf{\Delta}\hat{\mathbf{R}}_x$, $||\mathbf{\Delta}\hat{\mathbf{R}}_x||_F$, is given by

$$||\mathbf{\Delta}\hat{\mathbf{R}}_x||_F = \sqrt{\sum_{i=1}^{M}\sum_{j=1}^{M}|\mathbf{\Delta}\hat{\mathbf{R}}_x(i,j)|^2} \tag{36}$$

A threshold ζ can now be defined for $||\mathbf{\Delta}\hat{\mathbf{R}}_x||_F$ to determine the system's current operating condition as follows:

1. If $||\mathbf{\Delta}\hat{\mathbf{R}}_x||_F \leq \zeta$, it implies that the system operates under normal conditions and the signal parameters have varied slightly. In such conditions, the estimated phasors from previous and current data strings can be used for accurate estimation of frequency (see Section 6.1).
2. If $||\mathbf{\Delta}\hat{\mathbf{R}}_x||_F > \zeta$, it implies that the system is faced with an abnormal condition resulting in the signal parameters differing significantly from the previous data string. In such conditions, the proposed iESPRIT must be used for precise estimation of frequency, and subsequent correction of the estimated phasor (see Section 6.2).

The impact of ζ on the accuracy and computational burden of the proposed approach is investigated in detail in Section 8.6.

6. Accuracy Enhancement of Estimated Phasor and Frequency under Detected Conditions

The main goal of this section is to improve the accuracy of the estimated phasor (Section 3) and frequency (Section 4) based on the system operating condition (normal/abnormal) identified in Section 5.

6.1. Implementation of the Phasor Estimation Algorithm for Frequency Estimation under Normal Conditions

Let the estimated fundamental phasor obtained from the previous N-length data string with sample numbers $\{r, r+1,\cdots,r+N-1\}$ be denoted by $\hat{X}_{1old}$, which is calculated as follows at $\omega_{1old} = 2\pi f_{1old}$.

$$\hat{X}_{1old} = \frac{2}{N}\sum_{n=r}^{r+N-1} x[n]e^{-jn\omega_{1old}\Delta t} \tag{37}$$

Similarly, $\hat{X}_{1new}$ is obtained from the current N-length data string with sample numbers $\{r+1, r+2, \cdots, r+N\}$ at $\omega_{1new} = 2\pi f_{1new}$ as

$$\hat{X}_{1new} = \frac{2}{N} \sum_{n=r+1}^{r+N} x[n] e^{-jn\omega_{1new}\Delta t} \tag{38}$$

It is proved in Appendix C that, under normal conditions, the parameter $\acute{P}$ (defined in (12)) can be obtained as follows.

$$\acute{P} \approxeq \hat{X}_{1old} / \hat{X}_{1new} \tag{39}$$

Now, as $\acute{P}$ only depends on N, r, and $\Delta\omega_1$, for a pre-determined value of N, $\acute{P}$ can be calculated offline for different values of r and $\Delta\omega_1$ in the form of a look-up table. Then, in real-time, the $\Delta\hat{\omega}_1$ can be obtained with high accuracy from its corresponding $\acute{P}$ using the look-up table. Finally, the estimated frequency can be updated during normal conditions as

$$\hat{f}_{1new} = \hat{f}_{1old} + \Delta\hat{\omega}_1 / (2\pi) \tag{40}$$

The steps of the proposed frequency estimation algorithm are summarized in Algorithm 1.

Algorithm 1 Proposed Frequency Estimation Algorithm

Input: signal samples, $\hat{X}_{1old}$, $\hat{X}_{1new}$
Output: estimated frequency
Initialization: M and L
1: take L-length of input data
2: create $\mathbf{X}_f$, estimate $\hat{\mathbf{R}}_{x_{new}}$, and calculate $\Delta\hat{\mathbf{R}}_x$ using (22), (24), and (35), respectively
3: calculate $||\Delta\hat{\mathbf{R}}_x||_F$ using (36)
4: save $\hat{\mathbf{R}}_{x_{new}}$ as $\hat{\mathbf{R}}_{x_{old}}$ for the next estimation
5: **if** ($||\Delta\hat{\mathbf{R}}_x||_F \leq \zeta$) **then**
6: calculate $\acute{P}$ using (39)
7: find $\Delta\hat{\omega}_1$ using the lookup table
8: estimate frequency using (40)
9: go to line 18
10: **end if**
11: **if** ($||\Delta\hat{\mathbf{R}}_x||_F > \zeta$) **then**
12: calculate eigenvectors of $\hat{\mathbf{R}}_x$
13: find $\mathbf{Q}_S$ by (26), split it into $\mathbf{Q}_1$ and $\mathbf{Q}_2$ by (28)
14: estimate $\hat{\mathbf{\Psi}}$ using (31)
15: calculate eigenvalues of $\hat{\mathbf{\Psi}}$
16: estimate frequency using (34)
17: **end if**
18: **return** $\hat{f}_1$

6.2. *Implementation of the Proposed Frequency Estimation Algorithm for Phasor Estimation under Abnormal Conditions*

The first valid estimated frequency realized using Algorithm 1 is obtained after receiving the first L-length of input data; let this frequency be denoted by $\hat{f}_{1,L}$. It will be shown in Section 7.4 that the proposed iESPRIT is able to estimate the frequency precisely even when L is *lower than one-cycle-length of input data* (i.e., $L < N$), which is a significant achievement especially for protection applications. Therefore, it is possible to correct the estimated phasors by using $\acute{P}$ which is dependent on the accurate frequency $\hat{\omega}_1$ (see (9), (10) and (12)). To do so, we calculate the median (Mdn) of a set $\mathfrak{F}$ that contains the previous and current estimated frequencies (i.e., $\mathfrak{F} = \{\hat{f}_{1,L}, \hat{f}_{1,L+1}, \cdots, \hat{f}_{1,N}\}$) as shown in (41).

$$\hat{\omega}_1 = 2\pi.Mdn(\mathfrak{F}) \tag{41}$$

Employing the median brings additional robustness against noise. The value of the actual angular frequency $\hat{\omega}_1$, obtained from (41), can be used to correct the estimated phasors during abnormal conditions. Moreover, $\hat{\omega}_1$ can also be used to improve the performance of the proposed filter design (see Section 2) by replacing ω_{nom} with $\hat{\omega}_1$ in (5) and (6). In this regard, the filter's parameters can be calculated offline, for the pre-chosen filter order (see Section 7.1), for different values of ω_1 using (5) and (6), and stored in the form of a look-up table. In real time, the filter's parameters can be adjusted quickly using $\hat{\omega}_1$ and the look-up table.

Finally, note that the proposed frequency estimator itself provides strong immunity against noise and can be used for various power system applications that require accurate and fast frequency measurements, such as under-frequency relaying, power system stabilization/restoration, protection against loss of synchronism, and rapid synthetic inertia control.

Flowchart of the proposed approach for both phasor and frequency estimation is shown in Figure 3 in which normal and abnormal system operating conditions have been distinguished with green dashed and red solid arrows, respectively.

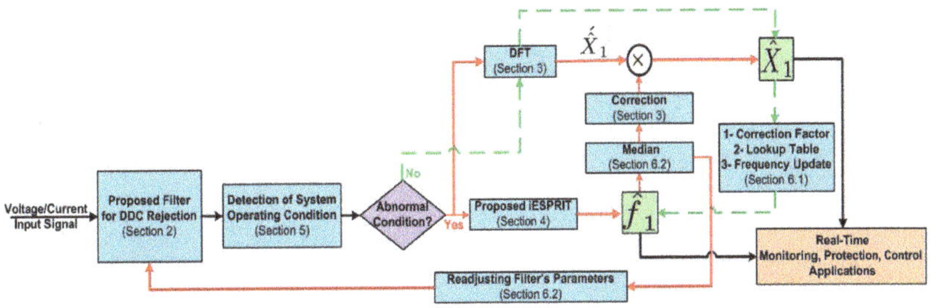

Figure 3. Flowchart of the proposed approach for phasor and frequency estimation.

7. Parameter Selection for Implementation

7.1. Choosing Optimal Value for the Filter Order R

To determine the filter order R of the complex gain filter (see Section 2), 100 synthetic test signals containing both fundamental and DDC components are generated. In the test signals, the D_0 and τ (DDC parameters) are randomly selected from a range of $[0, 1\ p.u.]$ and $[0.5, 5]$ (cycles), respectively [5]. Seven different values $\{3, 4, 6, 8, 10, 15, 20\}$ are considered for R. The input signals are passed through the proposed filter and DFT is then performed for estimation of the fundamental phasor. For each filter order, the maximum values of TVE [29,30] and settling time t_S [4], obtained over the 100 test signals are given in Table 1.

Table 1. Maximum values of TVE and t_S over 100 input test signals for different values of R.

R	3	4	6	8	10	15	20
Max. TVE (%)	0.76	0.67	**0.28**	0.22	0.14	0.04	0.01
Max. t_S (ms)	20.7	21.7	**26.9**	30.8	32.9	36.2	38.2

As can be seen from Table 1, by increasing R, the DDC is suppressed more effectively resulting in reduction of the maximum TVE at the cost of increasing the maximum t_S; i.e., there is a trade-off in choosing the optimal value of the filter order. Table 1 indicates that an acceptable value for the order of the proposed filter would be six since this provides a reasonable balance between DDC rejection (significantly lower maximum TVE than 1%) and response time.

7.2. Choosing Optimal Value for N

The impact of N (or equivalently f_s) on the accuracy of the estimated phasor by the proposed algorithm is investigated in this section. In this regard, a synthetic test signal containing a fundamental sinusoidal component that is distorted by DDC, harmonics, and noise is used as an input to the proposed algorithm. Six different values $\{16, 32, 128, 512, 1024, 16{,}384\}$ are considered for N. For each value, the fundamental phasor is estimated by the proposed algorithm, and the maximum TVE is calculated, which is given in Table 2.

Table 2. Impact of N on accuracy of the estimated phasor by the proposed algorithm.

N	16	**32**	128	512	1024	16,384
Max. TVE (%)	0.47	**0.43**	0.40	0.39	0.39	0.38

As can be seen in Table 2, the phasor estimation accuracy of the proposed algorithm converges to its limit value and is negligibly affected by increasing the number of samples per cycle, N (or equivalently the sampling frequency, f_s). On the other hand, increasing the sampling frequency causes (i) difficulties in hardware implementation of the phasor and frequency estimation algorithms in numerical relays, and (ii) increased computational burden, which is a major concern for the protection application. Based on Table 2, an appropriate value of N was found to be 32.

7.3. Choosing Optimal Value for M

The value of M plays a key role in computational complexity and robustness of the proposed iESPRIT against noise. As such, seven different values $\{3, 4, 5, 6, 8, 10, 15\}$ are considered for M. A synthetic test signal containing a sinusoidal component (with known frequency) which is distorted by noise is used as an input to the proposed iESPRIT. Note that the value of L (length of data string in (22)) is kept constant at N. The maximum absolute frequency estimation error (denoted by Max. $|\text{FE}|$) [30] and the execution time taken for estimation are presented in Table 3 for each of the seven values of M. Note that all of the simulations done in this paper have been carried out on a system with Core 2 Duo CPU (2.67 GHz) processor and 4 GB RAM.

Table 3. Maximum value of $|\text{FE}|$ and execution time for different values of M.

M	3	4	5	6	8	10	15
Max. $|$**FE**$|$ (mHz)	0.86	**0.41**	0.34	0.18	0.09	0.03	0.003
Exe. Time (ms)	0.12	**0.23**	0.47	0.81	1.74	3.05	9.7

It can be concluded from Table 3 that more robustness against noise and higher precision are achieved by choosing greater values for M, at the cost of increasing the execution time due to increasing complexity of calculations. Therefore, there is a trade-off between the precision of estimation and computational burden. From Table 3, an appropriate value of M is found to be four.

7.4. Choosing Optimal Value for L

The value for L (length of data string) plays a key role in convergence speed of the proposed iESPRIT. In this regard, six different values $\{N/4, N/2, 3N/4, N, 3N/2, 2N\}$ are considered for L. Likewise, as we did in Section 7.3, a synthetic signal is used while M is kept constant at four.

The Max $|\text{FE}|$ and settling time t_S are given in Table 4 for the six values of L. Considering optimal accuracy and convergence speed of the proposed iESPRIT, an appropriate value of L is found to be $N/2$.

Table 4. Maximum value of |FE| and t_S for different values of L.

L	$N/4$	$N/2$	$3N/4$	N	$3N/2$	$2N$
Max. |**FE**| (mHz)	1.03	**0.33**	0.25	0.04	0.007	0.0008
t_S (ms)	6.8	**11.4**	14.3	17.2	29.1	33.9

Note that the empirically obtained values of the parameters identified in Section 7 are kept constant for all the simulations performed in Section 8.

8. Simulation Results

In the following sub-sections, several static and dynamic test signals that comply with the IEEE and IEC Standards [29,30], as well as power system signals, are used to investigate the performance of the proposed approach. The above-mentioned standards only consider changes in the input signal parameters (i.e., magnitude, phase angle, and frequency). However, in this paper, the performance of the proposed algorithm is also evaluated in the presence of stray components, such as DDC and noise. Four recently published phasor and/or frequency estimation methods, applicable to P-class PMUs, have been selected for comparison purposes: (i) A phasor estimation method presented in [5], which is based on DFT and down sampling (called "DS-DFT"), (ii) a phasor and frequency estimation method presented in [13], which is based on LS and amplitude modulation modeling of the input signal (called "AM"), (iii) a phasor and frequency estimation method presented in [13], which is based on LS and phase modulation modeling of the input signal (called "PM"), and finally (iv) a subspace-based frequency estimation method inspired by the multiple signal classification (MUSIC) technique presented in [24] (called "A^{2}MUSIC"). The nominal frequency f_{nom}, sampling frequency f_s, and ζ are set to 60 Hz, 1920 Hz, and 10^{-4}, respectively.

8.1. Performance Evaluation in the Presence of DC Offset, DDC, Harmonics, and Noise in Both Nominal and Off-Nominal Frequency Conditions

The following test signal is used to investigate the performance of the proposed approach for phasor and frequency estimation in the presence of DC offset, DDC, harmonics, and noise, in both nominal and off-nominal frequency conditions

$$x_1(t) = A_0 + D_0 e^{-t/\tau} + \sum_{k=1}^{K} \frac{A_1}{k} \cos(2\pi k f_1 t + k\theta_1) + w(t) \tag{42}$$

where the total number of sinusoids K is set equal to 50 [29,30]. Five-hundred test signals are generated by selecting different values for A_0, f_1 (varied from 58 to 62 Hz [29]), D_0 (varied from 0 to 1 $p.u.$), τ (varied from 0.5 to 5 cycles), and signal-to-noise ratio (SNR) (varied from 10 to 80 dB) [24]. The fundamental phasor and frequency are estimated by the proposed approach and other understudied algorithms, and the mean, root mean square (RMS), and maximum values of maximum of TVE and |FE| are calculated. From the results shown in Table 5, it is clear that the proposed approach gives higher accuracy for both phasor (Max. TVE) and frequency (Max. |FE|) estimation in comparison to the other methods.

Table 5. Mean, RMS, and maximum values of the indices over the 500 test signals in the presence of undesirable components.

Algorithm	Max. TVE (%)			Max. |FE| (mHz)		
	Mean	RMS	Max.	Mean	RMS	Max.
DS-DFT [5]	1.05	2.09	4.86	N/A	N/A	N/A
AM [13]	1.23	2.54	5.78	8.23	11.4	32.4
PM [13]	1.06	2.38	4.97	6.51	9.21	34.6
A^{2}MUSIC [24]	N/A	N/A	N/A	2.28	3.74	6.63
Proposed Approach	0.58	0.67	0.79	0.56	0.69	1.01

8.2. Performance Evaluation for Dynamic Frequency Ramp Test

In this test, for the same test signals generated in Section 8.1, f_1 is ramped to $f_1 + 5$ (Hz) from 0 to 5 (s) at a rate of 1 Hz/s [3,29,30]. The mean, RMS, and maximum values of the two indices, namely, Max. TVE and Max. |FE|, for the 500 dynamic test signals are given in Table 6.

As seen in Table 6, the proposed approach gives better performance for both phasor and frequency estimation under frequency ramps in comparison to the other methods. Apart from the high estimation accuracy, the proposed approach also has faster response time under dynamic conditions which will be explored in more details in Section 8.4.

Table 6. Mean, RMS, and maximum values of the indices over the 500 dynamic test signals during the frequency ramp.

Algorithm	Max. TVE (%)			Max. |FE| (mHz)		
	Mean	RMS	Max.	Mean	RMS	Max.
DS-DFT [5]	1.32	2.59	5.66	N/A	N/A	N/A
AM [13]	1.35	2.67	5.93	9.14	12.6	36.21
PM [13]	1.13	2.52	5.07	6.63	9.42	35.06
A^{2}MUSIC [24]	N/A	N/A	N/A	2.69	4.17	7.12
Proposed Approach	0.67	0.78	0.89	0.66	0.80	1.07

8.3. Modulation Test

Transmission line faults/outages and bulk load switching often cause rotor-angle oscillations which may lead to power swings and subsequently mal-operation of the protective relays [32]. Generally, power swings are mathematically modeled as a modulated sinusoidal signal [13,29,30,32]

$$x_2(t) = X_m(1 + k_x \cos(\omega_m t)) \times \cos(\omega_{nom} t + k_a \cos(\omega_m t - \pi)) \quad (43)$$

where k_x and k_a denote amplitude and phase modulation coefficients, respectively, both of which are set to 0.1 [29,30]. In this test, the frequency of the input signal remains constant at 60 Hz, but the magnitude and phase-angle oscillate with an angular frequency of $\omega_m = 2\pi f_m$ where f_m changes from 0.1 Hz to 2 Hz in steps of 0.1 Hz as recommended by the Standards [29,30] for P-class PMUs. The maximum values of TVE and |FE| are given in Table 7 for different algorithms.

Table 7. Maximum TVE and |FE| for the modulation test.

Algorithm	Max. TVE (%)	Max. |FE| (mHz)
DS-DFT [5]	2.37	N/A
AM [13]	0.0096	2.3
PM [13]	0.0008	1.9
A^{2}MUSIC [24]	N/A	0.95
Proposed Approach	0.43	0.64

As can be observed from Table 7, the proposed approach gives better frequency estimation accuracy compared to the other algorithms. Since the algorithms presented in [13] are based on amplitude modulation (AM) and phase modulation (PM) modeling of the input signals, they have the highest phasor estimation accuracy (their Max. TVE is lowest). However, the proposed phasor estimation approach also satisfies the modulation test requirements for P-class PMUs specified in the Standards [29,30].

8.4. Step Change Test

Abrupt changes in the impedance and network configuration during faults may cause current/voltage transients which are mathematically expressed as [13,29,30,32]

$$x_3(t) = X_m(1 + k_x u(t)) \times \cos(\omega_{nom}t + k_a u(t)) \tag{44}$$

where k_x and k_a are the magnitudes of the unit step function $u(t)$ and set to 0.1 and $\pi/18$ rad, respectively, as recommended in the Standards [29,30]. Two well-known indices, namely, response time and maximum over/undershoot, which are defined in [30], are used to evaluate the performance of the different algorithms from the viewpoints of speed and accuracy. The results are given in Table 8 in which the response time and maximum over/undershoot are denoted by Rsp. and Max O/U.S., respectively. Based on the results obtained in Table 8, it is clear that the proposed approach has superior performance in terms of speed and accuracy, and its performance indices can comply with the requirements for P-class PMUs mentioned in the Standards [29,30].

Table 8. Performance indices for the step change test.

Algorithm	k_x	k_a	Phasor		Frequency
			Rsp. (ms)	Max O/U.S. (%) *	Rsp. (ms)
DS-DFT [5]	0.1	0	14.6	4.1	N/A
	0	$\pi/18$	15.5	4.6	N/A
AM [13]	0.1	0	15.3	5	41.3
	0	$\pi/18$	28.6	4	40.7
PM [13]	0.1	0	14.2	5	57.7
	0	$\pi/18$	30.9	4	57.1
A^2MUSIC [24]	0.1	0	N/A	N/A	12.4
	0	$\pi/18$	N/A	N/A	12.7
Proposed Approach	0.1	0	13.9	2.7	11.6
	0	$\pi/18$	14.4	3.3	12.1

* Max O/U.S. values are computed relative to the magnitude of the step [30].

8.5. Performance Evaluation Using Power System Signals

Power system signals of the IEEE 9-bus test system [13] are used to evaluate the performance of the proposed approach for both phasor and frequency estimation when severe oscillations in magnitude, phase angle, and frequency of the input signals take place. Two different scenarios are considered [13]:

Scenario 1: Outage of line 7-8 at $t = 0$ s. Voltage of bus 6 is used as the input signal.

Scenario 2: G_2 (connected to bus 2) trips at $t = 50$ ms. Current of line 6-4 is used as the input signal.

The maximum values of TVE and |FE| are given in Table 9 for the two scenarios. As can be observed from Table 9, the proposed approach provides higher accuracy compared to the other methods. It is also worth mentioning that the abnormal events in both the scenarios were detected by the proposed approach in less than one cycle.

Table 9. Maximum TVE (%) and |FE| (mHz) for each of the scenarios *.

Sc.	DS-DFT [5]	AM [13]	PM [13]	A^2MUSIC [24]	Proposed Approach
1	1.11 N/A	2.16 10.3	1.04 6.8	N/A 3.2	0.72 0.81
2	1.26 N/A	2.23 11.2	1.12 7.3	N/A 3.8	0.78 0.85

* For each cell, the first row is Max. TVE and the second row is Max. |FE|.

8.6. Computational Burden and Sensitivity Analysis

The threshold ζ has a significant impact on the accuracy and the computational complexity of the proposed approach (see Section 5). Hence, a sensitivity analysis is performed to investigate the effect of ζ on the accuracy of the estimated phasors and frequencies and the computational burden of the proposed approach. In this regard, Max. TVE, Max. |FE|, and average execution time (taken for each estimation) are given in Table 10 for different values of ζ for each of the test signals used in this section.

Table 10. Performance indices for different values of ζ.

Test	ζ	Max. TVE (%)	Max. |FE| (mHz)	Execution Time (μs)
Steady-State (Section 8.1)	10^{-2} 10^{-4} 10^{-6}	0.97 0.58 0.51	5.02 0.56 0.47	46 283 405
Frequency Ramp (Section 8.2)	10^{-2} 10^{-4} 10^{-6}	1.09 0.67 0.59	5.17 0.66 0.58	57 367 472
Modulation Test (Section 8.3)	10^{-2} 10^{-4} 10^{-6}	1.03 0.43 0.36	5.11 0.61 0.56	52 319 423
Magnitude Step Change (Section 8.4)	10^{-2} 10^{-4} 10^{-6}	1.12 0.69 0.60	5.09 0.59 0.50	54 326 446
Phase Angle Step Change (Section 8.4)	10^{-2} 10^{-4} 10^{-6}	1.14 0.71 0.63	5.15 0.63 0.57	56 339 468
Scenario 1 (Section 8.5)	10^{-2} 10^{-4} 10^{-6}	1.16 0.72 0.64	5.21 0.81 0.74	59 370 484
Scenario 2 (Section 8.5)	10^{-2} 10^{-4} 10^{-6}	1.25 0.78 0.71	5.26 0.85 0.77	63 391 510

As can be observed from Table 10, higher accuracy in estimation of phasor and frequency is achieved by choosing smaller values for ζ, at the cost of increasing the execution time (or equivalently computational burden/complexity) i.e., there is a trade-off in choosing the optimal value of the threshold ζ. Table 10 indicates that an acceptable value for ζ would be 10^{-4} since this provides a reasonable balance between the estimation accuracy and computational burden. The mean execution time of the proposed approach (over all of the test signals in this section) for each synchrophasor and frequency estimation by the adoption of 10^{-4} for ζ was around 0.34214 (ms), which makes it suitable for real-time applications.

8.7. Noise Propagation through the Proposed Algorithm

In this section, the following test signal is used to investigate the propagation of noise through the proposed algorithm:

$$x_4[n] = A_1 \sin(\frac{2\pi n f_1}{f_s} + \theta_1) + w[n] \quad \forall n = 1, 2, \ldots, N \tag{45}$$

Cramér Rao bounds (CRBs) that give lower bounds on the variances of estimated signal parameters (i.e., magnitude, phase angle, and frequency) are obtained for the above-mentioned test signal (45) as follows: [36].

$$\begin{aligned} CRB(A_1) &= \frac{2{\sigma_n}^2}{N} \\ CRB(\theta_1) &= \frac{4{\sigma_n}^2}{{A_1}^2}\frac{2N+1}{N(N-1)} \\ CRB(f_1) &= \frac{24{\sigma_n}^2}{{A_1}^2 N(N^2-1)}\left(\frac{f_s}{2\pi}\right)^2 \end{aligned} \tag{46}$$

In order to investigate the propagation of noise through the proposed algorithm, 100 test signals are generated by choosing different values for f_1 between 58 and 62 (Hz) [29,30]. The variance of the noise in the sinusoidal signal ${\sigma_n}^2$ is set to be 0.01^2 [36]. For each of the generated test signals, the signal parameters are estimated by the proposed algorithm. The mean square error (MSE) [37] as well as the CRBs are given in Table 11. As can be seen from Table 11, the values of MSE for all of the estimated parameters are very close to their corresponding CRBs.

Table 11. MSEs and CRBs of the estimated parameters by the proposed algorithm.

Parameter	CRB	MSE
A_1	6.25×10^{-6}	1.35×10^{-5}
θ_1	2.62096×10^{-5}	1.46×10^{-4}
f_1	6.8458×10^{-3}	9.23×10^{-3}

9. Conclusions

This paper proposes a fast, accurate, robust, and computationally-efficient approach for estimating phasor and frequency that can be applied to both voltage and current signals without making simplifying assumptions regarding the components present in the signal. The first step of the proposed methodology involves the removal of the DDC that may be present in the input signal using optimal filtering. Care is taken to ensure that the parameters of the fundamental component are left unaltered. In the next step, initial phasor and frequency estimates are obtained in parallel using DFT and the proposed iESPRIT, respectively. Finally, to further improve accuracy, the initial estimates of phasor and frequency are updated, based on the identified system condition.

The simulation results confirm that the proposed approach has high accuracy, fast response time, low computational complexity, and is robust against grid disturbances and off-nominal frequency conditions. Furthermore, the proposed approach fulfills the requirements recommended by the IEEE and IEC standards for P-class PMUs in both steady-state and dynamic conditions. This makes the proposed approach suitable for a variety of power system monitoring, protection, and control applications that need fast and accurate estimates of voltage and current phasors and system frequency.

To provide more robustness against stray components such as harmonics and inter-harmonics, particularly for measurement applications, the combination of more recent versions of DFT such as interpolated DFT (IpDFT) with iESPRIT will be explored in the future.

Author Contributions: Conceptualization, B.J. and A.P.; methodology, B.J. and A.P.; software, B.J.; validation, B.J. and A.P.; formal analysis, B.J. and A.P.; investigation, B.J.; writing—original draft preparation, B.J.; writing—review and editing, B.J. and A.P.; visualization, B.J. and A.P. All authors have read and agreed to the published version of the manuscript.

Funding: This research received no external funding.

Institutional Review Board Statement: Not applicable.

Informed Consent Statement: Not applicable.

Data Availability Statement: Not applicable.

Conflicts of Interest: The authors declare no conflict of interest.

Appendix A. P and Q Correction Coefficients

Appendix A.1. Variations of P and Q for Different Values of Δf and N

Variations of P and Q for different values of N and Δf are shown in Tables A1 and A2.

Table A1. Magnitudes and phase angles of correction coefficients for different frequency deviations and sampling rate of 24.

Δf (Hz)	$\lvert P \rvert$	$\angle P$ (°)	$\lvert Q \rvert$	$\angle Q$ (°)
−5	0.9886	−14.37	0.0434	29.37
−4.5	0.9908	−12.94	0.0390	27.94
−4	0.9927	−11.5	0.0346	26.5
−3.5	0.9944	−10.06	0.0302	25.06
−3	0.9959	−8.62	0.0258	23.62
−2.5	0.9972	−7.19	0.0215	22.19
−2	0.9982	−5.75	0.0171	20.75
−1.5	0.9990	−4.31	0.0128	19.31
−1	0.9995	−2.87	0.0085	17.87
−0.5	0.9999	−1.44	0.0042	16.44
0	1.0000	0	0	15
0.5	0.9999	1.44	0.0042	13.56
1	0.9995	2.87	0.0084	12.12
1.5	0.9990	4.31	0.0125	10.69
2	0.9982	5.75	0.0166	9.25
2.5	0.9972	7.19	0.0206	7.81
3	0.9959	8.62	0.0246	6.37
3.5	0.9944	10.06	0.0285	4.94
4	0.9927	11.5	0.0324	3.5
4.5	0.9908	12.94	0.0363	2.06
5	0.9886	14.37	0.0400	0.62

Table A2. Magnitudes and phase angles of correction coefficients for different sampling rates and frequency of 62 (Hz).

N	$\lvert P \rvert$	$\angle P$ (°)	$\lvert Q \rvert$	$\angle Q$ (°)
12	0.9982	5.5	0.0172	24.5
24	0.9982	5.75	0.0166	9.25
36	0.9982	5.83	0.0164	4.17
48	0.9982	5.876	0.0164	1.62
60	0.9982	5.9	0.0164	0.1
72	0.9982	5.92	0.0164	−0.92
84	0.9982	5.93	0.0164	−1.64
96	0.9982	5.94	0.0164	−2.19
108	0.9982	5.947	0.0164	−2.61
120	0.9982	5.95	0.0164	−2.95

Appendix A.2. Impact of Q on Phasor Estimation Accuracy

The following test signal is used to investigate the performance of the proposed algorithm with and without considering the effect of Q.

$$x_5(t) = \sum_{k=1}^{2} A_k \cos(2\pi k f_1 t + k\theta_1) \tag{A1}$$

where $A_2/A_1 = 0.1$. The above-mentioned test signal is applied to the proposed algorithm for estimation of the fundamental phasor. To investigate the impact of Q, seven different values of $\{0, \pm 0.5, \pm 1, \pm 2\}$ (Hz) are considered for the deviation of f_1 from its nominal value (60 Hz), and the maximum TVE is used for performance evaluation.

As can be seen from Table A3, the impact of Q on the accuracy of the estimated phasor by the proposed algorithm is small (in comparison to the TVE limit of 1% [29,30]) when the power system frequency varies in the range of -2 Hz and $+2$ Hz. Additionally, note that the frequency estimation algorithm improves the accuracy of the estimated phasor at off-nominal frequency conditions (see Section 6.2 and Figure 3).

Table A3. Impact of Q on accuracy of the estimated phasor by the proposed algorithm.

Δf_1 (Hz)	Max. TVE (%)	
	Without Q	With Q
−2	0.288	0.218
−1	0.223	0.168
−0.5	0.156	0.128
0	0.08	0.08
+0.5	0.152	0.126
+1	0.219	0.161
+2	0.285	0.217

Appendix B. Equality of the Matrices Q and U

By multiplying both sides of (23) by $\mathbf{X}_f{}^H$, we have

$$\mathbf{X}_f{}^H \mathbf{X}_f = \mathbf{X}_f{}^H \mathbf{L\Sigma U}^H \tag{A2}$$

By substituting (23) into (A2), we get

$$\mathbf{X}_f{}^H \mathbf{X}_f = \left(\mathbf{L\Sigma U}^H\right)^H \mathbf{L\Sigma U}^H = \mathbf{U\Sigma}^H \mathbf{L}^H \mathbf{L\Sigma U}^H \tag{A3}$$

Since the matrices **L** and **U** are unitary, (A3) is simplified as

$$\mathbf{X}_f{}^H \mathbf{X}_f = \mathbf{U\Sigma}^H \mathbf{\Sigma U}^H = \mathbf{U\Sigma}^2 \mathbf{U}^H \tag{A4}$$

By multiplying both sides of (A4) by $1/G$, we get

$$\frac{1}{G}\mathbf{X}_f{}^H \mathbf{X}_f = \frac{1}{G}\mathbf{U\Sigma}^2 \mathbf{U}^H \tag{A5}$$

Substituting (24) into (A5) yields

$$\hat{\mathbf{R}}_x = \frac{1}{G}\mathbf{U\Sigma}^2 \mathbf{U}^H \tag{A6}$$

By equating the right-hand sides of (25) and (A6), we get

$$\mathbf{Q\Lambda Q}^H = \frac{1}{G}\mathbf{U\Sigma}^2 \mathbf{U}^H \tag{A7}$$

It is clear from (A7), (23), and the (25) that the squared magnitudes of the data matrix singular values are equal to the eigenvalues of the $\hat{\mathbf{R}}_x$ scaled by a factor of G (i.e., $\mathbf{\Sigma}^2 = G\mathbf{\Lambda}$), and the columns of the matrix $\mathbf{U}$ are exactly equal to the columns of $\mathbf{Q}$ (i.e., $\mathbf{Q} = \mathbf{U}$). □

Appendix C. Proof of $\acute{P}$ Calculation Using (39)

The DFT of the previous data string in (37) is split as

$$\hat{X}_{1old} = \frac{2}{N}x[r]e^{-jr\omega_{1old}\Delta t} + \underbrace{\frac{2}{N}\sum_{n=r+1}^{r+N-1} x[n]e^{-jn\omega_{1old}\Delta t}}_{J} \tag{A8}$$

Similarly, (38) decomposes as follows for the current string:

$$\hat{X}_{1new} = \frac{2}{N}x[r+N]e^{-j(r+N)\omega_{1new}\Delta t} + \underbrace{\frac{2}{N}\sum_{n=r+1}^{r+N-1} x[n]e^{-jn\omega_{1new}\Delta t}}_{Z} \tag{A9}$$

According to (7), (8), (10) and (12), the following equation exists that relates J to Z:

$$J \approxeq \acute{P}Z \tag{A10}$$

By substituting (A8) and (A9) into (A10), we have

$$\hat{X}_{1old} - \frac{2}{N}x[r]e^{-jr\omega_{1old}\Delta t} \approxeq \acute{P}\left[\hat{X}_{1new} - \frac{2}{N}x[r+N]e^{-j(r+N)\omega_{1new}\Delta t}\right] \tag{A11}$$

Now, note the following assumptions (which are not violated under *normal operation* of power systems):

- Deviation of frequency from its nominal value is very small (hence, $e^{-jN\omega_{1new}\Delta t} \approxeq 1$).
- The input signal is approximately periodic (hence, $x[r] \approxeq x[r+N]$).
- Frequency varies negligibly (hence, $e^{-jr\omega_{1old}\Delta t} \approxeq e^{-jr\omega_{1new}\Delta t}$).

Based on these assumptions, (A11) can be simplified to obtain (39). □

References

1. Bollen, M.H.; Gu, I.Y. *Signal Processing of Power Quality Disturbances*; John Wiley & Sons: Hoboken, NJ, USA, 2006.
2. Amirat, Y.; Oubrahim, Z.; Ahmed, H.; Benbouzid, M.; Wang, T. Phasor estimation for grid power monitoring: Least square vs. linear Kalman filter. *Energies* **2020**, *13*, 2456. [CrossRef]
3. Zhan, L.; Liu, Y.; Liu, Y. A clarke transformation-based DFT phasor and frequency algorithm for wide frequency range. *IEEE Trans. Smart Grid* **2018**, *9*, 67–77. [CrossRef]
4. Jafarpisheh, B.; Madani, S.M.; Shahrtash, S.M. A new DFT-based phasor estimation algorithm using high-frequency modulation. *IEEE Trans. Power Deliv.* **2017**, *32*, 2416–2423. [CrossRef]
5. Jafarpisheh, B.; Madani, S.M.; Jafarpisheh, S. Improved DFT-based phasor estimation algorithm using down-sampling. *IEEE Trans. Power Deliv.* **2018**, *33*, 3242–3245. [CrossRef]
6. Jafarpisheh, B. Phasor estimation algorithm based on complex frequency filters for digital relaying. *IEEE Trans. Instrum. Meas.* **2018**, *67*, 582–592. [CrossRef]
7. Kang, S.H.; Seo, W.S.; Nam, S.R. A frequency estimation method based on a revised 3-Level discrete Fourier transform with an estimation delay reduction technique. *Energies* **2020**, *13*, 2256. [CrossRef]
8. Jin, T.; Zhang, W. A novel interpolated DFT synchrophasor estimation algorithm with an optimized combined cosine self-convolution window. *IEEE Trans. Instrum. Meas.* **2021**, *70*, 1–10.
9. Yu, H.; Jin, Z.; Zhang, H.; Terzija, V. A phasor estimation algorithm robust to decaying DC component. *IEEE Trans. Power Deliv.* **2021**, 1. [CrossRef]
10. Frigo, G.; Derviškadić, A.; Paolone, M. Reduced leakage synchrophasor estimation: Hilbert transform plus interpolated DFT. *IEEE Trans. Instrum. Meas.* **2019**, *68*, 3468–3483. [CrossRef]
11. Derviškadić, A.; Romano, P.; Paolone, M. Iterative-interpolated DFT for synchrophasor estimation: A single algorithm for P- and M-class compliant PMUs. *IEEE Trans. Instrum. Meas.* **2018**, *67*, 547–558. [CrossRef]

12. Fu, L. A multiple frequency Taylor model-based dynamic synchrophasor estimation algorithm. *IEEE Trans. Smart Grid* **2019**, *10*, 6870–6882. [CrossRef]
13. Vejdan, S. Accurate dynamic phasor estimation based on the signal model under off-nominal frequency and oscillations. *IEEE Trans. Smart Grid* **2017**, *8*, 708–719. [CrossRef]
14. Kim, W.J.; Nam, S.R.; Kang, S.H. Adaptive phasor estimation algorithm based on a least squares method. *Energies* **2019**, *12*, 1387. [CrossRef]
15. Rao, A.V.K. Accurate phasor and frequency estimation during power system oscillations using least squares. *IET Sci. Meas. Technol.* **2019**, *13*, 989–994.
16. Choqueuse, V.; Belouchrani, A.; Auger, F.; Benbouzid, M. Frequency and phasor estimations in three-phase systems: Maximum likelihood algorithms and theoretical performance. *IEEE Trans. Smart Grid* **2019**, *10*, 3248–3258. [CrossRef]
17. Castello, P.; Ferrero, R.; Pegoraro, P.A.; Toscani, S. Space vector Taylor–Fourier models for synchrophasor, frequency, and ROCOF measurements in three-phase systems. *IEEE Trans. Instrum. Meas.* **2019**, *68*, 1313–1321. [CrossRef]
18. da Silva, C.D.L. Phasor estimation in power systems using a neural network with online training for numerical relays purposes. *IET Sci. Meas. Technol.* **2015**, *9*, 836–841. [CrossRef]
19. Affijulla, S.; Tripathy, P. Development of phasor estimation algorithm for P-class PMU suitable in protection applications. *IEEE Trans. Smart Grid* **2018**, *9*, 1250–1260. [CrossRef]
20. Reddy, M.V.; Sodhi, R. An open loop fundamental and harmonic phasor estimator for single-phase voltage signals. *IEEE Trans. Ind. Inform.* **2020**, *16*, 4535–4546. [CrossRef]
21. Wang, L.; Suonan, J. A fast algorithm to estimate phasor in power systems. *IEEE Trans. Power Deliv.* **2017**, *32*, 1147–1156. [CrossRef]
22. Zamora-Mendez, A.; Paternina, M.R.A.; Vázquez, E.; Ramirez, J.M.; de Serna, J.A.L.O. Distance relays based on the Taylor–Kalman-Fourier filter. *IEEE Trans. Power Deliv.* **2016**, *31*, 928–935. [CrossRef]
23. De Apráiz, M.; Diego, R.I.; Barros, J. An extended Kalman filter approach for accurate instantaneous dynamic phasor estimation. *Energies* **2018**, *11*, 2918. [CrossRef]
24. Jafarpisheh, B. Power system frequency estimation using adaptive accelerated MUSIC. *IEEE Trans. Instrum. Meas.* **2018**, *67*, 2592–2602. [CrossRef]
25. Drummond, Z.D. An optimized subspace-based approach to synchrophasor estimation. *IEEE Trans. Instrum. Meas.* **2021**, *70*, 1–13. [CrossRef]
26. Xu, S.; Liu, H.; Bi, T. A novel frequency estimation method based on complex bandpass filters for P-class PMUs with short reporting latency. *IEEE Trans. Power Deliv.* **2020**, 1. [CrossRef]
27. Chen, L. Harmonic phasor estimator for P-class phasor measurement units. *IEEE Trans. Instrum. Meas.* **2020**, *69*, 1556–1565. [CrossRef]
28. Manolakis, D.; Ingle, V.; Kogon, S. *Statistical and Adaptive Signal Processing*; Artech: Norwood, MA, USA, 2005.
29. *IEEE Standard for Synchrophasor Measurements for Power Systems—Amendment 1: Modification of Selected Performance Requirements*; IEEE Std C37.118.1a-2014 (Amendment to IEEE Std C37.118.1-2011); 2014; pp. 1–25. [CrossRef]
30. *IEEE/IEC International Standard—Measuring Relays and Protection Equipment— Part 118-1: Synchrophasor for Power Systems—Measurements*; IEC/IEEE 60255-118-1:2018; 2018; pp. 1–78. [CrossRef]
31. Benmouyal, G. Removal of DC-offset in current waveforms using digital mimic filtering. *IEEE Trans. Power Deliv.* **1995**, *10*, 621–630. [CrossRef]
32. Affijulla, S.; Tripathy, P. Development of dictionary-based phasor estimator suitable for P-class phasor measurement unit. *IEEE Trans. Instrum. Meas.* **2018**, *67*, 2603–2615. [CrossRef]
33. Phadke, A.G.; Thorp, J.S. *Synchronized Phasor Measurements and Their Applications*; Springer: Berlin/Heidelberg, Germany, 2008; Volume 1.
34. Björck, Å. *Numerical Methods in Matrix Computations*; Springer: Berlin/Heidelberg, Germany, 2015.
35. Roy, R.H.; Kailath, T. ESPRIT–Estimation of signal parameters via rotational invariance techniques. *Opt. Eng.* **1990**, *29*, 296–313.
36. Singh, A.K.; Pal, B.C. Rate of change of frequency estimation for power systems using interpolated DFT and Kalman filter. *IEEE Trans. Power Syst.* **2019**, *34*, 2509–2517. [CrossRef]
37. Choqueuse, V.; Belouchrani, A.; Benbouzid, M. Phasor estimation using conditional maximum likelihood: Strengths and limitations. In Proceedings of the 2015 23rd European Signal Processing Conference (EUSIPCO), Nice, France, 31 August–4 September 2015; pp. 2251–2255.

MDPI AG
Grosspeteranlage 5
4052 Basel
Switzerland
Tel.: +41 61 683 77 34

Energies Editorial Office
E-mail: energies@mdpi.com
www.mdpi.com/journal/energies

www.ingramcontent.com/pod-product-compliance
Lightning Source LLC
Chambersburg PA
CBHW061442210926
51850CB00008B/1579